BADLAND GEOMORPHOLOGY AND PIPING

edited by

Rorke Bryan
and
Aaron Yair

ISBN 0 86094 113 2 cloth
0 86094 114 0 paper

Published by;
Geo Books (Geo Abstracts Ltd)
Regency House
34 Duke Street
Norwich NR3 3AP
England

Printed in Great Britain at the
University Press, Cambridge

Contents

1 Perspectives on studies of badland geomorphology — 1
R.B. BRYAN AND A. YAIR

2 The influence of material behaviour on runoff initiation in the Dinosaur Badlands, Canada — 13
W.K. HODGES AND R.B. BRYAN

3 The relationship of soil physical and chemical properties to the development of badlands in Morocco — 47
A.C. IMESON, F.J.P.M. KWAAD AND J.M. VERSTRATEN

4 Difference between "calanchi" and "biancane" badlands in Italy — 71
D. ALEXANDER

5 Spatial variations in infiltration, runoff and erosion on hillslopes in semi-arid Spain — 89
H. SCOGING

6 Sheetwash and rill development by surface flow — 113
J. SAVAT AND J. DE PLOEY

7 Hydraulic characteristics of a badland pseudo-pediment slope system during simulated rainstorm experiments — 127
W.K. HODGES

8 Experimental study of drainage networks — 153
R.S. PARKER AND S.A. SCHUMM

9 Application of "diffusion" degradation to some aspects of drainage net development — 169
Z.B. BEGIN

10 Sediment and solute yield from Mancos Shale hillslopes, Colorado and Utah — 181
J. LARONNE

11 Salt scalds and subsurface water: a special case of badland development in southwestern Australia — 195
A. CONACHER

12 Surface morphology and rates of change during a ten-year period in the Alberta badlands — 221
I.A. CAMPBELL

13 Quaternary evolution of badlands in the south-eastern Colorado Plateau, USA 239
S.G. WELLS AND A.A. GUTIERREZ

14 How old are the badlands? A case study from south-east Spain 259
S.M. WISE, J.B. THORNES AND A. GILMAN

15 Long term denudation rates in the Zin-Havarim badlands, northern Negev, Israel 279
A. YAIR, P. GOLDBERG AND B. BRIMER

16 Piping in the Big Muddy badlands, southern Saskatchewan, Canada 293
D.P. DREW

17 The occurrence of piping and gullying in the Penticton glacio-lacustrine silts, Okanagan Valley, B.C. 305
O. SLAYMAKER

18 The role of piping in the development of badlands and gully systems in south-east Spain 317
A. HARVEY

19 Throughflow and pipe monitoring in the humid temperate environment 337
M. ANDERSON AND T.P. BURT

20 Experimental studies of pipe hydrology 355
J.A.A. JONES

References 371

1 Perspectives on studies of badland geomorphology

R. Bryan and A. Yair

Although true badlands are of restricted occurrence a number of important geomorphic concepts have been formulated or tested there, and many relevant papers have been published. In summarizing the salient aspects of badland research we have not attempted an exhaustive review and accordingly have cited references very sparingly. Virtually all important references, and the basis for many statements in the introduction may be found in the papers which form the body of the volume.

The term 'badlands' was originally used to describe intensely dissected natural landscapes where vegetation is sparse or absent and which are useless for agriculture. They were generally considered to be typically of fluvial origin, characterized by very high drainage densities, V-shaped valleys and short, steep slopes, often fringed by gently sloping planar surfaces referred to as pediments or miniature pediments. With time, and increasing knowledge, application of the term has been expanded to include:

1. areas where piping, tunnel erosion and mass-wasting processes combine with fluvial processes to produce a rugged, hummocky and dissected topography where the dense dendritic drainage network of 'classical' badlands is obliterated. In such conditions integration of the drainage system is often incomplete making stream network analysis difficult, and sometimes inappropriate.
2. large areas, particularly in semi-arid regions, where a fragile natural equilibrium has been disturbed by ill-advised land use practices causing the delicately balanced system to move swiftly into extensive badland degradation. Although this is often triggered by man, the processes and resultant landforms are very similar to those in naturally-developed badlands.

In parts of Australia the term 'badlands' has been further extended to include flat, unvegetated surfaces incised by a rill network of low density. These develop as a result of land use changes which cause water-tables to rise carrying soluble salts into topsoil, changing soil properties, eliminating vegetation and agricultural potential. Although these features share some properties with badlands, they lack the diagnostic characteristics of highly dissected topography and rapid erosion rates. Extension of the term in

this way is problemmatic for if lack of vegetation and salinized regolith are used as defining criteria, extensive arid and semi-arid areas would be included, regardless of lithologic or topographic character.

BADLAND CHARACTERISTICS

Numerous factors and processes influence badland development to produce a complex diversity of landforms at various scales. Nevertheless most areas in which natural badland development is extensive share certain lithologic and climatic characteristics.

Lithologic Conditions

Extensive badland development is usually associated with unconsolidated or poorly cemented materials. The most widespread and well-developed examples occur in shales, marls or silty-clay formations, but they may also develop on sandy conglomerates with clay matrices, weakly cemented sandstones and even on thick sandy weathering mantle derived from granite.

Because of poor consolidation the role of structural control is usually limited, but this does not prevent marked topographic variations over short distances related to changes in mechanical, chemical or mineralogical properties of the regolith. This may be seen in the Zin badlands, Israel, where there is marked topographic variation between the Upper and Lower Taquiya shales. The upper shales with 80-100 percent montmorillonite have long spurs with limited rilling, while the lower shales, with 40 percent kaolinite and 10 percent illite, have short spurs and dense rilling. In the Alberta badlands, Canada, similar differences exist between densely-rilled silty sandstones and interbedded montmorillonitic shales which have a sparse or non-existent rill network.

Although differences in physical resistance to particle size and mineralogical composition are responsible for many, perhaps most, topographic variations in badlands, they do not explain all differential developments observed. In some areas chemical properties are most significant, and the balance between the exchangeable sodium percentage (ESP) of the regolith and the electrical conductivity (E^c) of the regolith solution may determine the occurrence and threshold of geomorphic processes. The behaviour of clay minerals, particularly swelling, dispersion and flocculation, is strongly affected by variations in this balance. This applies particularly to montmorillonite which has an extremely high intrinsic swelling capacity which, in sodium-rich environments like the Alberta badlands, is an important control on geomorphic processes. In the Zin badlands, in different conditions of regolith chemistry, however, the swelling capacity of montmorillonite is almost entirely suppressed, with pronounced impact on processes.

Climatic conditions

Numerous examples of man-induced badlands exist in humid areas such as the Perth Amboy badlands described in Schumm's classic paper which developed where precipitation exceeded 1000 mm yr^{-1}. Apart from the Hong Kong badlands where precipitation averages 1899 mm yr^{-1} all natural badlands appear to be characterized by limited precipitation and prolonged hot, dry weather. Areas of natural occurrence may be broadly classified into two groups, both of which show marked seasonal variation in geomorphic activity:

(a) Mediterranean semi-arid and arid areas of the sub-tropical climatic belt where most rainfall occurs in high-intensity storms of brief duration during the winter and little geomorphic activity occurs during the long, hot, dry summers.

(b) Continental areas in higher latitudes where very cold conditions prevail during the winter. Frost action becomes geomorphologically significant, particularly in more mountainous regions like western Colorado, and much fluvial development may take place during seasonal snowmelt. Rainstorms are confined to the summer when evaporation rates are high and surfaces dry very quickly.

In Mediterranean and sub-tropical badlands erosional processes are almost entirely confined to the winter or the wet season. Erosion by overland flow is dominant but considerable local variations in frequency, duration and effectiveness occur, reflecting climatic and regolith characteristics. More systematic regional variations may appear related to differences in average annual rainfall. As rainfall decreases leaching of salts from the regolith is reduced and so a transition may exist between the more humid areas where regolith is thoroughly leached through an intermediate zone where calcium dominates the regolith to the most arid sector where sodium dominates. These variations will be reflected in the swelling, flocculation or dispersion of clay-rich regolith, and therefore in infiltration characteristics and response to rainfall. Although threshold conditions necessary for runoff show extreme local variability it may be possible to identify more general regional trends in relation to regolith chemistry. In the intermediate zone dominated by calcium and by flocculation, infiltration capacities would be generally higher, requiring larger rainfall amounts to generate equivalent runoff and erosion. This may also lead to more extensive pipe development and some cases to a predominance of piping over surface runoff and erosion.

Within any of the zones identified microclimate also strongly influences geomorphic development. In the Zin badlands differences in temperature, and consequently in moisture regime, between adjacent north and south-facing slopes on homogeneous shales were responsible for quite different surface characteristics. These in turn resulted in an approximate fourfold difference in the amount of rainfall necessary to generate runoff.

Badland development in continental mid-latitude regions appears to be even more complex than in the sub-tropics. Some geomorphic activity continues at all seasons but there is marked seasonal variation in the processes active. During prolonged severe winters fluvial action is limited but substantial amounts of material are moved by frost creep on steep, unstable shale slopes. This is particularly important in mountainous areas where freeze-thaw cycles are numerous, and in the mountains of western Colorado creep rates well above those from humid areas have been reported. Frost action also affects aggregate size and stability and may render shale surfaces particularly vulnerable during snowmelt. This is particularly important in badlands of the plains and in the Alberta badlands appears to be the dominant geomorphic event of the year providing the only complete saturation of shale surfaces. So far precise monitoring has been prevented by the erratic, unpredictable occurrence of snowmelt.

In summer the dominance of processes changes. Creep is still active as a result of rainsplash impact and swelling-induced instability, but rates appear to be lower than in winter. In general rainsplash, sheetwash, rilling and piping all became more significant. Although all can occur at any time between snowmelt and freeze-up there may be some seasonal variation in significance.

PROCESSES OF BADLAND DEVELOPMENT

Several authors have suggested that badlands form ideal field 'laboratories' for testing landform evolution hypotheses. This is related to three basic assumptions: that badland landforms develop rapidly, facilitating monitoring; that badland landforms are analogous, except in scale, to major fluviatile landforms; that geomorphic processes, dominated by overland flow, are comparatively simple and easily studied. The assumption of rapid development has been verified by many studies but does not apply to all badlands. While the morphologic analogy of landforms is tempting, data on nature and rates of processes are so scant that attempts to draw geomorphic analogies are certainly premature. The most difficult problem is the assumed simplicity of processes, for many studies shown that badland development is both complex and varied including a wide range of surface and subsurface processes. The occurrence and relative importance of these vary between badlands and within any area with time. At all scales such variability and the landforms produced appear to relate primarily to the physical and chemical behaviour of regolith material. A brief review of the most important processes and controlling factors follows.

Rainsplash erosion

Rainsplash erosion should be favoured by the absence of vegetation, friable bedrock and apparently intense rainfall in badland areas, and in many badland areas evidence of its significance is widespread (for example, in the ubiquitous hoodoos of the Dinosaur badlands, Alberta). Unfortunately quantitative data on rainsplash in badlands have not yet been published. Field observations in the Dinosaur badlands show that rainsplash is an important agent on montmorillonite-

rich sandstones, and a trigger for particle instability on steep 'popcorned' shales, but ineffective on crusted shale surfaces. A similar result was found on crusted shales in the Zin badlands. This can be explained by the high mechanical strength of dry shales and the low rate of water absorption by montmorillonite. When wetting is extremely prolonged as in sequential rainstorms or snowmelt, shale moisture contents approach the liquid limit, strength is greatly reduced, and rainsplash becomes an effective erosional agent. These conditions, though infrequent, do occur in temperate badlands, but probably not in Mediterranean or subtropical badlands where rainstorms are typically very short-lived.

Overland flow

The extremely high drainage densities of badland areas are usually regarded as evidence of dominance of overland flow, but actual measurements are scarce and information on factors controlling runoff generation is limited. The few data available are somewhat conflicting indicating considerable complexity in the nature and frequency of overland flow.

Generation of runoff

It is often assumed that a combination of scant vegetation, steep slopes and impermeable regolith in badlands will result in widespread Hortonian runoff. In fact regolith infiltration capacities are frequently high though variable indicating considerable spatial variation in runoff generation and although Hortonian runoff can occur it is probably no more frequent than in other areas. Problems with application of the Hortonian runoff-infiltration model in semiarid areas have been suggested by Scoging and Thornes (1979) and Scoging (this volume) following field experiments in which rapid runoff response was observed at rainfall intensities below measured infiltration capacities. This was attributed to the limited storage capacity of the surface soil layer rather than profile control of infiltration, although it is difficult to follow the logic of this distinction. It is possible that limited storage capacity of the surface regolith layer is responsible for rapid runoff response from the smooth pseudo-pediments of the Dinosaur badlands where infiltration into the lower profile is restricted by large vesicles caused by air entrappment during sediment deposition.

While rapid runoff response can result from restricted storage capacity it more frequently reflects low infiltration capacities. In the Dinosaur badlands this occurs both on montmorillonitic sandstones and on rilled portions of shales. On the sandstones infiltration capacity drops very swiftly on wetting due to dispersion of the immediate surface layer, possibly caused by high exchangeable sodium concentrations. Rapid runoff from rills on shales reflects the compact structure of a localized depositional layer of silt and fine sand. The importance of surface structure has also been shown, by studies in Israel. On talus-mantled slopes in Sinai runoff generation is limited to incised gullies where large blocks are concentrated and where roughness and surficial storage

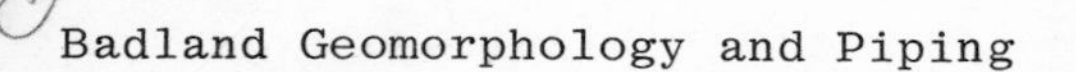

are high. No runoff developed on adjacent smoother interfluves. This apparently confusing observation stems from a compact fine-grained layer in which the large blocks of the gully are embedded, while interfluves are uncompacted. In the Zin badlands more rapid runoff at higher rates occurs on rough but compacted north-facing slopes than on smooth but uncompacted south-facing slopes. It therefore seems that a negative relationship between detention storage capacity and runoff generation is far from general.

On shale interfluves runoff generation is complex and variable. Although the shale material itself has a very low infiltration capacity, this is greatly increased by a dense network of desiccation cracks. The first stage of runoff generation is localized shedding of water from shale aggregate units into cracks. Gradual sealing of cracks from the base leads to crackflow, and later to complete surface sealing and general runoff. This could be interpreted as Hortonian runoff, though uniform widespread runoff occurs only after prolonged wetting.

It is clear from this brief review that a number of different processes of runoff generation are active in badlands. Field experimental data indicate that these can vary over very short distances on a single slope, closely reflecting changes in soil moisture content and surface structure, and also at any location with time.

Sheet wash

True unconfined sheetwash seems to occur only in those parts of badland areas which produce runoff swiftly and uniformly even in low intensity rainstorms. Observations of such flow come from two environments: (1) montmorillonitic sandstones in Dinosaur badlands where there is a quick transition from sheetwash to confined flow in deeply incised rills; (2) gently sloping smooth pseudo-pediments developed at the edge of retreating scarps, composed of successive thin layers of silt and fine sand deposited during flow recession. As described above, rapid runoff response relates to compaction and vesicular layers. In both the environments noted wetting depth at the end of rainfall is only a few millimetres. Flow depths are very shallow and flow is laminar or transitional, disturbed greatly by raindrop impact. On pseudo-pediments this flow pattern is interrupted by the development of ephemeral rills which are obliterated by sedimentation during flow recession.

Concentrated flow

Concentrated flow in badlands may occur as channel flow in rills and gullies or, on desiccated shale surfaces, as crackflow. It is not yet clear if these differ only in scale or also in hydraulic characteristics. In some areas flow types are sequentially linked and, for example, crackflow leads to rill flow, but in other areas no such linkage is apparent.

Crackflow appears in shale badlands as the initial stage of interfluve flow, which may or may not precede channel flow. Initially cracks form a highly efficient infiltration route, but as rain continues they start to close from the

bottom. The crackflow thus initiated persists until cracks seal completely and general sheetwash starts. The time required varies with clay mineralogy, antecedent moisture and rainfall conditions. In most badlands, particularly in more arid regions, rainfall is seldom sufficiently prolonged for complete sealing and crackflow persists as the dominant interfluve flow contribution. Most crack patterns are irregular and tortuous, and of low hydraulic efficiency, so flow is typically discontinuous and hydraulically varied. Occasionally, if rainfall is very intense crack capacity is exceeded and over-topping leads to localized, discontinuous sheetwash.

Much of the surface flow in badlands is in shallow rills and in shale badlands these usually show the most rapid runoff response due to compact silt and sand deposition. Except in extreme rainstorms or during snowmelt, flow is confined to the immediate vicinity of rills. Due to the relatively high frequency and magnitude of rill flow these form an integrated, efficient flow delivery system on shale surfaces. Although this applies to all badlands, comparative studies in the Dinosaur badlands, Canada, and the Zin badlands, Israel, reveal basic differences between the 'arid' and 'humid' badlands. At Dinosaur where mean annual precipitation is 325 mm, once rill flow is initiated it progresses downstream, is continuous in time and occurs simultaneously in adjacent rills. In the Zin badlands, with mean annual rainfall of 90 mm, rill flow during continuous simulated rainfall is intermittent and pulsatory, and usually does not occur simultaneously in adjacent rills. Although this may partly reflect regolith properties, it also suggests climatic control leading to more rapid flow response and better integration of rill flow and sediment delivery in more humid badlands.

The basic problem of rill initiation has been addressed by Savat and De Ploey (this volume) who suggest from an extensive literature review that:

(1) permanent rills most commonly develop on unconsolidated loess or silts
(2) rills typically form on slopes between 2 and 12 degrees, regardless of climate or regolith, and on steeper slopes more incised gullies are typical
(3) rills can start very close (<1m) to divides

In laboratory flume experiments rills formed on very wet cohesionless loess at Froude Numbers of 1.2 and in drier conditions between 2 and 3. Critical Froude Numbers were reached at slope angles of 2 - 4 degrees and did not vary appreciably with unit discharge. As changes in slope length or rainfall intensity affect unit discharge, but not the Froude Number, this explains the initiation of rills with very low discharge close to divides. If, as for rivers, Reynolds and Froude Numbers determine bedforms, undulations must be created by supercritical flow which will deepen to form rills, even on an initially smooth surface. Clearly irregular microtopography in the field may accelerate flow concentration and rill initiation.

The field data reviewed by Savat and De Ploey were mostly collected in cultivated or cleared areas of relatively low slope angle. Field observations from natural badlands

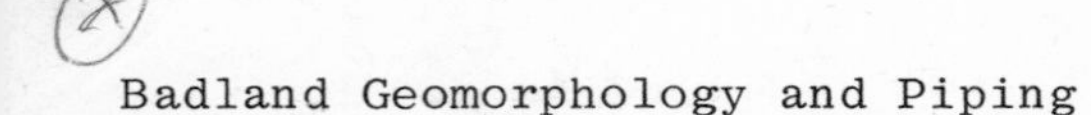

show that permanent shallow or incised rills which start close to divides persist on both shales and sandstones at slope angles of 45° and even, in some cases, near vertical. Slope angle cannot, therefore, always form a limiting factor in rill development. The hydraulic conditions governing formation and persistence of rills on very steep slopes have not yet been determined.

Pipeflow

Although tunnel erosion and piping were first described in semi-arid regions most recent reports concern piping in humid areas, and the geomorphic and hydrological role of pipeflow in arid and semi-arid badlands is still rather unknown. Virtually all the available experimental data come from humid upland peat moors of north-western Europe and extrapolation to semi-arid regions is very uncertain. The following synthesis with field observations from these regions is therefore tentative.

Piping can develop in many different materials and circumstances, the only essential requirement being juxtaposition of a steep hydraulic gradient and a steep free face. Many early technical descriptions come from inadequately designed earth-fill dams. Under natural conditions piping and the larger, more permanent forms of tunnel erosion apparently develop most readily where high infiltration capacities occur on material of low intrinsic permeability. This somewhat paradoxical combination allows ready, but concentrated, infiltration, yet prevents rapid lateral diffusion of water within the material. Suitable conditions are uncommon and are usually caused by intense desiccation cracking related to the high shrinkage capacity of materials such as montmorillonitic shales or *Sphagnum* peat. Pipe development is therefore encouraged by factors which enhance cracking, such as high sodium adsorption on clays, or marked seasonal variation in rainfall. Sometimes cracking is not related to shrinkage, and the large pipes and tunnels developed in glaciofluvial silts in British Columbia, Canada, appear to have been triggered by glacial unloading.

While tunnel erosion and piping are often associated with desiccation cracking, it is not essential. They can also be caused by subsurface flowline convergence, and engineering literature would distinguish between this as true piping and tunnel erosion related to cracking. Flowline convergence can occur in numerous circumstances, but in natural conditions it is frequently caused by surface or bedrock topography. Such topographically-controlled piping has been described from peats and many other regolith materials in humid areas, but no reports from arid or semi-arid areas have been found. In humid areas pipes are typically almost horizontal but in arid areas major vertical pipes can also be found, suggesting tectonic joint control and possibly solution processes.

The information available suggests that pipe and tunnel networks are extremely diverse. Individual pipes range from several millimetres (macropores) to about 3 metres in diameter and may be aligned at any angle from horizontal to vertical. The largest pipes are usually associated with an

indurated layer or cap-rock giving some protection collapse and assuring more prolonged development. saur badlands, for example, massive localized lent sandstone often forms the cap-rock. The absence of parable layer in comparatively homogeneous peat dep explain the apparent limiting diameter of about 30 cm. In any given regolith pipe size presumably reflects permanence and the amount of water passing through the pipe, and very large pipes are accordingly more widespread in semi-arid than in arid regions.

It is extremely difficult to map subsurface drainage networks and information on the morphometry and integration of pipe networks is still quite deficient. Observations from numerous field areas show that pipes may occur at any point on the slope, at one or several levels beneath the surface, and in diverse, complex combinations of size, shape and alignment. Few attempts have been made to relate pipe and surface drainage networks but initial experimental data from Dinosaur badlands indicate that although sometimes direct and immediate, the relationship is often obscure and remote.

The best available indication of pipe integration is the duration of pipeflow after rainfall. In semi-arid badlands such as those of Alberta and Saskatchewan, and in the pipe networks of the British peat uplands, prolonged duration of flow at low discharge indicates a high degree of integration. In more arid badlands, as in Israel, this is much less common, indicating poorer integration and probably higher absorption losses within the network. Climatic characteristics clearly influence flow persistence and network integration.

Frequency and duration of rainfall are obviously influential, and in temperate badlands snow accumulation and persistence may also be important. In Alberta, for example, snow sometimes persists in deep pipes until July, encouraging rapid and prolonged flow response even to quite minor rainstorms.

It is not yet possible to make a general statement on the hydrologic and geomorphic significance of piping in badland systems. It seems clear from visual impressions in heavily piped badlands that subsurface flow can provide the dominant denudational process in some badlands. In the Saskatchewan badlands Drew (this volume) has demonstrated the high geomorphic capacity of pipeflow, while experimental data from Dinosaur badlands indicate that in high intensity rainfall pipeflow is comparable in response time and erosional capacity to surface flow, and in low intensity rainfall may be both more frequent and more prolonged. All pipe flow in badland areas is ephemeral, but in the peat uplands of Britain where most hydrological data on pipeflow have been obtained, both ephemeral and perennial pipes occur. The hydrologic characteristics of these pipes vary considerably with examples of both rapid and greatly delayed response, and varied response from the same pipe depending on storm characteristics. In general, however, they are of considerable hydrologic significance and Jones (this volume) reports that some pipe systems can produce up to 60 per cent of direct runoff.

RATES OF BADLAND DEVELOPMENT

It is clear from the preceding discussion that badlands are much more diverse in their origins and development than early descriptions would indicate. It is not surprising therefore that the increasing volume of research in badland areas also demonstrates a considerable range in rates of development. Some badlands are currently undergoing rapid erosion and evolution. The Alberta badlands developed over an area of approximately 800 km^{-2} with a maximum incision of 100 metres during approximately 12-14 000 years since the last glacial recession. If an average incision of 50 metres is used, the calculated average rate of surface lowering is just over 4 mm yr^{-1}. Clearly such a figure has little precise meaning over such a large and diverse area, although it is remarkably close to rates of lowering observed experimentally in the Steveville and Dinosaur badlands in recent years. The chief significance is that several independent lines of enquiry point to a very rapid rate of erosion sustained over quite a long period of time. Comparable denudation rates are found in the badlands of south-eastern Colorado, and even higher rates have been recorded for short periods in the South Dakota badlands (17.93 mm yr^{-1}, Schumm, 1956b) and the Hong Kong badlands (17.36 mm yr^{-1}, Lam, 1977). Although such rates are very rare in 'natural' geomorphic systems they may be matched in rapidly eroding agricultural lands in the tropics.

Although badlands developed around the Mediterranean basin appear to be morphologically identical to some of the rapidly eroding badlands noted, several studies have indicated that they are eroding at very much lower rates. In the Zin badlands Yair *et al.*, (this volume), calculated long term (70 000 years) denudation rates of only 0.45 mm yr^{-1} while archaeological evidence from the Guadix badlands of southeastern Spain confirms the stability of apparently highly active badlands (Wise *et al.*, this volume). Although the figures quoted suggest a forty-fold discrepancy, the long term figures for Zin should not be compared directly with short-term rates for Dakota and Hong Kong. A more realistic comparison is with long-term rates for Alberta and Colorado which indicates a discrepancy closer to ten-fold. It is not yet possible to attempt a complete explanation of such discrepancies between areas that appear morphologically similar. Unquestionably climatic variations are important, higher erosion rates being found in semi-arid continental areas of higher latitudes and in the humid tropics than in semi-arid and arid Mediterranean areas. In continental high latitude areas fluvial processes during the summer typically dominate, but frost action in winter and slumping, mudflows, surface liquefaction and other flow processes engendered by snow melt are also very important. In humid tropical badlands fluviatile erosion is active at all seasons. In Mediterranean areas, in contrast, it is largely limited to the winter period during which long dry spells between storms increase the threshold precipitation necessary to generate runoff.

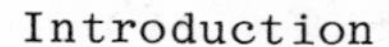

While broad climatic variations explain some of the discrepancies in erosion rates observed, differences in the lithological characteristics of bedrock and regolith also result in dramatic variations within one climatic region. This has been demonstrated particularly by experimental work in the Dinosaur badlands, Alberta, but also occurs in many other badland areas. In the Zin badlands, Israel, on the other hand, variations in erosion rates in areas of homogeneous lithology and macroclimate are determined by subtle variations in microclimate related to local small-scale topography.

Local topography is an important factor in all badland areas, which by definition have highly accidented relief. The effect on runoff generation and erosion is complex, however, and all recent experimental studies show that slope angle *per se* has little effect on runoff generation. In most badlands the most significant loci of rapid runoff generation are low angle depositional areas where compacted layers of silt provide much lower infiltration capacities than adjacent steep, cracked clay slopes.

Finally it is noteworthy that despite the extreme complexity of badland morphology, studies of badland geomorphology and hydrology at catchment scale indicate that badland catchments generally conform to partial and variable source area concepts of runoff generation. In humid areas the typical pattern of runoff generation in valley bottoms close to drainage lines is determined essentially by groundwater patterns. In badlands, the pattern of runoff generation is much more erratic, being controlled primarily by surface lithology. To the extent that depositional areas and pseudo-pediments tend to concentrate close to drainage lines, the resulting pattern tends to resemble the standard pattern for humid regions.

CONCLUSIONS

Although some badlands are comparatively simple others are extremely complex and, despite increasing attention from geomorphologists and hydrologists, still present formidable theoretical and technical problems. Unquestionably these will continue to stimulate attention to badland geomorphology. Inevitably, however, the significance of this focus will be questioned. Natural badlands are very widespread but they do not cover large areas and are in a sense geomorphic curiosities. It sometimes seems that the attention they have already received from geomorphologists is disproportionate. While many major geomorphic concepts can be applied to or tested in badlands, it is often difficult to apply the results of badland research to other, more areally significant, geomorphic systems.

Although the role of badland research in geomorphology may be questioned it has certainly gained new significance in land-use management. Several contributions to this volume stress the importance of human disturbance which can extend badland development far beyond its natural climatic or lithologic range. Any disturbance of vegetation over highly erodible soil or regolith can trigger intense, rapid dissection

and badland development. The most common causes have been ill-advised agricultural practices and, on an increasingly large scale, industrial atmospheric pollution.

Badlands triggered by agricultural practice can develop anywhere but appear to be most common in areas with marked seasonal climatic variations which cause soil cracking. They are particularly widespread in densely-settled, intensely-weathered clay-rich soils of Africa and the Mediterranean basin where prolonged excessive land-use has destroyed vegetation, exposing deep, highly-erodible soils. Many of these badlands are of recent origin and have not yet reached equilibrium, and accordingly show some of the most rapid development rates, but lithologically and morphologically they are virtually indistinguishable from natural badlands. Badlands triggered by industrial pollution are controlled by industrial location but often develop most rapidly in areas of comparatively high rainfall, sometimes on regolith materials which under natural circumstances are not vulnerable to such development.

While research on badland geomorphology is interesting and significant, it will become infinitely more important if the results can be used to predict areas of severe land degradation, their nature and rate of development, and to provide a basis for rehabilitation. It is hoped therefore that the collection of papers in this volume, by providing a current summary of the status of research in badland geomorphology will lead not only to advances in geomorphic research, but to significant benefits in land-use management.

2 The influence of material behavior on runoff initiation in the Dinosaur Badlands , Canada

W.K. Hodges and R.B. Bryan

Extensive badlands covering some 800km^2 flank the Red Deer river in Alberta, Canada, for the last 300km of its course above the Saskatchewan border, and form the major source contributing sediment to the river (Campbell, 1977a, 1977b). Various processes are involved in movement of sediment from badland slopes to valley flats, but during summer sediment delivery to the Red Deer invariably coincides with flow in ephemeral tributaries following sporadic rainstorms. The relationship between rainfall and sediment movement is complex, reflecting spatial and temporal rainfall fluctuations and the varied surface response (Bryan and Campbell, 1980).

The Dinosaur badlands lie in a semi-arid area with a mean annual moisture deficit of 600mm and mean annual precipitation of 330mm, 70% of which occurs between April and September. Annual variability is about 35% (Longley, 1968, Heywood, 1978) and spatial variability is extreme. Although no long-term rainfall records exist for the badlands, short-term observations (Bryan and Campbell, 1980; Hogg, 1978) show that most rain falls in isolated sporadic storms of rather low intensity. Since 1976, recorded intensities have reached 20-30mm hr^{-1} once or twice each year. Rainfall data from Brooks, 50km to the south, for 1965-76 provide the closest indices of intensity, duration and frequency (Table 1). Although the same storm rarely covers both areas, the rainfall pattern is probably broadly similar.

Badland development varies in character and intensity but reaches maximal expression in the Deadlodge Canyon secton of Dinosaur World Heritage Park (see Campbell, this volume) where the river is incised 100m below the prairie surface. The badlands are formed in Upper Cretaceous strata of the Belly River group which includes several marine and non-marine units but only the Oldman formation outcrops in the immediate area. This consists of intercalated lagoonal shales and sandstones of variable character. All shales are marked by desiccation cracking and most have shallow, moderately dense rill networks. The sandstones include highly indurated cross-bedded arkoses and friable muddy sandstones with thin shale or silt partings, most being densely rilled and, like the shales, piped. Resistant concretionary ironstone fragments add further diversity.

Table 2.1 Rainfall intensity, duration and frequency[1]: Brooks horticultural station

Storm duration	Average total rainfall (mm) per storm Return Period (years)						Return period rainfall rates as intensity (mm hr^{-1}) with 50% confidence limits Return Period (years)					
	2	5	10	25	50	100	2	5	10	25	50	100
5 min.	4.3	6.3	7.7	9.3	10.6	11.8	52.0±3.9	76.0±7.6	91.8±10.7	111.8±14.8	126.7±17.9	141.4±21.4
10 min.	6.4	9.6	11.7	14.4	16.3	18.3	38.3±3.1	57.5±6.1	70.1± 8.6	86.1±11.8	98.0±14.3	109.8±16.8
15 min.	8.2	13.3	16.7	21.0	24.2	27.3	32.8±3.3	53.3±6.5	66.8± 9.2	83.9±12.7	96.6±15.3	109.2±18.0
30 min.	11.8	19.9	25.2	32.0	37.0	42.0	23.5±2.6	39.7±5.1	50.4± 7.2	64.0±10.0	74.0±12.1	84.0±14.2
1 hr.	15.6	27.4	35.2	45.1	52.4	59.7	15.6±1.9	27.4±3.8	35.2± 5.3	45.1± 7.3	52.4± 8.8	59.7±10.4
2 hr.	19.8	32.2	40.5	50.9	58.6	66.3	9.9±1.0	16.1±2.0	20.2± 2.8	25.4± 3.9	29.3± 4.7	33.1± 5.4
6 hr.	27.0	37.9	45.0	54.1	60.9	67.6	4.5±0.3	6.3±0.3	7.5± 0.8	9.0± 1.1	10.2± 1.4	11.3± 1.6
12 hr.	31.4	45.6	55.0	66.9	75.7	84.4	2.6±0.2	3.8±0.2	4.6± 0.5	5.6± 0.7	6.3± 0.9	7.0± 1.0
24 hr.	36.9	58.0	71.9	89.5	102.6	115.6	1.5±0.1	2.4±0.3	3.0± 0.4	3.7± 0.5	4.3± 0.7	4.8± 0.8

[1]all data based on 11 year record (1965-1976). Source: Atmospheric Environment Service, Canadian Climate Centre, 1978.

The regional dip is 1-2° N.E., but detailed stratigraphy is complex with numerous lenticular beds. Parts of the surface retain grassed glacial till and alluvial valley flats are also vegetated. Elsewhere bedrock dominates the surface, and initiation of runoff and sediment movement is closely controlled by lithologic variations, as shown by simulated rainfall tests (Bryan, *et al.*, 1978). Sandstones responded with almost instantaneous Hortonian flow and rapid sediment entrainment from the complete catchment whereas shales showed delayed response with surface flow initiation primarily in rills. In general the tests suggested surface erosion on sandstones in virtually every rainstorm, but on shales only several times per year.

These results were unequivocal, but concerned only a limited lithologic range. An extended study was initiated to provide more detailed monitoring of response on a wider range of lithologies supported by laboratory study of the behaviour of selected materials.

Experimental design

Ten experimental microcatchments were selected, including four used in the initial study. These covered 33 lithologic units and examples of most small-scale topographic forms. Detailed geological and topographic maps were drawn (eg., Fig. 1, 2, and Hodges, this volume). Site and unit characteristics are shown in Table 2a, 2b.

Simulated rainstorm experiments (SST runs) were completed at all sites, using a simulator with two pole-mounted spray units (Plate 1) and a fall-height of 5m, giving an estimated 80% kinetic energy reproduction. Rain was applied at a constant rate, but wind turbulence caused some intensity variation (Table 3a, 3b). Due to wind conditions paired wet and dry antecedent moisture tests could be completed at only 8 sites. Average intensity was 29mm hr^{-1} and usual duration 30 min., conditions which recur in the area every 2-5 years (Table 1). Regolith moisture content was determined before and after each test, and in most cases at commencement of runoff on rills and interfluves. During tests, location, time and pattern of runoff initiation were observed. Hydraulic and sediment transport parameters were also measured (see Hodges, this volume).

Sixty-six samples were collected from different lithologic units for laboratory testing, which included particle size analysis, saturated hydraulic conductivity and permeability, Atterberg consistency limits, slaking, dispersion and vane shear strength. Clay mineral analysis by X-ray diffraction and cation adsorption by atomic spectrophotometer were completed for some samples. Some bulk samples were immersed for prolonged periods to determine saturated moisture holding capacity, and then dried by air and sunlamp, simulating natural conditions. Some were also tested under simulated rainfall to determine patterns and rates of crust development.

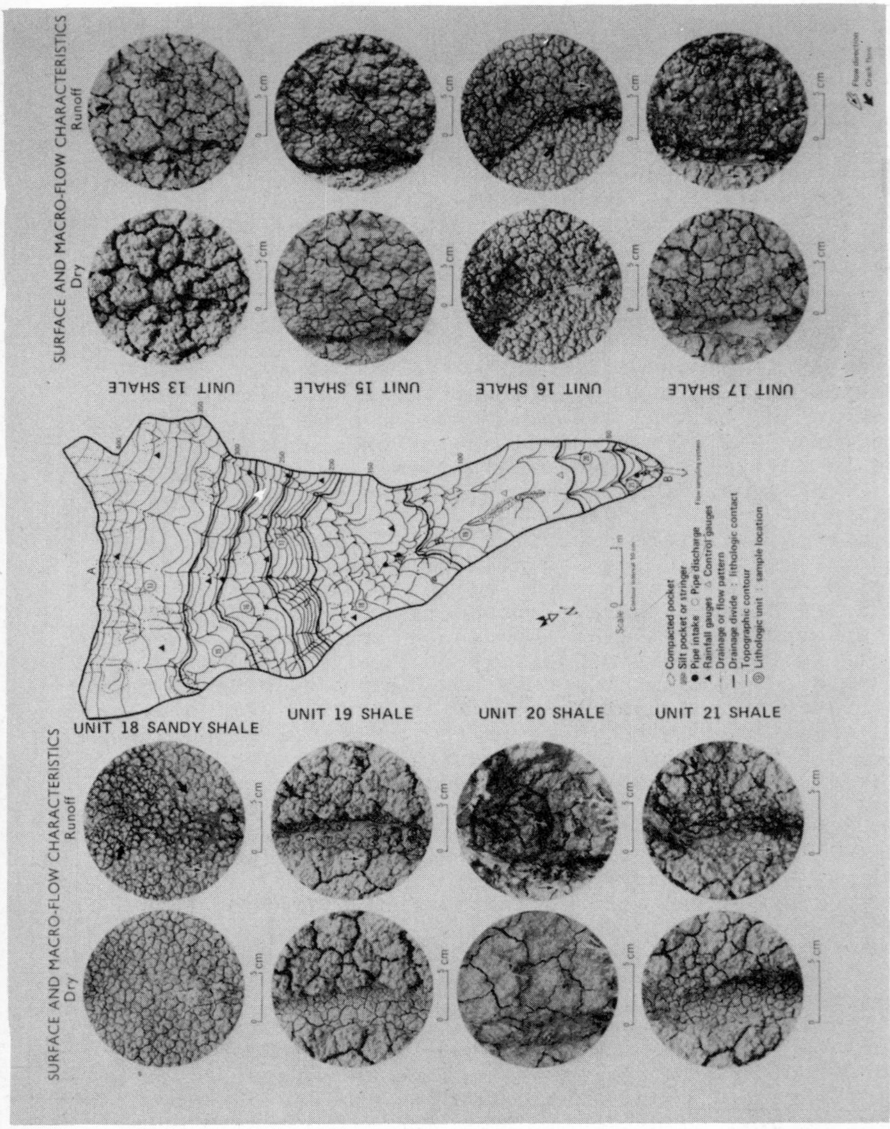

Figure 2.1. Catchment 6: surface characteristics

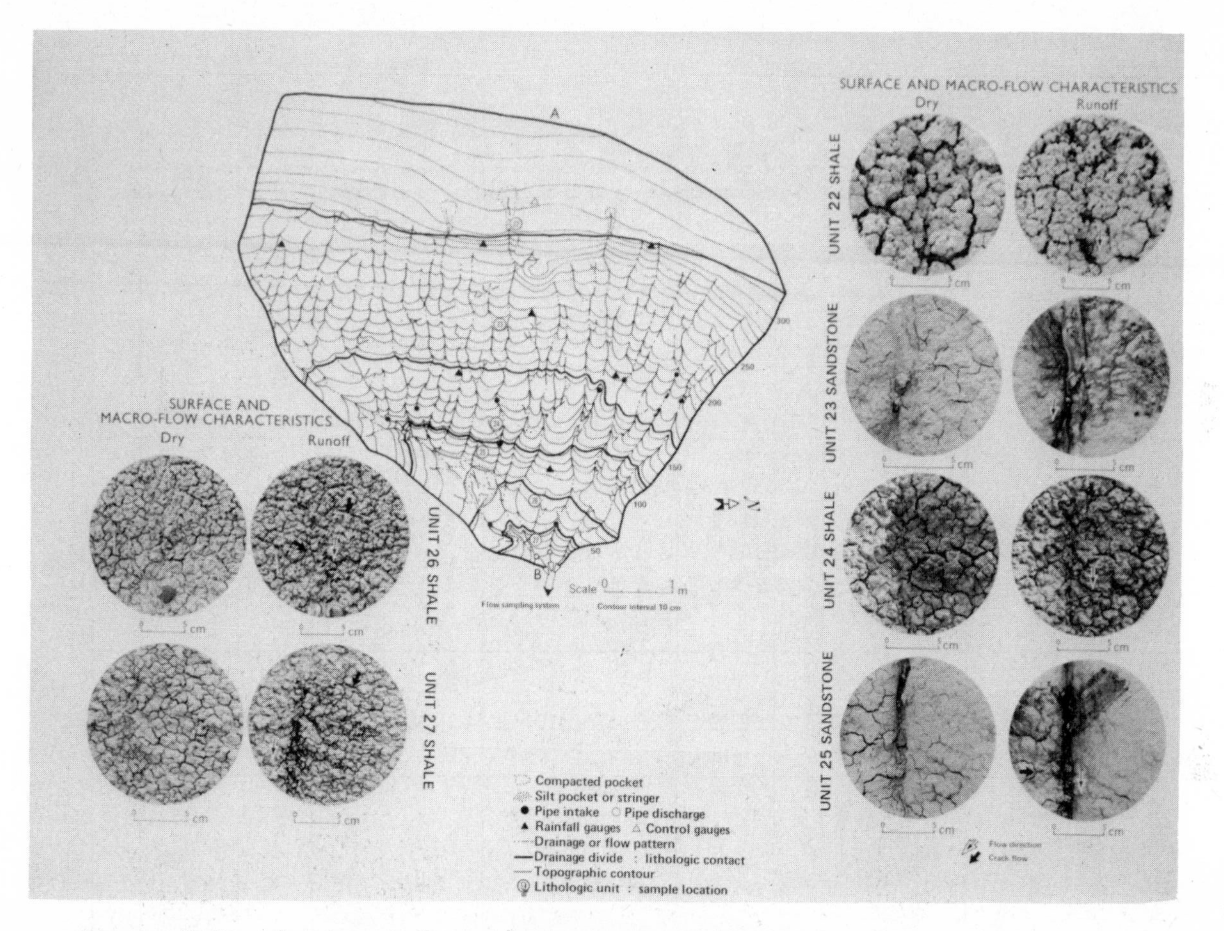

Figure 2.2. Catchment 7: surface characteristics

Table 2.2a Lithologic unit characteristics

Micro-Catchment No.	Unit No.	Lithology	Avg. Slope Angle (°)	Unit Thick-ness (m)
1	1	claystone	17.5	0.1
	2	sandyshale	21.0	0.5
	3	claystone	20.0	0.2
	4	sandstone	19.0	1.7
	5	claystone	52.0	0.4+
2	deb-ris	shale reg.	39.5	N.A.
3	6	shale	12.5	0.5+
	7	shale	8.5	0.4
	8	shale	23.0	0.2
	9	shale	40.0	0.4
4	10	shale	14.0	0.3+
	11	shale	6.5	1.0+
5	12	shale	7.5	0.6+
	13	shale	8.0	0.2
	14	sandstone	19.5	0.3
	14p	sandy-silt	4.5	0.3+
6	13	shale	29.5	1.2+
	15	shale	40.5	0.4
	16	shale	35.5	0.4
	17	shale	47.0	0.6
	18	shale	25.0	0.8
	19	shale	9.5	0.6
	20	shale	21.5	0.3
	21	shale	37.5	0.4
7	22	shale	16.5	0.8+
	23	sandstone	32.5	1.3
	24	shale	33.5	0.6
	25	sandstone	37.0	0.5
	26	shale	20.5	0.2
	27	shale	46.5	0.5
	28	shale	45.0	0.3+
8	29	shale	14.5	1.3+
	30	shale	25.5	1.2
	32	sandstone	N.A.	N.A.
9	29	shale	30.0	0.8+
	30	shale	27.5	0.6
	31	sandyshale	23.5	0.7
	32	sandstone	20.0	2.1
10	33	loam	22.5	N.A.

EXPLANATION:

1
s. = surface
sb. = subsurface

2
I. = interfluve
r. = rill

Regolith thickness[1] (cm)		Bulk density	Bearing Cap.[2] (kNm^{-2} $.10^3$)		Crack density
s.	sb.	$g\ cm^{-3}$	I.	r.	($cm\ cm^{-2}$)
N.A.	N.A.	N.A.	N.A.	N.A.	N.A.
1-2	N.A.	-	-	-	N.A.
0.5-1	N.A.	-	7.2	N.A.	-
0.5-1	N.A.	0.7	15.0	-	N.A.
N.A.	N.A.	N.A.	N.A.	N.A.	N.A.
5-8	20+	0.4	0.3	N.A.	0.7
2-3	9+	0.7	0.7	10.0	1.3
0.2-0.5	2-3	0.9	2.9	5.5	1.9
2-3	4-6	0.8	-	-	-
1-2	-	-	-	-	-
0.3-0.5	2+	0.4	0.3	N.A.	1.0
1.5-2	13+	0.5	1.0	7.9	1.1
1	1-1.5	0.7	0.8	4.4	1.3
1.5-2.5	3+	0.6	0.4	6.9	1.4
0.1-0.5	N.A.	-	12.0	15.9	1.6
0-30+	N.A.	1.4	2.8	N.A.	N.A.
1.5-2	5-7	0.9	0.8	4.2	1.4
1-1.5	N.A.	-	0.9	4.7	1.3
0.5-0.8	N.A.	0.8	10.2	13.7	2.0
1-1.5	N.A.	0.8	0.3	3.6	1.2
0.3-0.8	2+	1.2	6.5	11.5	2.6
2-3	5-7	1.2	1.9	8.5	0.8
0.5-1	3-4	0.7	0.2	3.5	0.6
2-3	3-5	1.5	5.1	7.8	0.8
2-4	15+	0.6	0.1	-	1.7
0.2-0.5	N.A.	-	11.1	24.2	1.2
0.5-0.8	N.A.	0.4	6.3	10.1	1.3
0.2-1	N.A.	-	21.8	26.4	-
0.5-0.8	N.A.	0.8	2.2	14.8	1.8
0.5-1	N.A.	1.1	3.1	14.0	2.3
N.A.	N.A.	N.A.	N.A.	N.A.	N.A.
1.5-3	6-15	0.8	3.1	6.5	1.0
0.5-1.5	3-5	1.1	0.6	8.8	1.3
N.A.	N.A.	N.A.	N.A.	N.A.	N.A.
1.5-3	6-15	-	-	-	-
1-2	3-5	1.0	0.7	13.0	1.6
0.2-1.5	N.A.	0.7	16.0	26.7	1.7
0.2-1	N.A.	1.6	17.0	30.4	-
0.1-0.3	-	-	0.4	N.A.	N.A.

Table 2.2b Experimental catchment and simulated rainstorm test characteristics

Micro-Catchment No.	No. of Units	Slope Aspect	Drainage Area (M^2)	SST Duration (hr)		SST Intensity (mm hr^{-1})	
				dry	wet	dry	wet
1	5	N.15°E	20.1	0.58	N.A.	40.7	N.A.
2	1	S.65°W	15.7	0.62	0.53	24.0	14.0
3	4	N.	36.8	0.46	0.47	28.6	28.2
4	2	S.35°W	44.3	0.44	0.44	20.2	25.5
5	4	N.32°W	31.5	0.42	0.45	26.5	24.2
6	8	N.12°E	26.0	0.41	N.A.	21.9	N.A.
7	7	S.87°E	34.2	0.50	0.40	19.3	40.4
8	3	N.32°W	10.6	0.30	0.38	47.0	34.8
9	4	N.60°E	30.2	0.23	0.46	23.7	23.4
10	1	N.84°E	14.9	0.42	0.38	38.0	43.5

Plate 2.1 Experimental instrumentation, catchment 7

Surface and material properties

Most surfaces fall into two classes, governed by bed rock characteristics. Shales typically develop a shallow regolith of one or two layers above fractured bed-rock 'shards'. The surface layer may consist of loose, puffy aggregates ('popcorn') or of a compact layer broken by desiccation cracks (Plates 2a, 2c), while the subsurface varies from a dense amorphous crust to 'welded' shards. Sandstones develop no regolith, but only a thin weathering rind (Plate 2d). Remaining surface types include surfaces armoured by ironstone fragments, alluvial fans and pseudo-pediments, and depositional sites stabilized by vegetation.

Most surfaces show rill development but its character and permanence is very variable. On shale debris slopes and on silty alluvial fans the development is ephemeral and changing, while on sandstones, rills are dense and deeply incised. Many units are also piped in a complex manner, but pipe development varies, apparently responding to subtle changes in lithology.

Basic properties of each unit are shown in Table 2a. Most shale units are fine grained with an average sand content in surface layers of 7.4% slightly higher than both subsurface and bedrock. The few sandstone units studied all had a rather high clay content, averaging 22.6%. Virtually all the 21 units analyzed for clay mineral content showed strong domination by montmorillonite, except unit 33 from vegetated catchment 10 which contains little clay (12.7%), and approximately equal intermixture of chlorite, illite and quartz. Duplicate analysis was completed for several samples but no other departure from montmorillonite dominance appeared; this was also true of the one unit (6) for which the complete regolith was analyzed.

Cation adsorption analysis shows dominance by sodium and calcium in the few units tested:

	Cation adsorption Meq/100g			
	Ca	Na	Mg	K
Catchment 3 Unit 6				
surface	26.25	13.98	6.0	1.92
subsurface	31.25	11.83	6.0	1.28
shard	28.75	12.90	6.0	1.28
Catchment 5 Unit 14p	11.25	4.30	2.0	0.64
sandstone	17.50	10.75	4.0	0.64

Bulk density (Table 2a) also changes with lithology and surface type, with the greatest variations occurring on the shales, due to differences in compaction, regolith thickness and cracking density. Although data are available only for surface layers, density clearly increases beneath the surface, approaching the bedrock.

Table 2.3a Simulated rainstorm, 'dry' antecedent test, moisture parameters

Catchment No.	Unit	Total rain: month before SST dry (mm)	Total rain in previous storm (mm)	Drying time to SST run (hr)
1	2 3 4	37.5	20.5	96
2	deb.	24.0	12.0	96
3	6 7 8	34.5	3.0	144
4	10 11	25.5	8.0	360
	12 13 14 14p	33.5	8.0	96
6	13 15 16 17 18 19 20 21	32.0	20.5	168
7	22 23 24 25 26 27	31.0	16.5	72
8	29 30	27.0	14.0	96
9	29 30 31 32	32.0	3.0	48
10	33	34.5	1.5	72

Material behaviour during and following rainstorms

Water transfer

Although clay-rich soils and regolith can seal quickly and become impermeable (McIntyre, 1958) this does not necessarily occur, particularly in badland clays which often retain high infiltration capacity (Schumm, 1956b, Yair *et al.*, 1980). This behaviour is governed by desiccation cracking, and material absorption characteristics, which also affect

Moisture conditions, SST dry runs (θdw, %)							
Initial			Runoff		Final		
s	sb	r	s	r	s	sb	r
7.2	8.1	N.A.	26.2	N.A.	23.3	8.0	N.A.
-	-	N.A.	-	N.A.	-	-	N.A.
6.0	7.2	N.A.	23.7	N.A.	22.2	8.2	N.A.
9.2	12.0	N.A.	-	N.A.	76.7	14.6	N.A.
8.2	8.9	7.1	86.5	77.5	75.9	27.2	75.5
8.3	9.2	9.1	59.2	-	53.9	9.3	50.5
-	-	-	N.A.	-	-	-	-
5.7	7.1	N.A.	N.A.	N.A.	45.2	6.9	N.A.
5.1	5.5	5.0	80.8	71.5	63.3	7.6	64.0
5.8	6.5	3.3	N.A.	69.1	52.0	12.9	63.7
3.6	6.6	3.9	74.4	50.5	70.0	18.1	-
2.5	-	4.0	-	38.6	50.3	-	50.2
0.6	8.5	N.A.	36.0	N.A.	37.8	6.0	N.A.
7.3	12.8	-	-	49.7	45.0	18.5	-
5.6	7.3	-	-	24.7	36.4	7.8	-
4.7	5.0	-	-	41.0	30.6	8.4	-
5.4	6.6	-	-	28.8	31.7	7.3	-
5.6	7.5	-	-	40.1	36.6	9.5	-
5.6	9.6	5.6	-	48.9	44.3	13.7	38.7
6.1	11.3	-	-	41.9	35.9	10.3	-
6.1	8.9	-	-	41.6	39.9	11.3	-
14.0	28.3	18.2	46.6	-	61.8	32.8	43.8
2.4	N.A.	2.1	35.6	41.9	46.1	N.A.	39.0
4.2	7.2	6.0	38.1	36.0	42.5	8.4	38.6
1.5	N.A.	2.0	-	-	27.4	N.A.	36.6
2.6	5.8	2.2	N.A.	-	-	-	-
4.6	9.1	3.7	N.A.	-	-	-	-
11.2	20.1	9.7	N.A.	57.3	65.4	31.5	59.3
5.4	7.1	12.0	N.A.	43.8	49.3	34.2	40.2
-	-	-	N.A.	N.A.	-	-	-
8.0	7.1	6.8	N.A.	N.A.	44.3	6.8	42.7
7.0	7.7	8.0	26.6	-	43.6	5.4	16.5
4.5	N.A.	3.8	52.9	47.8	42.5	N.A.	38.9
3.0	3.5	N.A.	54.9	N.A.	30.6	17.0	N.A.

sealing rates, permeability and saturation capacity. Laboratory results (Table 4) show a wide range of saturation capacities (gravimetric basis) with most materials capable of holding more than 150% water, and some over 200%. Apart from sandstones most samples show positive relationship between clay content and saturation capacity, but changes in clay content or character are insufficient to explain the range of saturation capacities.

Most badland materials have low permeability when fully hydrated and when desiccation cracks are sealed, water

Table 2.3b Simulated rainstorm, 'wet' antecedent test, moisture parameters

Catchment	Unit	Total rain in previous storm + SST dry (mm)	Drying time to SST run (hr)	Moisture conditions, SST Wet runs (θdw.%)							
				Initial			Runoff		Final		
				s	sb	r	s	r	s	sb	r
	2										
1	3	N.A.	N.A.	N.A.	N.A.	N.A.	N.A.	N.A.	N.A.	N.A.	N.A.
	4										
2	deb.	15.9	24	23.1	10.9	N.A.	70.7	N.A.	50.8	11.1	N.A.
	6			43.9	35.3	31.0	56.6	40.0	62.8	40.4	51.5
3	7	36.2	72	26.1	13.8	19.2	49.0	47.4	44.2	20.7	53.7
	8			-	-	-	-	-	-	-	-
	10			10.7	11.0	N.A.	48.2	N.A.	38.8	10.3	N.A.
4		16.9	72								
	11			11.8	8.8	11.0	36.6	49.7	61.1	9.0	56.3
	12			9.0	10.3	11.1	85.9	74.8	81.0	20.5	75.0
	13			30.8	18.6	9.2	53.6	57.6	54.2	42.9	54.3
5	14	11.1	24	8.6	N.A.	6.9	-	49.7	52.8	N.A.	60.0
	14p			3.0	7.2	N.A.	32.7	N.A.	39.8	7.2	N.A.
	13										
	15										
	16										
	17										
6	18	N.A.	N.A.	N.A.	N.A.	N.A.	N.A.	N.A.	N.A.	N.A.	N.A.
	19										
	20										
	21										
	22			34.5	26.4	36.1	55.0	55.7	48.1	42.0	45.7
	23			5.5	N.A.	4.8	42.9	44.1	27.3	N.A.	34.1
	24			5.2	7.8	6.0	41.6	42.1	30.2	10.5	36.5
7	25	9.7	24	3.7	N.A.	14.2	45.7	50.2	33.4	N.A.	40.1
	26			4.9	6.9	4.7	31.0	42.3	47.7	9.4	33.8
	27			3.9	6.0	5.6	19.9	17.9	34.7	7.9	36.3
8	29	14.1	24	13.1	35.5	15.8	50.5	45.8	56.4	22.4	53.3
	30			13.9	16.7	15.8	48.5	49.8	57.5	32.8	51.5
	29			-	-	-	N.A.	N.A.	-	-	-
9	30	5.5	24	19.0	7.1	14.1	40.2	39.7	55.3	13.1	53.8
	31			16.8	1.8	13.3	41.5	36.5	47.9	7.6	43.7
	32			18.4	N.A.	15.4	36.9	41.1	37.8	N.A.	34.8
10	33	16.0	24	21.2	9.9	N.A.	45.0	N.A.	25.4	17.2	N.A.

transfer capacity is minimal. The prolonged wetting necessary to reach this state is probably rather rare. Even at lower moisture contents hydraulic discontinuities in the regolith greatly affect transfer patterns, particularly between the surface layer and denser subsurface crust. Saturated hydraulic conductivities measured in the laboratory with disturbed samples are typically substantially

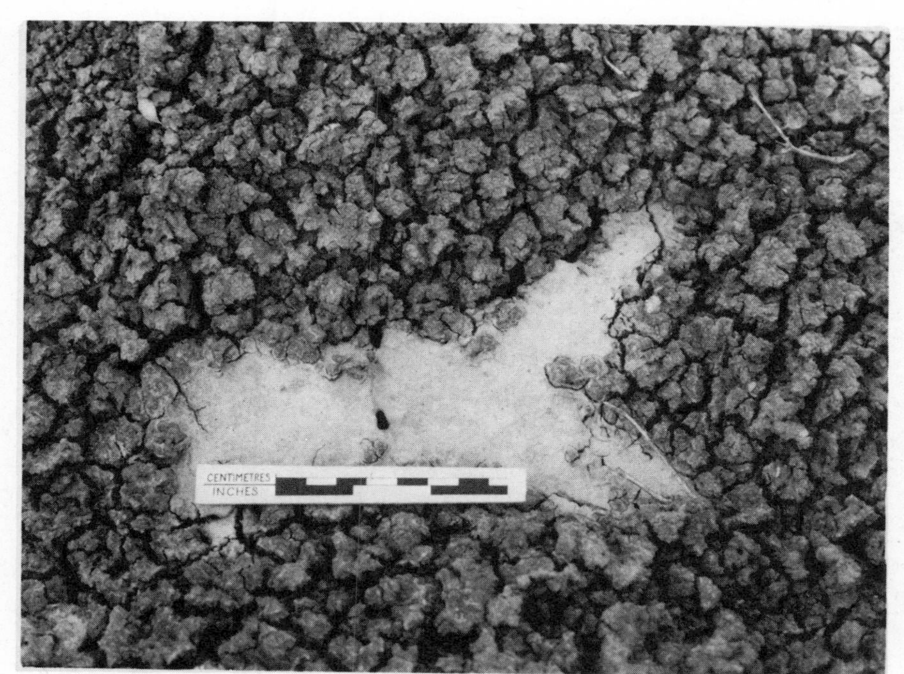

Plate 2.2a Shale regolith with 'popcorn' surface and compact silt pocket

Plate 2.2b Compact surface layer of shale regolith broken by coarse desiccation cracks: silt laminae on surface

Plate 2.2c Typical shale regolith profile illustrating popcorn, crust and bedrock shard

Plate 2.2d Raindrop pitted weathering rind developed on sandstone

Table 2.4 Water transfer characteristics of materials

Catch- ment No.	Unit	Saturation Capacity (% θdw)			Sat. hydraulic Conductivity (cms^{-1}.10^{-6})			Permeability (cm^2.10^{-10})		
		s	sb	r	s	sb	b	s	sb	b
1	2	-	86	N.A.	-	N.A.	6.2	-	N.A.	0.62
	3	-	-	N.A.	-	N.A.	-	-	N.A.	-
	4	131	-	N.A.	0.8	N.A.	-	0.08	N.A.	-
2	deb	-	146	N.A.	95.7	23.4	N.A.	9.80	2.40	N.A.
3	6	213	223	241	1.0	0.3	-	0.11	0.03	-
	7	-	172	-	0.3	1.7	-	0.03	0.17	-
	8	-	-	-	-	-	-	-	-	-
4	10	137	97	N.A.	-	-	-	-	-	-
	11	214	237	235	6.7	1.3	-	0.67	0.13	-
5	12	234	-	206	-	-	-	-	-	-
	13	-	194	-	-	8.5	-	-	0.85	-
	14	-	-	-	-	N.A.	-	-	N.A.	-
	14p	-	-	N.A.	22.1	-	N.A.	2.21	-	N.A.
6	13	153	-	-	6.9	4.3	-	0.69	0.43	-
	15	119	-	-	29.7	-	21.8	2.97	-	2.18
	16	163	-	-	2.4	-	14.0	0.24	-	1.40
	17	115	-	-	12.2	-	17.4	1.22	-	1.74
	18	-	183	-	0.3	0.8	-	0.03	0.08	-
	19	114	153	-	47.7	13.3	-	4.77	1.33	-
	20	-	77	-	899.4	1494.7	-	89.94	149.47	-
	21	116	189	-	30.9	16.3	-	3.09	1.63	-
7	22	130	237	-	-	6.6	-	-	0.66	-
	23	166	N.A.	-	-	N.A.	-	-	N.A.	-
	24	244	-	-	0.6	N.A.	0.7	0.06	N.A.	0.07
	25	-	N.A.	-	-	N.A.	-	-	N.A.	-
	26	234	-	-	-	N.A.	0.7	-	N.A.	0.07
	27	215	-	-	0.8	N.A.	-	0.08	N.A.	-
8	29	225	209	237	1.0	0.2	-	0.10	0.02	-
	30	248	214	-	-	1.0	-	-	0.10	-
9	29	-	-	-	-	-	-	-	-	-
	30	-	-	-	1.8	0.3	-	0.18	0.03	-
	31	198	N.A.	-	-	-	-	-	-	-
	32	190	N.A.	-	2.5	N.A.	-	0.25	N.A.	-
10	33	55	-	N.A.	-	2209.5	N.A.	-	220.95	N.A.

lower for the subsurface. The depths of both surface and subsurface layers are positively related to saturated hydraulic conductivity (figure 3) up to a threshold at about 100cm^{-1} 10^{-6}. Bryan *et al.*, (1978) and Yair *et al.*, (1980) showed a generic relationship between subsurface crust development and water penetration in montmorillonite-rich materials.

The data in Table 4 indicate very low water movement rates, and show that surface layers have limited capacity to transmit water during storm events. Most added water is

Table 2.5 Strength and stability of materials

Catchment No.	Unit	Atterberg Consistencies (% θdw)		
		Liquid Limit (Ll)	Plastic Limit (Lp)	Plasticity Index (Ip)
1	2	-	-	-
	3	-	-	-
	4	52.5	6.3	46.2
2	deb.	59.0	25.9	43.1
3	6	89.5	12.6	76.9
	7	78.0	12.6	65.4
	8	99.0	19.1	79.9
4	10	63.0	19.5	43.5
	11	107.0	22.0	85.0
5	12	93.5	7.0	86.5
	13	89.5	27.1	62.4
	14	84.5	8.6	75.9
	14p	27.0	21.4	5.6
6	13	76.0	19.3	56.4
	15	54.5	9.4	45.1
	16	63.0	20.5	42.5
	17	45.0	8.5	37.0
	18	92.5	11.5	81.0
	19	50.5	5.4	45.1
	20	36.5	17.6	18.9
	21	53.0	15.7	37.3
7	22	35.0	24.7	10.3
	23	76.0	16.3	59.7
	24	28.2	15.0	13.2
	25	-	-	-
	26	62.0	21.15	40.5
	27	81.5	26.1	55.4
8	29	110.5	33.6	76.9
	30	57.0	21.7	35.3
9	29	-	-	-
	30	83.0	23.1	59.9
	31	91.0	27.3	63.7
	32	79.5	20.3	59.2
10	33	22.0	19.3	2.7

EXPLANATION:

1. Vane shear strength at runoff during SST runs uses moisture content of material at incipient flow (see Table 3a-3b). Estimates for missing data use final moisture content.
2. Materials are:

   S = surface; sb = subsurface; b = bedrock; r = rill

Vane Shear Strength, τs[1] (kNm^{-2})				Disperson index[2] (% wt. loss/% wt. stable)			
Runoff antecedent dry	Runoff antecedent wet	At Ll	At Lp	s	sb	b	r
-	N.A.	-	-	0.09	N.A.	0.11	-
-	N.A.	-	-	-	N.A.	-	-
20.0	N.A.	7.0	115.0	0.07	N.A.	-	-
3.5	4.3	4.6	59.0	0.62	0.10	N.A.	N.A.
7.8	11.6	7.2	24.0	0.40	0.06	0.44	0.07
12.0	15.0	2.6	1000+	0.14	0.06	0.08	0.14
-	-	2.4	290.0	0.10	0.08	0.55	-
N.A.	18.0	14.0	160.0	0.38	0.24	0.19	N.A.
10.5	24.0	1.1	270.0	5.63	0.12	0.45	0.16
17.0	16.3	14.5	52.0	0.12	-	0.10	0.11
13.5	8.0	1.2	210.0	0.09	0.07	0.06	0.13
13.5	8.4	3.2	210.0	0.05	N.A.	-	0.14
4.8	5.6	8.0	12.0	5.43	N.A.	N.A.	N.A.
9.9	N.A.	2.3	250.0	0.10	0.05	0.29	0.10
71.0	N.A.	4.4	1000+	0.12	N.A.	0.60	-
5.6	N.A.	4.2	8.8	0.13	N.A.	0.09	-
16.2	N.A.	7.6	120.0	0.12	N.A.	0.12	-
13.5	N.A.	8.1	29.0	0.06	0.05	0.24	-
3.8	N.A.	3.3	1000+	0.10	0.05	0.73	-
8.2	N.A.	10.5	41.0	3.21	0.40	0.19	-
9.4	N.A.	4.8	150.0	0.06	0.04	0.08	-
19.0	7.5	32.0	71.0	0.14	-	-	-
8.4	8.0	3.8	31.0	0.04	N.A.	-	-
15.5	9.0	36.0	330.0	0.07	N.A.	0.06	-
-	-	-	-	-	-	-	-
-	18.0	10.5	51.0	0.25	N.A.	0.07	-
-	79.0	5.2	40.0	0.13	N.A.	0.06	-
12.0	28.0	1.7	58.0	0.13	0.13	0.15	0.10
35.0	31.5	28.0	70.0	0.10	0.08	0.45	0.08
N.A.	N.A.	-	-	-	-	-	-
N.A.	25.0	11.0	48.0	0.10	0.16	0.09	0.11
32.0	21.0	3.0	39.0	0.06	-	0.06	0.06
7.5	8.8	4.4	16.0	0.03	N.A.	-	-
2.0	2.5	5.2	6.1	-	-	N.A.	N.A.

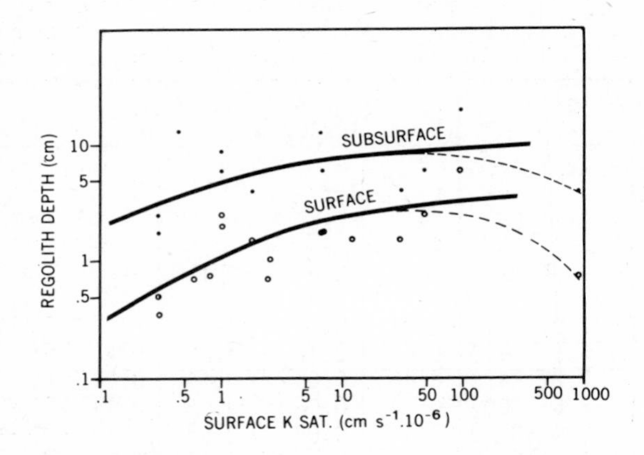

Figure 2.3 Regolith depth as a function of surface saturated hydraulic conductivity

taken up in hydration, swelling and sealing near the surface, so that subsurface layers are effectively inactive during rainstorms. This is not true where micropiping occurs, which can transmit water rapidly to the subsurface crust and shard layers (Bryan *et al.*, 1978). Although the functioning of micropipe systems is not yet fully understood, penetration of water into the main body of the crust through crack and micropipe walls is restricted and throughflow is thereby enhanced. At Dinosaur, micropipes are typically wetter than the surrounding regolith for a long time after rainfall. As most pipes are directly integrated with surface drainage nets much surface flow must originate in this manner. Micropipe flow frequently precedes surface runoff and may persist longer after rainfall.

Strength and stability changes on wetting

Badland materials show considerable changes in strength and stability on wetting which markedly affect the character and rate of geomorphic processes. These are shown by Atterberg consistency limits (Table 5); in general large differences between the limits reflect the high plasticity associated with high clay and montmorillonite contents. As the material passes from a plastic to a highly viscous liquid state infiltration capacity drops markedly and runoff may suddenly start. Disturbance of the regolith structure as the liquid limit is passed may also reduce pore space, generating positive pore water pressures and triggering instability (Bryan *et al.*, 1978). Consistency limits vary greatly even amongst units otherwise similar in character, suggesting considerable spatial variations in the initiation of some geomorphic processes, even on surfaces of apparently uniform lithology.

Material Behaviour and Runoff Initiation

Changes of state with increasing moisture content affect runoff generation and material resistance to entrainment. Coherent, massive surfaces like the sandstone and some shale units, resist fluid stress and raindrop impact through shear strength, but this drops markedly with increasing moisture content (Table 5). Curves relating shear strength and moisture content were drawn for all samples and by reference to simulated storm data used to estimate shear strength at runoff initiation. At the start of rainfall surfaces in the 'wet' antecedent moisture tests were invariably wetter and weaker, but the situation at runoff initiation was variable as runoff on some units started at lower moisture contents in 'wet' than in 'dry' tests.

When surfaces consist of loose aggregates and particles, like most shale units, different forms of resistance and entrainment are involved, at least until surfaces become fully hydrated and start to behave coherently. On wetting, aggregates are subjected to dispersive stresses which reflect the interaction of initial moisture content, wetting rate, and the amount and rate of swelling. All depend in part on rainstorm characteristics and in part on aggregate and surface fabric which changes markedly during wetting, particularly on some shales. Field observations using micrometer gauges showed high swelling, on 'popcorn' surfaces, but negligible change on compact shale surfaces. More detailed swelling tests in the laboratory were unsuccessful, due to structural disruption during sample transport and preparation, or interaction between the sample and container wall. Tests with individual aggregates showed swelling ranging from 10 to 40% of dry volume.

Although the dispersive stresses associated with wetting are easily distinguished conceptually, measurement and assessment of relative significance in the field during rainstorms is much more difficult. Because of climatic conditions slaking stress is clearly significant. Slaking (Yoder, 1936) results from air compression ahead of a wetting front, which can burst aggregates. It is maximal where wetting is rapid and aggregated material is dry. Although rainfall in Dinosaur is typically of moderate intensity, most falls on rather dry surfaces. Rainstorm patterns are erratic, but Bryan and Campbell (1980) showed an average of 4.7 days between rainstorms in 1968 and 3.9 in 1976. A drying time of approximately 96 hours is enough to reduce moisture contents to very low levels, typically ranging from 2 to 8% (Table 3A, figure 4b). In these circumstances slaking at the start of rainfall should approach a maximum.

Swelling stress is most effective in dispersion when it is differential. This may be caused by wetting from one direction, which is typically the case during rainstorms. Field observations in Dinosaur show that this can be sufficiently disruptive to cause widespread surface instability and mass movement on steep slopes. Differential swelling may also be associated with material of heterogeneous mineralogy, which is probably of less importance in Dinosaur because of the dominance of montmorillonite, but substantial variations in swelling rates and amounts can still be associated with the intermixture of coarse particles with clay, or with regolith chemistry.

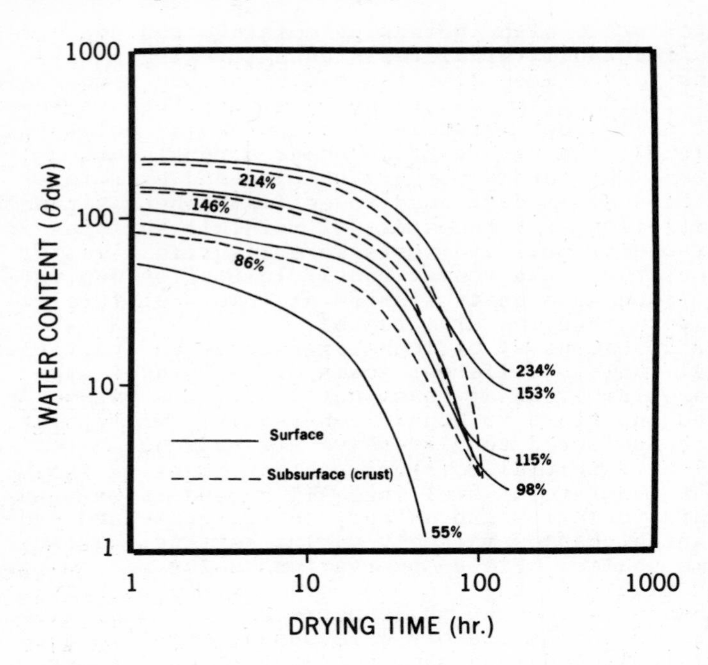

Figure 2.4a Summary of drying experiments for materials at specific saturation capacities

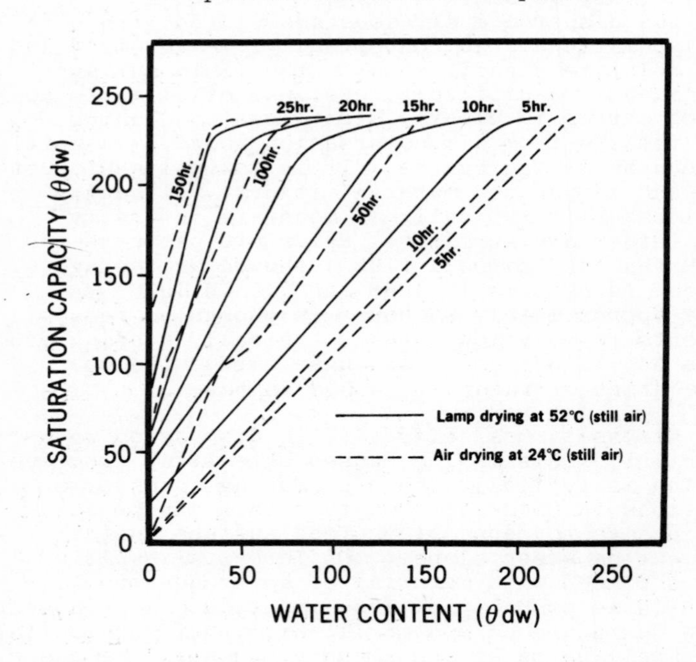

Figure 2.4b Drying rates for materials of different saturation capacities

The effect of adsorbed cations on swelling has been shown in numerous studies (cited in e.g. Baver *et al.*, 1972), and high sodium adsorption, in particular, is associated with high swelling capacity (Arulanadan *et al.*, 1975). The limited data available show high, though variable sodium adsorption. The varied swelling noted in field measurements is presumably related, at least partially, to differences in adsorbed sodium, and these would also indicate some dispersion related to differential swelling.

A related dispersive process of potentially great significance is peptisation where high exchangeable sodium percentages are associated with low electrolyte concentrations in the regolith solution (Imeson *et al.*, this volume). Available data are insufficient to judge its significance in the Dinosaur badlands, but the character of the system is such that sudden reduction of electrolyte concentrations caused by rainfall could destabilize aggregates from units with high adsorbed sodium percentages. The behaviour of the weathering rind on sandstone units which loses strength and disperses almost instantly on wetting is probably caused by peptisation.

Liquefaction has been observed on some shale units but the precise mechanism is not clear. As the moisture content increases following rainfall a point is reached (presumably close to the liquid limit) at which the complete aggregated structure starts to collapse. This must be associated with reduction in pore space, and probably positive pore water pressures, greatly reducing the surface resistance to fluid stress or raindrop impact.

The collective effect of different dispersive stresses is alteration of the regolith surface which strongly influences runoff generation. As material disperses and falls into desiccation cracks (Plate 3a) the surface becomes a small-scale layer of 'intra-pedonic turbation' (Yaalon and Kalmar, 1978). Relatively impermeable micro-depressions are formed and selective sediment transport leads to silt and fine sand 'pockets' (Plate 2a). After prolonged wetting shale surfaces tend to be hydraulically smoothed, but the sandstone weathering rind increases in roughness due to raindrop impact (Hodges, this volume).

Laboratory tests to determine slaking and dispersion were carried out for most units. These involved flood-wetting samples on a 1.00mm sieve in distilled water for one minute and recording the ratio of material loss: retention as a dispersion index. Because of rapid wetting and low electrolyte concentrations this would overestimate field dispersion but nevertheless some interesting relationships emerged. The subsurface crust shows lower dispersion than either interfluve or rill surface layers, or than bedrock. The crust is formed partly by physical densification during wetting: drying cycles, but the low dispersion may also reflect lower subsurface exchangeable sodium percentages as shown by data for unit 6 above. As a relatively stable feature the crust affects the depth and intensity of 'popcorn' surface layer development, thin surface layers being associated with dense resistant crusts. This in turn affects infiltration characteristics and runoff generation. High dispersion indices are also related to low surface bearing

Plate 2.3a Deposition of silts and fine sands within desiccation cracks

Plate 2.3b Silt stringers deposited in rills

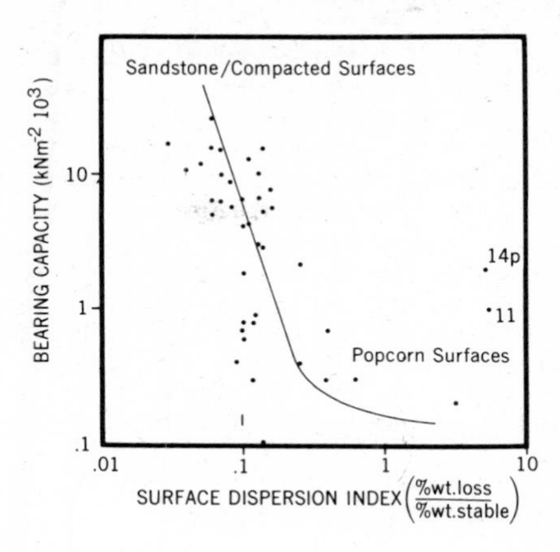

Figure 2.5 Bearing capacity as a function of surface dispersion index

capacities as shown in figure 5. If the surface is not disturbed this may be insignificant, but if disturbance occurs, compact 'pockets' form which fill with silt or fine sand and become preferred loci for runoff initiation (Bryan *et al.*, 1978). Once compacted areas are formed, either by subsurface crust development or by disturbance of a wet 'popcorn' surface, the compacted depressions are difficult to disperse or erode and become quasi-permanent landscape features, controlling runoff initiation and rill development (Plate 3b).

Behaviour of materials during drying

The data discussed show that resistance to dispersion is minimal in dry materials, while shear strength is minimal at very high moisture contents. This accords with Grissinger and Asmussen's (1963) observation that maximum stability of cohesive materials occurs between 10 and 25 per cent moisture content, and Grissinger's (1966, 1972) observations on the relationship between stability, aging and antecedent moisture content. Resistance to entrainment during storm events is therefore closely related to behaviour on drying and to the interval between storms. The samples used in saturation capacity tests (Table 4) were also subjected to sequences of air and infrared lamp drying to simulate diurnal drying patterns. The results are summarized in drying curves in figure 4a. Starting from the ordinate with saturation capacity, intersection with the appropriate lamp or air drying curve allows determination of the moisture content of any sample after a given drying period, on the abscissa. So a sample with saturation capacity of 150% will dry to 115% in 5 hrs. by lamp drying or 145% in 5 hrs. by air drying. Both rates are lower than in the field as wind influence was not included. Figure 4b shows the combined effects of lamp and air drying for surface and subsurface samples from selected units. These show that most materials with saturation capacities above 100% take between 70 and 100 hours to dry to about 10% moisture content.

The data presented show that when rainstorms are separated by more than about four days, antecedent moisture contents will be similar, regardless of lithology. When storms occur at shorter intervals there will be major differences in moisture content which will affect all controls on erosive effectivenss including infiltration capacity, runoff generation, surface dispersion and material shear strength. The interval between storms is therefore a critical factor controlling their erosive capacity and, therefore, their geomorphic significance.

Evolution of surface and profile features

The preceding discussion has shown the geomorphic significance of material behaviour in response to wetting and drying, and the importance of differences in surface and profile characteristics. Little research has been carried out on the evolution of regolith profiles in badland areas, but the data presented cast some light on the processes involved.

Crust development

Crusting on soils results from interaction of raindrop impact, dispersion, swelling, densification on drying, institial straining of fines and in some cases, clay seal formation (Duley, 1939, McIntyre 1958a, 1958b, Hillel, 1960, Tackett and Pearson, 1964, Hillel and Gardner, 1970, Bryan, 1973, Farres, 1978). It produces surface horizons which are denser and often less permeable than those beneath. In the Dinosaur badlands true surface crusting is confined to residual soils (e.g. unit 33), and the term 'crusting' as applied to shale regolith in the preceding discussion applies to subsurface crusts. These form by alteration of the shale 'shard' layers, which are fissile unloading features probably resulting from glacial overconsolidation. Crusting is always beneath a surface layer, usually 'popcorn', but in some cases no crust development occurs.

Surface soil and subsurface regolith crusts have some structural similarities and both cause decline in permeability, but whereas the soil crust reduces infiltration, the subsurface crust tends to cause water concentration at the surface: subsurface interface. All the crusting processes may be involved in subsurface crust formation, except raindrop compaction and clay seal formation, which are strictly surface phenomena. Observations during natural and simulated rainstorms and crusting experiments in the field and laboratory (Plate 4) show the following development sequence.

When rain starts, water initially infiltrates rapidly through desiccation cracks, but is held at the crust interface where few cracks penetrate. The cracks start to close from the bottom isolating the crust from the water supply, except where micropiping occurs. Crust wetting usually does not occur for several hours after rainfall. Penetration of

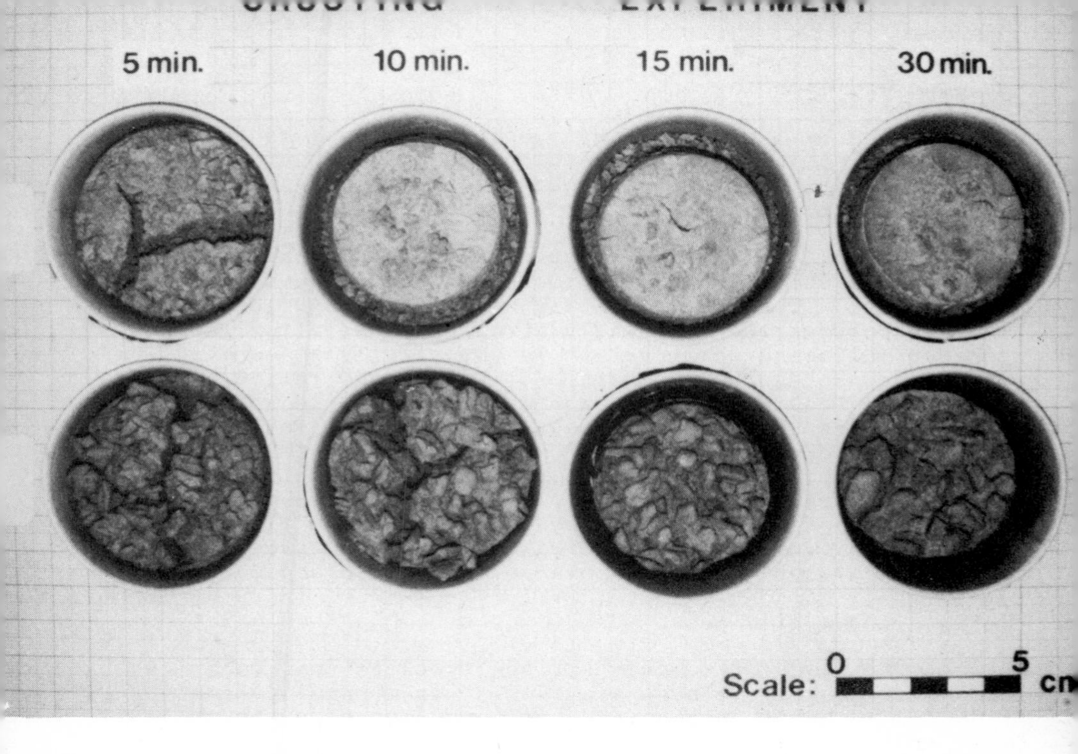

Plate 2.4 Laboratory crusting experiment with unit 6 (A) and unit 7 (B): changes for different durations at 50mm hr^{-1} rainfall intensity

the wetting front into the crust is slow and may persist for a long time (see also Yair *et al.*, 1980), varying with the gravitational potential of the wet layer and the combined matric and osmotic potential of the dry layer.

The rate of crust development depends on the amount and frequency of water penetration to the crust: shard interface. Adequate water to cause saturation is essential for slaking and dispersion of the shards. Comparison of field measurements of crust depth with dispersion and saturated hydraulic conductivity show generally positive relationships, but in the latter case this extends only to a certain 'equilibrium depth' which appears to represent the maximum depth to which water from the 'average' rainstorm can penetrate. Once crust development reaches this depth further saturation of the upper shards by rainfall must be infrequent and increase in crust depth negligible. At this stage any significant expansion is probably caused by snowmelt.

'Popcorn' and rind development

It is not clear from available data why some shale units develop a loose, puffy 'popcorn' surface and some a dense platelet structure. It seems that it must relate to colloidal characteristics but all the shales are clay-rich and dominated by montmorillonite, and no systemmatic variations in, for example, silt: clay ratios were found. The most likely controlling variable, currently being examined, is the exchangeable sodium percentage which, as pointed out above, affects the rate, amount and pattern of swelling and contraction. In some cases frost action may also contribute to structural evolution.

Although obviously coarser than the shale units, sandstone units still contain 20 to 25% clay, mainly sodium montmorillonite. The main cementing agent is calcium carbonate. The weathering rind apparently forms by dissolution and dispersion following wetting. It is essentially an equilibrium profile whose depth is governed by water penetration; during rainstorms sealing is rapid and wetting depth is about 3mm. This, combined with rapid removal of the dispersed sediment by runoff and rainsplash ensures that the weathering rind is very thin, though if it is ever stripped completely it must reform rapidly. In some places it reaches up to 15mm in depth; because of greater penetration due to microjointing, snowmelt (when penetration is not retarded by raindrop impact), or reduced removal of dispersed sediment.

Surface compaction

The significance of compact surface 'pockets' has been mentioned previously. These are similar in morphology and influence on runoff initiation, but have diverse origins. Some start as random depressions where water accumulates. These undergo enhanced dispersion and densification as well as deposition of transported silt and sand and, as the water film dries, formation of an oriented clay seal (Bryan, 1973). Pockets can also form by disturbance of weak saturated shale surfaces by mechanical loading, producing compaction and porewater expulsion. The results of loading by animal and human tracks has been observed (Plate 5a), but more frequent random loading is probably associated with minor landslides and slumps such as those shown in Plate 5b.

Plate 2.5a Compact silt pockets produced on shale surface by animal tracking

Plate 2.5b Extensive shallow slumping on steep (50°) shale slope resulting from, hydration, swelling and instability of a thin surface layer

Subsurface water flow along the surface: crust interface can also cause compaction. This was observed on several steep slopes such as plot 2. It causes liquefaction and instability (Plate 5c), followed by slumping or mudflows, which extends loading and disturbance downslope. In this way a compacted layer of oriented clay particles can produce linear features leading to rill formation. Initially these are fragile and vulnerable to seasonal disturbance, but ultimately they may become 'armoured' by silt deposition (Plate 3b). Schumm and Lusby (1963) noted incision of rills on Mancos shale in Colorado by summer runoff and their elimination by winter frost action. In Dinosaur a reversed sequence appears as many rills are initiated during snowmelt and eliminated by desiccation cracking and disturbance during the summer. Under average conditions recovery of the surface with full development of cracking is rapid, occurring within 3 to 4 days (Plate 5d), but recovery after more extreme rainfall can be prolonged.

Plate 2.5c Very liquid slump produced by 27 minutes of rainfall on Unit 2

Plate 2.5d Slump scar on Unit 2 after 24 hours drying

Vesicular alluvial fan surfaces (peri-pediments)

Silty miniature alluvial fans (peri-pediments) behave somewhat differently from the sandstone and shale surfaces described. They change imperceptibly between storms, but during storms, ephemeral rills are scoured, then eliminated by deposition in receding flow (Hodges, this volume). Immediately below the deposited material is a vesicular layer, which impedes water percolation and enhances runoff. Engelen (1973) suggested formation by ice crystal action in the Dakota badlands, but in fact vesicular layers are a common feature of reg soils in many deserts (e.g. Springer, 1958) and result from air entrappment during rapid sedimentation, rather than frost action. Vesicular layers recur cyclically with depth, and it may be possible to use them to identify the absolute incidence of runoff events required to form the alluvial fans.

Runoff generation

The simulation experiments showed several types of runoff occurring simultaneously due to lithologic differences. Overland flow includes crack flow on shales, and rill and sheetflow, primarily on sandstones and silts. Subsurface flow on shales may originate at the surface in crack flow or rillflow, or as interflow either at the top of crust (plot 2) or at the crust: shard interface. This is believed to occur extensively in most shale regolith, but is not readily observed unless it generates slumping, as on plot 2, or leads to micro-pipe flow. Pipeflow is widespread in the area and pipes are well-developed in shales, sandstones and alluvial fill, varying in diameter from millimetres up to about 3m. All the flow types can occur simultaneously or sequentially within a single microcatchment if appropriate lithologic variation occurs. As the lithology is very heterogeneous, this is commonly the situation, and the overall pattern of runoff generation is therefore extremely complex, particularly as rainfall is seldom uniform in incidence or intensity over more than a very restricted area.

Point generation of runoff is governed by complex and variable interaction of rainfall (or snowmelt) with surface and subsurface materials. The timing and contributing area for microcatchment runoff, however, is governed by the frequency and location of the most active units. These are the sandstones and peri-pediment silts which generate flow almost immediately after rainfall starts, regardless of antecedent moisture conditions, often less than one minute. Runoff from shale units is more variable, and can be delayed from 9 to 23 minutes on a dry surface, and 3 to 15 minutes on wet surfaces depending on lithology. Units with slow response can be influenced by discharge from adjacent upslope 'active' units which can promote sealing and runoff generation where it might not otherwise occur. Surface flow on shale units is generally confined to desiccation cracks (figure 1,2) or to compacted depressions and silt pockets. The high infiltration and storage capacity of 'popcorn' interfluves precludes runoff, except in extreme or prolonged rainstorms.

Badland Geomorphology and Piping

Depressions and silt pockets usually occur randomly and so therefore does runoff generation on shales. Subsurface flow on crust or shard layers requires saturation of much of the regolith profile and must be infrequent, occurring only during intense rainstorms, and with a lengthy lag time. Subsurface crack and micropipeflow, which does not require saturation, is more significant and appears to occur at least as frequently as surface flow. Mapping of the three-dimensional micropipe systems has not yet been successfully accomplished, but many appear to be integrated and carry a substantial portion of catchment discharge. Any given system may intersect the surface several times before the main surface drainage network is reached.

Changes in runoff parameters between rainstorms

The moisture regime of a lithologic unit or a complete microcatchment is the critical factor determining the incidence, timing and magnitude of runoff response to rainfall. This reflects both material behaviour during a rainstorm and antecedent moisture conditions, which show great temporal and spatial variations, in response to lithology, to topography and microtopography and to the location and characteristics of, and interval since, preceding rainfall. Moisture regime dynamics were assessed by monitoring natural and simulated rainstorms, by moisture sampling before, during and after simulation tests, and by laboratory analysis of drying characteristics (Table 3a, 3b, figure 4a, 4b).

The data in Table 3a show considerable differences in natural rainfall on study catchments. Most units except 22 and 29 were very dry prior to the 'dry antecedent' simulated rainstorms. Most units dried by at least 50% during the 24 hours between 'dry' and 'wet' tests (Table 3a, 3b) aided in part by the fact that in no case did final moisture contents in the 'dry' test approach saturation capacity. These rates agree closely with laboratory drying data (figure 4a, 4b). On some units, particularly sandstones and silts, there was little difference between initial moisture contents in the two tests.

Moisture regimes for 'wet' and 'dry' tests are shown in figure 6. Runoff generally occurs on sandstone or silt interfluves at 37% moisture content in 'dry' tests and 41% in 'wet' tests. Comparable figures for rills are 43 and 46%. These differences reflect variations in weathering rind thickness. Unrilled interfluves on shale generate runoff at 59% in 'dry' tests and 49% in 'wet' tests, while comparable figures for rills are 48% and 46%. Although individual units differ, these data indicate that the influence of antecedent moisture conditions on runoff generation is greatest on shales, particularly on interfluves where the critical threshold moisture content drops by almost 20% in wet antecedent conditions.

Comparison of data on moisture contents at runoff (Table 3a, 3b), saturation capacities (Table 4) and liquid limits (Table 5) indicates conditions at runoff. Runoff on most units is generated at moisture contents well below saturation capacity. This suggests that true Hortonian runoff is not common on these badlands slopes, but moisture

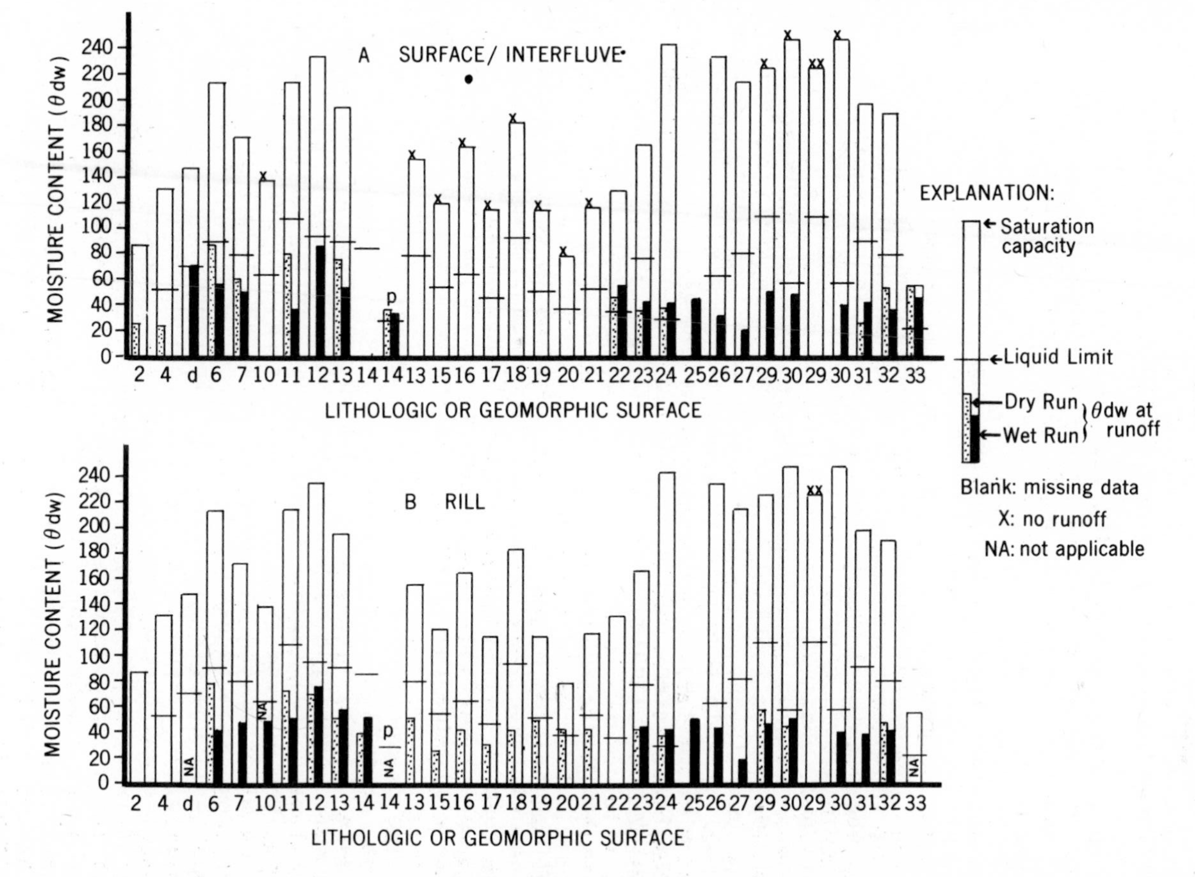

Figure 2.6 Comparison of moisture regimes at runoff for interfluves and rills

samples taken as runoff started were deeper than the immediate surface layer, and a very thin saturated layer may have existed at runoff. This was certainly the case on sandstone units where the wetted layer is often only 1mm and almost instantaneous Hortonian runoff occurs (Bryan *et al.*, 1978). In many cases runoff in the 'wet' test was generated at lower moisture contents than in the 'dry' test, and runoff on rills was generated at lower moisture contents than on interfluves.

In general liquid limits were reached neither at runoff nor by the end of the tests. The only exceptions in both 'wet' and 'dry' tests were units 14p; 22 33 and the debris unit of plot 2. Unit 33 is the vegetated glacial soil, and the data suggest that this would suffer rapid erosion if it were not stabilized by vegetation. On unit 14p, the alluvial fan, infiltration is impeded by the vesicular layer described. The heavy scouring observed (Hodges, this volume) is influenced by the achievement of liquid limit. Likewise the surface of the debris unit on plot 2 overlies an impermeable crust where lateral water flow occurs; again the achievement of liquid limit is reflected in instability and mudflows (Plate 5c). Unit 22, the crest unit on plot 7, is an extreme 'popcorn' unit on which the aggregated surface collapses and liquifies during rainfall. Apart from these units the data would suggest that units do not fail from liquefaction during rainfall. Again this must be taken reservedly for liquefaction of thin surface layers unquestionably occurs, perhaps due to fluid stress or raindrop impact at moisture contents slightly below liquid limit.

Stages in runoff generation

Runoff generation in the Dinosaur badlands is complex, and depends ultimately on lithologic variations. Sandstones and silts respond swiftly to rainfall with Hortonian flow from the complete surface, though flow may start on interfluves before rills due to differences in weathering rind characteristics.

The desiccation crack system is critical to runoff on shales, providing surface and subsurface micro-channel drainage nets (Haigh, 1978), which develop into pipes. The stages of runoff generation on shales are shown in figure 7. The initial stage (B) follows hydration and partial sealing by clays. The wetting front advances through the surface layer sealing minor cracks in compacted areas and generating rill flow. Some flow is diverted into pipes, with little loss through wall absorption because of impermeable clay linings. As rainfall continues, the wetting front penetrates to the surface: crust contact and accumulates to cause saturation (C). This reflects the generally lower hydraulic conductivity of the crust, though on some units, particularly those susceptible to piping, the crust hydraulic conductivity is actually higher. As the surface approaches saturation, aggregates swell and disperse, secondary cracks disappear and primary cracks start to close from the base.

Primary cracks may collect some seepage from the saturated zone at the surface: crust contact, but effective crackflow does not start until desiccation 'cells' seal and

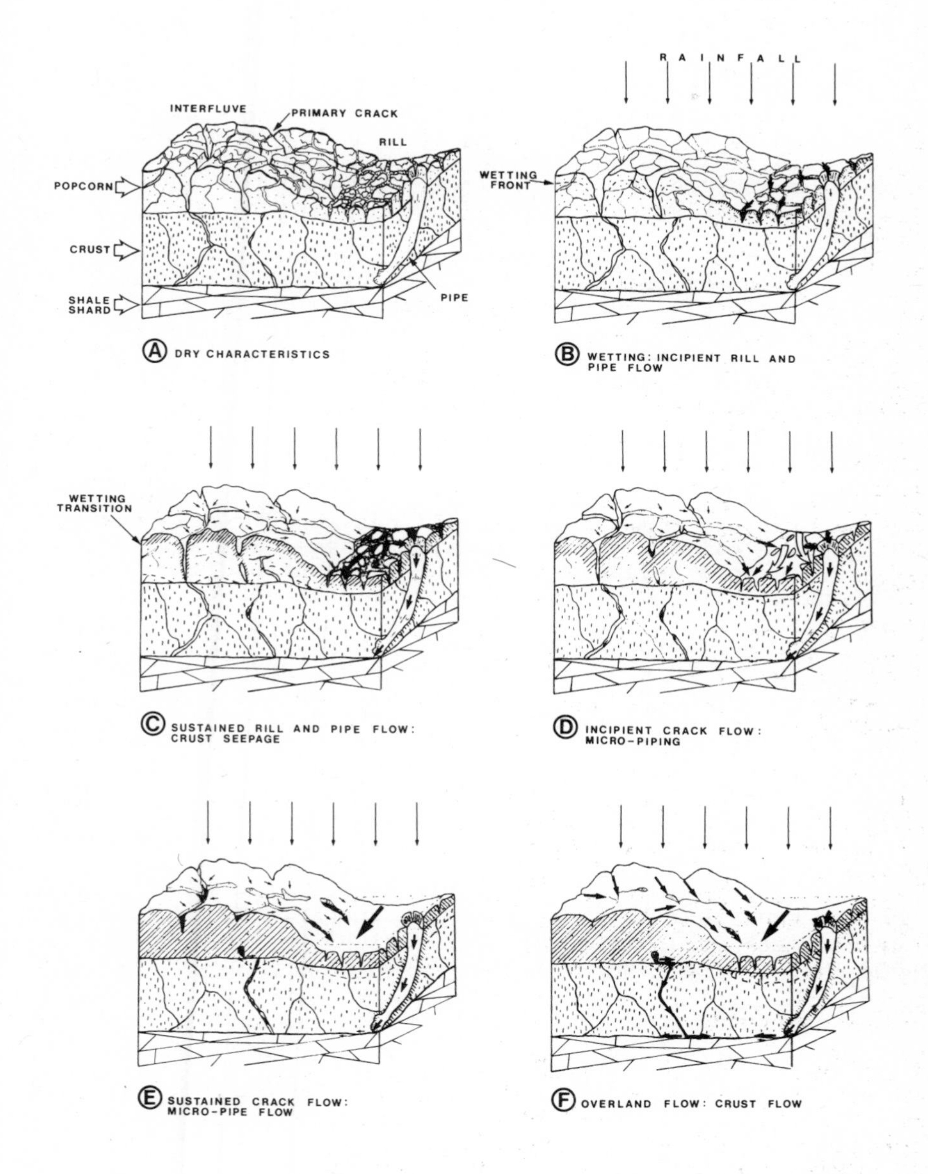

Figure 2.7 Stages in the initiation and development of runoff from shale slopes

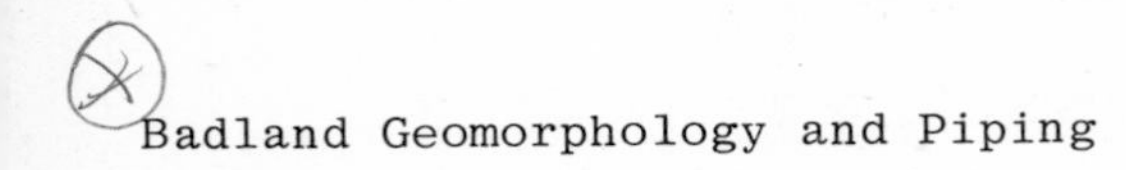

surface water flows into cracks (D). By this stage rill and pipeflow is sustained and discharge increases as the source area expands into 'popcorn' interfluves. Continued rainfall leads to flow from most of the surface, sealing of primary cracks (E), and true sheetflow begins to exceed confined channel crackflow. Within the crust, flow penetrates through cracks to the shard layer, with little absorption into the crust matrix, causing slaking and dispersion at the shard contact, and development of a second saturated zone.

If rainfall is intense or prolonged all surface fractures may seal, and rill flow reaches maximum discharge with the complete surface contributing (F). Pipeflow is also maximal with flow routed through pre-existing large pipes, ephemeral micropipes in the surface layer and through pipes developed by slaking and dispersion at the crust: shard interface.

The sequence outlined shows that true overland flow from virtually the complete surface occurs on shale units in extreme rainfall, which recurs only once or twice per year. Stages C to E are more frequent and rill and crackflow represent the most common response to rainfall. These cause high drainage densities in intermediate stages which decrease as true sheetflow is approached.

Conclusion

Observation of the behaviour of different lithologic units under natural and simulated rainfall has shown a more complex response pattern than is usually assumed for semi-arid badlands. Runoff development in the Dinosaur badlands has been compared with partial area and variable source area models (Bryan *et al.*, 1978) but with recognition of lithology as the dominant factor controlling contributing areas. More precisely runoff is controlled by the behaviour of surface materials and can expand or contract much more swiftly than envisaged in the standard variable source area model. Existing runoff models (reviewed in Kirkby, 1978) do not appear to be well-adapted to the complex, heterogeneous lithology of the Dinosaur badlands. During most rainstorms runoff comes almost entirely from sandstones and silts, but in extreme events, which appear to be of greater geomorphic significance, the varied response of shale units becomes critical. Development of a variable source area model to cover the behaviour of these units is complicated, involving a three-dimensional hierarchy of contributing areas which become fully integrated only in extreme rainfall events.

Acknowledgements

The field work in Dinosaur World Heritage Park was made possible by the Alberta Provincial Parks Branch, and the assistance of J. Stomp, Chief Ranger at Dinosaur Park, is particularly appreciated. C. Wilson-Hodges assisted with field work, and S. Kistoth, W. Leonard and R. Gorman with laboratory analysis. Research was funded by a grant to Professor Bryan from the National Sciences and Engineering Research Council, Canada.

3 The relationship of soil physical and chemical properties to the development of badlands in Morocco

A.C. Imeson, F.J.P.M. Kwaad and J.M. Verstraten

In the neighbourhood of Beni Boufrah, in the Rif Mountains of northeast Morocco (figure 1), three forms of gullying have produced contrasting badland areas. On old red colluvial non-saline/non-sodic soils on moderately steep north-facing slopes, V-shaped gullies have developed into badlands. Nearby two types of badlands have developed from U-shaped gullies, one on gently sloping marine deposits with very saline-sodic soils and another on fine, stratified wadi sediments, which are also partly saline and completely sodic, where gullies have been mainly formed by piping. This paper examines the influence of various soil properties on these different developments.

Many soil properties have been studied in attempt to explain various aspects of gully and pipe erosion (Heede, 1971; Crouch, 1976). Particular attention has been given to properties related to dispersion. Soils having high exchangeable sodium percentages (ESP) readily disperse in solutions with electrolyte concentrations below a certain threshold (Quirk and Schofield, 1955). Apart from the ESP and the related sodium adsorption ratio (SAR) of the soil solution, dispersion has also been investigated with a number of easily measured or calculated indices (e.g. Arulanandan and Heinzen, 1977; Crouch, 1978; Loveday and Pyle, 1973). Dispersed soil systems are characterised by high bulk densities, low permeabilities and infiltration rates, and high dry and low wet mechanical strength (Russell, 1973).

The chemical composition of the soil solution also influences the swelling and shrinking of the soil mass (e.g. McNeal *et al.*, 1966; McNeal, 1970; Rhoades and Merrill, 1976; Rowell, 1963, 1965; Rowell *et al.*, 1969). The importance of swelling for pipe erosion has been stressed by Parker (1963). In fact swelling and dispersion are closely related, dispersion generally being preceded by swelling.

Chemical conditions favouring swelling and/or dispersion are not essential for badland development. A highly erodible saprolite may be sufficient (Ireland *et al.*, 1936), particularly if it contains low amounts of free sesquioxides and organic matter. In such cases soil properties influencing the rate of water acceptance of a soil are important, particularly aggregate stability, since these determine whether

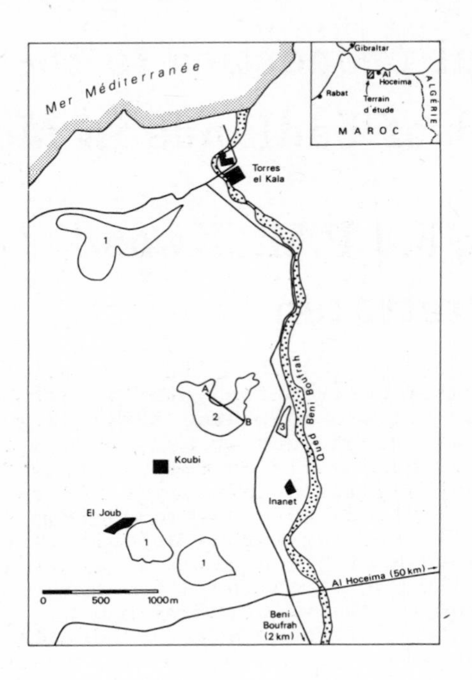

Figure 3.1 Location of the badland areas in the fieldwork area in northern Morocco. The numbers 1, 2 and 3 refer to the badland types described in the text. The line A-B gives the location of the slope profile shown in figure 4

sufficient runoff will be generated and concentrated to erode through more resistant surface horizons into the weak saprolite below (Ireland *et al.*, 1936).

Some of the soil properties related to gullying and badland development are more difficult to determine than commonly suggested or indicated. Consequently relationships with erosional phenomena occasionally appear contradictory. For example, in saline soils exchangeable sodium is analytically extremely difficult to determine so that the accuracy of ESP values reported when these are not related to SAR values in the saturation extracts of the soil, may sometimes not be very great. For this reason some details of the analytical procedures used are provided.

THE FIELD AREA

A number of physical and social factors make the Beni Boufrah region particularly susceptible to soil erosion. Long steep slopes have been produced by recent uplift. The flysch rocks weather rapidly producing silt-rich sediments and both these and former marine and wadi sediments found near the coast contain large amounts of soluble salts which in places lead to a deterioration of the soil structure. The climate, with a highly variable average annual precipitation of about 300 mm, results in too little organic matter being produced to stabilise and maintain the soil structure when the land is cultivated. Apart from the few remaining areas of maquis, almost the entire area is cultivated and exposed to erosion with virtually no cover between June and February. This includes steep slopes (up to 30°) which are tilled at least twice a year. Only the low rainfall erosivity and small runoff volumes prevent erosion from making cultivation impossible. Badland development is thought to have started in the 1920's, and it probably increased considerably in area in the 1950's when badly designed and maintained terraces were introduced to protect against sheet erosion and to conserve water for the establishment of almond trees.

The types of badlands in the Beni Boufrah area

The three different types of badland can be considered with respect to the simple classification of gully types proposed by Imeson and Kwaad (1980a), summarised in table 1. In this scheme, gullies are classified according to the erosion processes responsible for gully development, the materials in which the gullies are formed and their position in the landscape.

Badlands produced by V-shaped gullies on moderate to steep slopes with non-saline/non-sodic soils (type 1 badlands)

Badlands of this type (plate 1) are very extensive in this part of the Rif Mountains. They are composed of networks of V-shaped gullies cut into remnants of old colluvial red soils and flysch weathering products found on north-facing slopes. The red soils, which date from the Villafranchian, and the gullies associated with them, are most extensive in amphitheatre-shaped embayments. On south-facing slopes only remnants of a former cover of red non-saline/non-sodic, $EC_e < 4$ mS cm^{-1}, ESP < 15, soils are now found. The soil profiles developed in the red slope deposits are typified by profile 1 (Table 8).

The steep slopes on which these badlands have formed appear to have been cleared for cultivation only relatively recently (since 1920). There is little field evidence to indicate association with a soil environment which favours dispersion. Surface sealing and high soil erodibility, resulting partially from structural deterioration induced by cultivation and over-grazing, are considered important in their development.

Table 3.1 Sets of conditions characterising particular gully types (after Imeson and Kwaad, 1980a)

Gully type	Gully cross section	Position in landscape	Principal source of runoff
Type 1	V-shaped	Anywhere, except valley bottoms, where runoff becomes concentrated	Overland flow
Type 2	U-shaped	Anywhere in landscape except valley bottoms	Overland flow with a contribution of subsurface water of lesser importance; occasional seep caves at headcut
Type 3	U-shaped	Anywhere, but usually on pediments and gentle lower slopes	Subsurface flow predominates, as is apparent from piping
Type 4 (arroyos)	U-shaped	Valley bottoms	Overland flow, mainly from tributary gullies, and subsurface flow

Badlands formed by gullies in marine sediments (type 2 badlands)

These badlands (plate 2) are restricted to an area of saline-sodic soils (EC_e > 4 mS cm^{-1}, ESP > 15) near the village of Koubi (figure 1). The gullies are generally more U-shaped than plate 2 indicates and the white salt efflorescence seen in the channel bottom is only sometimes present. These soils occur both on and downslope of an outcrop of marine sediments presumably of Pliocene age. Samples were collected for analyses from the catena illustrated in figure 2, one profile of which is described in table 8. It is thought that badlands have developed on the relatively gentle slopes of this area due to the saline soil environment, and in contrast to the V-shaped badlands on the red soils, swelling and dispersion are important.

Material in which gully is developed	Favourable conditions
Relatively resistant weathering products of impermeable parent materials, B-horizons of deep soil profiles. The resistance of the material does not decrease with depth	Intense rainfall, poor soil structure, steep slopes, poorly built terraces and tracks
Relatively little resistant weathering products or slope deposits which do not increase in resistance with depth	Dispersive soil materials, sub-humid climate with pronounced wet and dry seasons
Weathering products and slope deposits as for type 2	Dispersive soil materials are essential; further as type 2
Alluvial and slope deposits	Semi-arid climate, lack of valley bottom vegetation, dispersive materials

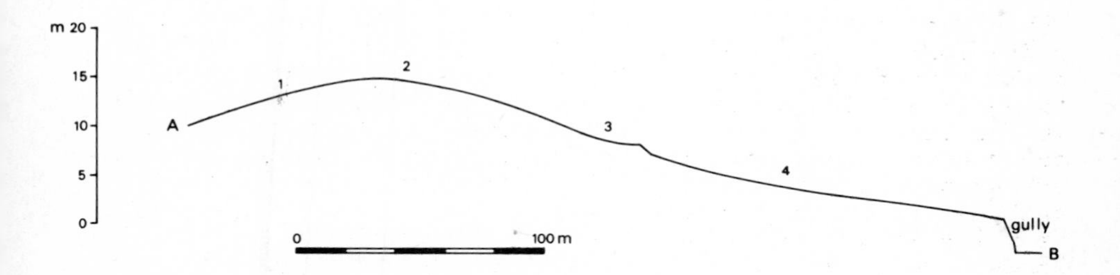

Figure 3.2 Locations of sampling points in the area of type 2 badlands described in the text

Plate 3.1 Badlands of type 1 near El Joub, Beni Boufrah, Morocco

Badlands formed by piping in wadi sediments (type 3 badlands)

These badlands (plate 3) are confined to a narrow zone of subrecent wadi deposits along the Beni Boufrah wadi south of Jnanet (figure 1). Unlike the soils on the former marine sediments, some soil horizons are not saline ($EC_e < 4$ mS cm^{-1}), but the entire soil is sodic (ESP > 15) and the parent material is texturally stratified (table 8). A relatively pure fine sand deposit above an impermeable subsoil and soil chemistry which promotes swelling and dispersion are thought to be important in the development of these badlands.

FIELD AND LABORATORY METHODS

Many soil profiles were described in the field according to the FAO Guidelines for soil profile description (1968). Soil moisture determinations were made with a calcium carbide soil moisture meter. To investigate the infiltration characteristics of field soils, a portable rainfall simulator was used to establish sorptivity values (Talsma, 1969; Imeson and Kwaad, 1980b), and the ponding times of rainfall of different intensities (Smith, 1972; Dunin, 1976). Amounts of rainfall required to pond soils were further estimated from

Plate 3.2 Badlands of type 2 in marine sediments near Koubi, Beni Boufrah, Morocco

the sorptivity measurements using a relationship described by Smith and Parlange (1978).

Routine laboratory analyses included grain size distribution (without a pre-treatment to remove carbonates), pH, carbonates, organic matter and saturated hydraulic conductivity. The pH was measured using a 1:2.5 soil 0.01 M $CaCl_2$ solution or H_2O ratios. Non-oriented powder specimens of the fraction < 2 μm were prepared and exposures obtained with a Guinier De Wolff camera using CoK α radiation. Quantities were estimated visually.

The resistance of aggregates to the impact of falling water drops was used as a measure of aggregate stability. For this purpose the procedure described by Low (1954) was slightly modified. The number of impacts from 2.7 mm water drops, allowed to fall 1 m, required to break 4-5 mm diameter aggregates sufficiently for them to pass through a 3 mm wire mesh was counted.

As a measure of dispersion a modified version (Jungerius and van der Wusten, 1980) of the Emerson (1967) test described by Loveday and Pyle (1973) was used. The soil erodibility factor (k) of the Universal Soil Loss Equation (Wischmeier and Smith, 1978) was estimated with the Wischmeier *et al.* (1971) nomogram. The procedures used to measure the shrinkage limit, shrinkage ratio and volumetric shrinkage were taken from Singh (1967).

Table 3.8 Soil profile descriptions

BADLANDS OF TYPE 1

Profile 5: classification, Calcic Luvisol (FAO-Unesco, 1974); profile dry throughout

Ap	0-15	2.5 YR 3/6 (dark reddish brown), slightly stony, silt loam, slightly sticky, slightly hard, weak coarse angular blocky to structureless, few fine pores, few fine roots, abrupt and smooth to
B2t	15-30	10 R 3/4 (dark red) silt loam, slightly sticky, slightly hard, moderate to strong fine to medium angular blocky, patchy cutans on ped faces or as lining in pores, common fine and medium pores, rounded caliche and fresh sandstone gravel, common fine roots, abrupt and smooth to
BCca	30-50+	5 YR 5/8 (bright reddish brown) silt, medium weak subangular blocky to structureless, parent material mainly powdery caliche, no gravel or stones, few fine roots

Profile 9: classification: Calcic Luvisol (FAO-Unesco, 1974); profile dry throughout
Similar red soil outside of badland area under maquis

A1	0-10	2.5 YR 2/3 (black reddish brown) gravelly and stony silt loam, weak angular blocky, soft to slightly hard, non sticky, few fine pores, frequent fine and medium roots, gradual and smooth to
B1t	10-20	5 YR 3/4 (reddish brown) clay loam, fine to medium strong angular blocky, hard, continuous cutans on ped faces, few fine pores, common fine and medium roots, gradual and smooth to
B2t	20-40+	2.5 YR 4/8 (reddish brown) silt loam, strong fine and medium angular blocky, hard, cutans on ped faces, few fine pores, few fine roots.

The cation exchange capacity (CEC) was determined by Polemio and Rhoades' (1977) method, which is very suitable for (non-saline) calcareous and gypsiferous soils and can also be used for saline soils if samples are initially prewashed with distilled water.

Exchangeable bases were determined according to Bower and Hatcher (1962), and also for some profiles, after washing the soil six times with a 60% ethanol solution of 0.4 N $LiNO_3$/0.1 N LiOac, pH 8.2. Calcium, magnesium, sodium, potassium, chloride and sulphate were determined in the extract. Exchangeable sodium and calcium were calculated by

BADLANDS OF TYPE 2

Profile 1: classification: Orthic Solonetz (FAO-Unesco, 1974); profile dry throughout

Ap	0-6	10 YR 7/4 to 6/4 (grey yellowish orange to strong yellowish brown) slightly gravelly, clay, very sticky, hard, strong 'coarse' crumb breaking to medium crumb, many fine and medium pores, no roots, abrupt and smooth to
B2tca	6-30	7.5 YR 4/4 to 4/6 (brown, clay, very firm, strong prismatic breaking to medium and coarse angular blocky, continuous cutans on ped faces, few to common fine pores, few to common diffuse light yellow brownish grey (10 YR 8/2) soft powdery calcareous concretions, no roots, gradual to
B3tca	30-50+	mottled 2.5 Y 4/4 (dark yellowish brown) 7.5 YR 4/4 to 4/6 (brown) and 5 Y 5/1 (yellowish grey) clay, moderately coarse angular blocky breaking to medium and fine angular blocky, firm to very firm, broken cutans on ped faces, common very fine pores, no roots

BADLANDS OF TYPE 3

Profile 7: classification: Orthic Solonchak (FAO/Unesco, 1974); profile dry throughout

1 (Ap)	0-10	5 YR 4/4 (reddish brown), gravelly loam, 'coarse' crumb, hard, many fine and medium pores, no roots, abrupt and smooth to
2	10-40	10 YR 5/6 (yellowish brown) sand, weak coarse angular blocky to structureless, few fine pores, no roots, abrupt and smooth to horizon 3
clay lens	13-14	2.5 Y 5/2 (yellowish brownish grey) angular blocky, many pores, no roots
3	40-90	7.5 YR 5/2 (brownish grey) common clear medium distinct reddish brown (r YR 4/8) nodules, silt loam, sticky hard, coarse weak prismatic to structureless, common fine pores, no roots, abrupt and smooth to
4	90-120+	consists of a laminated deposit of horizons 2 and 3, each lamina a few cm thick, a few roots.

---

correcting their extractable amounts with equivalent amounts of chloride and sulphate (Verstraten, 1980).

Water soluble salts were determined from saturated pastes made according to the procedure outlined in Richards (1954). ESP values were calculated from the practical SAR values (SARp) of the saturation extracts using the relationship also reported in Richards (1954). Real SAR values (SAR), corrected for ion pair association and ionic strength, were calculated with the computer program SOLMNEQ (Kharaka and Barnes, 1973).

Since ESP, SAR and SARp values are positively linearly related for values of ESP $\leq$ 40 (Richards, 1954; Sposito, 1977;

Plate 3.3 Badlands of type 3 associated with piping near Jnanet, Beni Boufrah, Morocco

Sposito and Mattigod, 1977; Oster and Sposito, 1980, and the analyses reported here) only SAR and SARp values will be referred to in the following discussion.

RESULTS

Differences in parent material mean that the results presented below usually fall into three groups corresponding to each badland area. Results are accordingly shown separately in figures and tables for each area.

The mechanical analyses (table 2) indicate that most soils are slightly gravelly with 5-15% > 2 mm. Only the deeper horizons of the partly saline-sodic soils (badlands 2 and 3) contain very little gravel. In all of the soils there is very little material coarser than fine sand. The soils in which the V-shaped gullies have developed (type 1 badlands) have fine earth fractions with textures in the clay, clay loam and loam classes. The saline-sodic soils have a variable textural composition reflecting contrasts in parent material. The most outstanding textural feature is differentiation in the soils affected by piping (profile 7), particularly the contrast between horizons 2 and 3 (table 8).

Table 3.2 Particle size distribution, $CaCO_3$, and humus content, pH and k factor (USLE) for soils from the three different badland areas

	profile number	horizon	% material[1] > 2 mm	material < 2 mm (%)[2] 0-2	2-50	50-2000 µm	$CaCO_3$[2] %	humus[2] %	pH H_2O	$CaCl_2$	k(USLE)
	5	Ap	4	31	31	38	5.0	1.5	7.7	7.3	0.21
	5	B2t	5	37	27	36	5.4	2.0	7.8	7.2	0.17
Type 1 badlands	5	BCca	13	44.5	25.5	30	14.5	0.8	7.6	7.3	0.13
	A1	Ap	22	36	44	20	5.0	1.5	7.9	7.3	0.24
	A4	Ap	6	26.5	46	27.5	3.6	1.0	7.4	7.0	0.24
	A6	Ap	18	23	31	46	1.7	0.8	7.9	7.4	0.33
	1	Ap1	12	41	42	17	31	1.0	7.7	7.4	0.22
	1	Ap2	11	44.5	37	18.5	34	1.0	7.9	7.8	0.17
	1	B2tca	1	52	41.5	6.5	34	0.7	7.9	7.9	0.18
	1	B3tca	2	54.5	40	5.5	26	0.6	7.9	7.8	0.16
Type 2 badlands	3	Ap1	14	25.5	45.5	29	29	0.7	7.9	7.8	0.29
	3	Ap2	27	27	43.5	29.5	25	0.8	7.9	7.9	0.28
	3	C	2	43.5	52.5	4	6	0.5	8.4	8.1	0.37
	4	Ap1	6	13.5	41	45.5	23	0.8	7.5	7.3	0.21
	4	Ap2	8	27.5	34	38.5	23	1.2	7.8	7.7	0.25
	4	B2t	2	26	29.5	44.5	24	0.3	8.2	7.8	0.27
	7	1	5	25	30	44	6.1	0.5	7.7	7.5	0.24
Type 3 badlands	7	2	0	8	11.5	80.5	0.9	0.1	8.8	7.7	0.28
	7	3	1	31	31	38	3.2	0.3	8.2	8.0	0.28
	7	4	1	30	24	46	6.6	0.3	8.7	7.9	0.25

[1] air dry soil; [2] abs. dry fine earth fraction.

Badland Geomorphology and Piping

The calcium carbonate contents of all soils are relatively high (table 2), particularly in the saline-sodic soils where it may reach 34% of the fine earth fraction. To some extent these high contents explain the high silt percentages of these soils, but not always. The organic matter content of all soils is low.

The main clay mineral (table 3) in all of the soils is illite with chlorite being a major component of the soils on the marine sediments. Kaolinite is found above trace amounts in all soils except in wadi sediments where piping occurs. There is little recognisable variation in clay mineralogy with depth. In spite of high values of volumetric shrinkage, no smectites (montmorillonite) could be detected.

Aggregate stability, swelling and dispersion

The aggregate stability tests show that all soils are highly erodible (table 4). The test was discontinued when an aggregate had received 100 water drop impacts. The number of aggregates, out of the total test population of 25, which survived the test is indicated in table 4, together with the mean and median number of impacts survived. Differences between the three badland areas exist but, with the exception of sample 17 and 19 from the area of piping, are not large enough to imply anything other than a very high erodibility.

To facilitate comparison with other studies USLE k factor values are shown in table 4. These are not corrected for structure and permeability since the uncertainty of this correction can be avoided if soil loss is not being estimated. As all soils develop crusts and structure is generally weak, corrected values would be in the order of 0.12 units higher. No trends emerge concerning soil erodibility as expressed by this parameter.

The dispersion test results seem to indicate that this is only likely to be important in the badlands affected by piping. Dispersion of aggregate from marine deposits on immersion in distilled water might have been expected, but did not occur reflecting the limitation of this test for some saline soils. Soluble salt content was so high that dispersion was repressed by the high electrolyte content rapidly obtained by the distilled water in the closed system. The threshold for dispersion with respect to the SAR and total dissolved solids concentration was reached (figure 5) only in horizons 2 and 4 in the soils subject to piping.

This is also evident from the chemical composition of the saturated paste extracts (table 5). Here, however, three distinct groups of samples, corresponding to the three badland areas, can be distinguished. Type 1 badlands have very low SARp values (< 1) and relatively low electrolyte concentrations (EC_e 0.3 - 0.5 mS cm^{-1}) resulting in chemically non-dispersive environments. Type 2 badlands have very high SAR values (SARp 20-35) and relatively high electrolyte concentrations (EC_e 10-26 mS cm^{-1}), which also result in a non-dispersive chemical environment, as indicated by the dispersion test above. In contrast soils from the badlands affected by piping contain two non-saline horizons (2 and 4, profile 7, table 8) with relatively low electrolyte

Table 3.3 Mineralogical composition of the fraction < 2 μm; the intensity of the X-ray reflections is indicated by the symbol +, except for quartz which is given in weight percentages

	Profile	Sample no.	Horizon	chlorite	illite	kaolinite	quartz	feldspar	anatase
Type 1 badlands	5	1	Ap		++++	+	5-8	tr	tr
	5	2	B2t		+++(+)	(+)	5-8	tr	tr
	5	3	BCca		++++	+	5-8	tr	tr
	A1	4	Ap						
	A4	5	Ap						
	A6	6	Ap	tr	+++	(+)	4-6		
Type 2 badlands	1	7	Ap1	++	+++	+	4-6		
	1	8	Ap2	++	+++	+	4-6		
	1	9	B2tca	++	+++	+	4-6		
	1	10	B3tca	++	+++	+	4-6		
	3	11	Ap1	+	+++	+	3-5		
	3	12	Ap2	(+)	++(+)	+	3-5		
	3	13	C	(+)	++(+)	+(+)	3-5		
	4	14	Ap1	tr	++(+)	(+)	3-5		
	4	15	Ap2	tr	++(+)	(+)	2-4		
	4	16	B2t	tr	++(+)	(+)	2-4		
Type 3 badlands	7	17	1	tr	++++	tr	5-8	tr	tr
	7	18	2	tr	++++	tr	5-8	tr	tr
	7	19	3	tr	++++	tr	5-8	tr	tr
	7	20	4	tr	++++	tr	5-8	tr	tr

Table 3.4 Dispersion index and aggregate stability measurements: columns A and B show respectively the average and median number of drop impacts survived by the aggregates tested. In column C the number of aggregates surviving 100 drop impacts is indicated

	profile no.	sample no.	horizon	Dispersion Index	Aggregate stability A	B	C
Type 1 badlands	5	1	Ap	3	17	13	0
	5	2	B2t	2	18	8	0
	5	3	BCca	2	18	12	1
	A1	4	Ap	5	15	10	0
	A4	5	Ap	3	11	9	0
	A6		Ap				
Type 2 badlands	3	11	Ap1	2	6	6	0
	3	12	Ap2	4	8	7	0
	3	13	C	0	5	7	0
	4	14	Ap1	0	15	14	0
	4	15	Ap2	0	14	10	1
	4	16	B2t	0	22	16	0
Type 3 badlands	7	17	1	0	30	36	6
	7	18	2	9	5	4	0
	7	19	3	5	27	23	3
	7	20	4	14	6	5	0

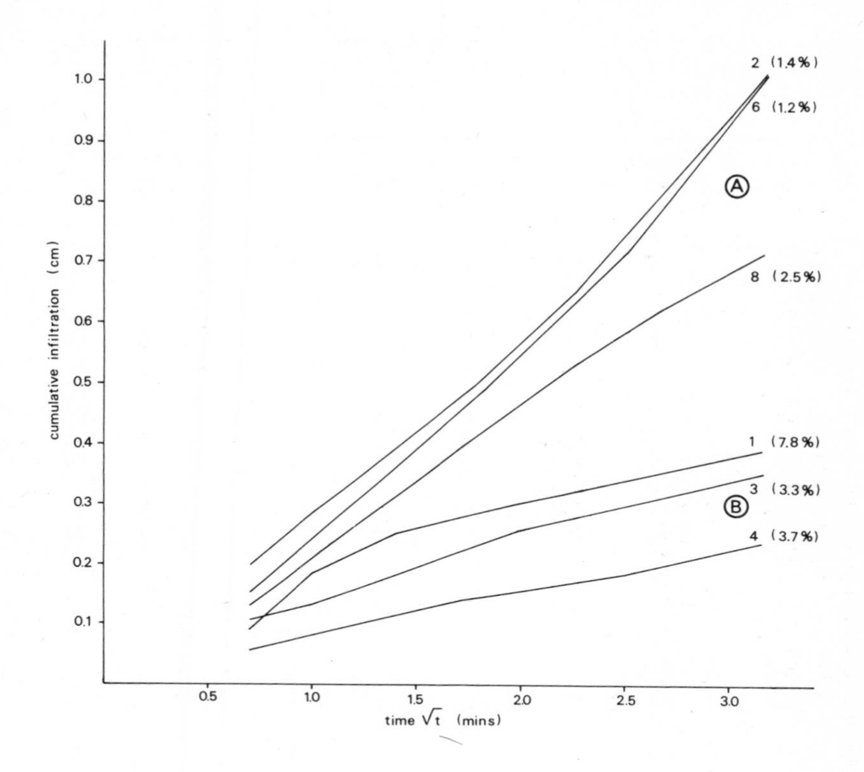

Figure 3.3 Cumulative infiltration measured at sites in badland areas 1 (A) and 2 (B). The figures shown in parenthesis refer to the soil moisture contents at the start of the experiment

concentrations (EC_e 2.0 - 2.5 mS cm^{-1}) and very high SAR values (SARp 25 and 29) which produce highly dispersive environments. The other two horizons are more or less similar to those in soils from area 2, except that the SAR value for horizon 3 is relatively high with respect to the dissolved solids concentration.

Shrinkage test results for badland areas 2 and 3 are given in table 6. The shrinkage limit is defined as the maximum water content at which a reduction in water content will not cause a further decrease in volume of the soil. Of particular interest are the high values of volumetric shrinkage measured for the saline-sodic soils and the differences which occur between the various soil horizons where piping occurs. These results obviously reflect soil texture and the relationship between SAR values and electrolyte concentrations discussed above.

Table 3.5 Water soluble salts and electrical conductivity at 25°C of the saturated pastes; exchangeable bases and cation exchange capacity at pH 8.2

	profile no.	sample no.	horizon	water soluble salts (saturated paste) pH	Na	K	Ca	Mg	Cl	NO_3
							meq/l			
Type 1 badlands	5	1	Ap	8.4	0.63	0.34	2.46	0.73	0.46	0.34
	5	2	B2t	8.3	0.57	0.15	1.73	0.56	0.39	0.27
	5	3	BC_{ca}	8.2	1.13	0.13	2.82	0.78	0.55	0.98
	A1	4	Ap	8.3	9.3	0.61	1.38	0.04	5.40	1.34
	A4	5	Ap	8.1	1.02	0.29	2.04	0.08	0.82	0.98
Type 2 badlands	1	7	Ap1	7.6	218	1.41	82	70	306	5.24
	1	8	Ap2	7.8	113	1.15	36	27	105	1.10
	1	9	B2t	7.8	176	1.15	25	44	209	3.02
	1	10	B3t	7.6	178	1.15	39	42	198	2.10
	3	11	Ap1	8.0	184	2.30	49	25	180	2.88
	3	12	Ap2	7.5	228	2.43	52	37	226	2.64
	3	13	C	7.6	77	0.56	5.7	5.7	60	0.19
	4	14	Ap1	7.9	108	2.30	37	22	148	0.05
	4	15	Ap2	7.7	184	2.43	56	49	212	2.17
	4	16	B2t	7.9	61	0.80	11.3	10.9	64	0.08
Type 3 badlands	7	17	1	7.7	89	1.28	24.8	11.7	93	1.91
	7	18	2	8.3	21	0.29	0.92	0.08	16.4	0.43
	7	19	3	7.8	133	2.24	11.6	15.7	132	3.37
	7	20	4	8.3	19	0.45	1.18	0.08	10.6	0.38

1) Alk = alkalinity ≈ HCO_3^-; 2) abs. dry base

Rainfall acceptance and infiltration

The soil properties described above influence the amount of water a soil can accept before ponding and overland flow begin. Infiltration measurements were made so that ponding times (or amounts of rainfall) could be estimated for the different soils for a given rainfall intensity. The results of a number of infiltration measurements are plotted cumulatively against $\sqrt{t}$ (mins) in figure 3. The slope of this relationship gives an approximation of the sorptivity which can be used in an expression given by Smith and Parlange (1978) to estimate ponding times. Such estimates are given in table 7 together with actual measured ponding times for a few soils. Differences between measured and calculated rates are most pronounced at high rainfall intensities because of measurement errors associated with the small amounts of water involved and because ponding is not just a function of sorptivity and permeability. For the first few minutes of the infiltration runs used to measure ponding times, water entered the many cracks which were present before these had time to swell. Usually at least 0.15 cm of water could

water soluble salts (saturated paste) SO$_4$	Alk[1]	EC$_e$ mS cm^{-1}	SAR	SARp	moisture content %[2]	exchangeable bases Na	K	Ca	Mg	Σcat	CEC pH 8.2 meq/100g
								meq/100g			
0.48	2.67	0.43	0.55	0.50	63	0.3	0.6	12.4	2.6	15.9	15.2
0.49	1.88	0.31	0.58	0.53	65	0.2	0.3	12.3	2.8	15.6	14.0
0.68	3.06	0.52	0.93	0.84	93	0.2	0.3	14.8	2.8	18.1	18.1
1.34	2.77	1.20	12.60	11.0	66	n.d.	n.d.	n.d.	n.d.	n.d.	20.2
0.74	2.54	0.58	0.78	0.70	44	n.d.	n.d.	n.d.	n.d.	n.d.	15.5
49	0.86	25.8	32.9	24.9	75	3.6	0.1	8.5	2.1	14.3	14.3
68	0.97	13.1	27.5	20.3	79	3.3	0.5	8.5	2.6	14.9	15.1
31	0.77	20.7	38.4	29.8	92	5.6	0.1	5.8	5.7	17.2	18.7
57	1.58	22.1	36.8	27.8	94	6.4	0.2	8.8	4.8	19.5	22.1
61	0.78	12.6	40.8	30.4	62	5.1	0.2	8.1	2.0	15.4	16.8
73	0.80	21.0	46.3	34.3	62	5.2	0.3	7.4	2.6	15.5	15.8
19.3	1.32	26.2	41.1	32.3	96	8.1	0.2	7.5	6.4	22.2	25.5
14.1	2.77	15.2	25.3	19.8	39	2.6	0.2	5.0	1.7	9.5	11.5
65	0.90	22.7	34.1	25.4	61	4.2	0.2	8.4	2.8	15.6	15.7
18.2	1.09	7.5	23.2	18.3	62	3.1	0.2	6.3	2.9	12.5	15.2
23.1	0.76	12.0	26.8	20.8	67	2.2	0.4	4.2	2.0	9.0	8.6
1.87	2.11	2.01	37.9	29.0	44	1.1	0.2	1.7	0.9	3.8	3.6
27.5	1.70	15.4	46.4	35.9	95	3.0	0.9	2.8	3.0	9.6	9.0
0.38	2.16	2.49	30.5	24.5	80	2.1	0.6	2.9	2.9	8.5	7.6

infiltrate in this way, even though according to the calculations, ponding should have occurred much earlier.

The great differences in the amounts of water soils from badlands 1 and 2 can accept, are of particular interest. In area 2, swelling causes infiltration to drop rapidly to a very low rate. The difference in infiltration rates in area 1, between crusted and non-crusted soils, is also of interest. The infiltration measurements reported here were made during dry antecedent conditions so that ponding times might be expected to be at a maximum.

DISCUSSION

The three badland areas have in common soils with low rates of water acceptance and high erodibility. Rainfall intensities which occur several times a year produce ponding and overland flow on most soils. Nevertheless, differences exist in the relative importance of the soil properties examined, due in part to environmental or historical factors.

Table 3.6 Results of shrinkage test determinations for soils from badlands of types 2 and 3.

	profile no.	sample no.	horizon	(1) shrinkage limit (% volumetric soil moisture)	(2) shrinkage ratio	(3)* volumetric shrinkage %	(4) original moisture content
Type 2 badlands	1	7	Ap1	18.2	1.8	107	78
	1	8	Ap2	17.8	1.8	125	86
	1	9	B2tca	15.6	1.8	200	123
	1	10	B3tca	16.6	1.8	186	118
	3	11	Ap1	15.8	1.9	128	83
	4	14	Ap1	17.4	1.9	116	79
Type 3 badlands	7	17	1	16.2	1.7	64	53
	7	18	2	32.2	1.5	25	
	7	19	3	18.2	1.9	103	73
	7	20	4	19.7	1.7	45	46

*) The volumetric shrinkage refers to the shrinkage occurring during the drying of the saturated paste (Singh, 1967). The original moisture content of the paste is indicated in column 4

Table 3.7 Amounts of rainfall (cm) required to pond the soil at the indicated intensity and saturated hydraulic conductivity

		k	Amounts of rain (cm) required to pond soil at indicated intensity (cm hr^{-1})									
		cm hr^{-1}	0.5	1.0	1.5	2.0	2.5	3.0	3.5	4.0	5.0	6.0	7.0
Type 1 badlands	Site												
non crusted	F1	1			3.5	2.2	1.6	1.3	1.0	0.91	0.7	0.57	0.48
	F1*	-								0.73	0.71	0.73	0.73
	F1	1.5				2.9	1.0	1.4	1.2	0.99	0.75	0.6	0.5
	F1	0.5		4.4	2.5	1.8	1.4	1.15	0.97	0.84	0.66	0.55	0.47
Type 1 badlands	F4	0.25	1.8	0.75	0.47	0.35	0.27	0.23	0.19	0.17	0.13	0.11	0.09
crusted surface	F4	0.5		0.9	0.53	0.37	0.29	0.24	0.2	0.25	0.25		
	F4*	n.d.						0.22	0.23	0.25	0.25	0.25	0.23
	F6	0.2	6.1	2.6	1.7	1.2	0.96	0.8	0.67	0.59	0.47	0.39	0.33
Type 2 badlands	3	27×10^{-4}	0.63	0.32	0.21	0.16	0.13	0.11	0.09	0.08	0.06	0.06	0.05
	4	n.d.	0.36	0.17	0.11	0.08	0.07	0.05	0.05	0.04	0.03	0.03	0.02
	J8*	n.d.		0.17		0.10		0.11		0.13	0.14	0.14	0.12

*) The values are calculated from a relationship described by Smith and Parlange (1978), except for those marked by an asterisk which were measured in the field

Badlands produced by V-shaped gullies on non-saline/ non-sodic soils (type 1 badlands)

Several factors other than the presence of suitable materials have favoured the development of these badlands. These include the moderately steep gradients and the collection and concentration of runoff along roads and terraces. Red soils similar to those in badland areas occur in places under maquis but here the organic matter may reach 4%, soil aggregates are very stable, and infiltration rates and permeability are high. It would seem that the present high erodibility and low infiltration rates are related to the decline in organic matter which has accompanied cultivation (de Mas and Jungerius, 1980). Field infiltration measurements have consistently demonstrated the effect of crusting on water acceptance, and all of the red soils in the badland areas develop crusts after rainfall.

The soil chemistry does not suggest that crusting is associated with the chemical dispersion of the soil, but this can not be ruled out under conditions of ponded infiltration (Quirk, 1978). It is possible that at certain times atmospheric salts might accumulate on the soil surface so that conditions favouring dispersion may exist temporarily at the onset of rainfall, but there is no direct evidence of this. Beneath the crust the soil is always well aggregated probably reflecting the high free iron and carbonate contents and very low SAR values. The most likely explanation is that crusting is primarily a result of slaking and dispersion produced mechanically by raindrop impact. However, observations by Bryan *et al.* (1978) and Yair *et al.* (1980) in Canada and Israel indicate that surficial crusting can also be obtained by depositional processes producing thin varved layers.

Once runoff occurs and is concentrated by terraces or pathways, the soil on the relatively steep slopes offers little resistance to erosion. Gullies incise until underlying calcrete or flysch rocks are reached.

Badlands in the marine deposits (type 2 badlands)

Gullying in these saline-sodic soils is probably again related to the low rates of water acceptance and poor soil structure but soil properties promoting these conditions differ from the type 1 badlands. Although slopes are more gentle (figure 2), soil conditions are worse.

The high clay and salt contents promote very large volumetric changes in the soil mass (Table 6). When dry the soils have an irregular hummocky microtopography with a close network of desiccation cracks, sometimes > 1 cm wide. Rapid swelling of the soil during the early stages of infiltration appears to produce very low rates of water acceptance. Infiltration rates may be fairly high during the first 3 or 4 minutes but soon become very low (figure 3). Ponding might be expected to occur at least 5-10 times per year under the present rainfall conditions.

Particularly marked in these soils are the very low organic matter and very high silt contents. These properties,

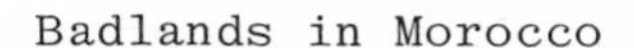

together with swelling, help to explain the extremely low aggregate stability. Another factor influencing the structure is the repeated growth and solution of salt crystals during drying and wetting cycles. The low aggregate stability occurs throughout the profile and the soil can offer little resistance to erosion once water becomes concentrated. Flow concentration may be favoured by the microtopographic depressions which form where cracks close upon swelling or along furrows formed by ploughing.

Soil dispersion is hindered by the high salt content of the soil solution. Should the area experience a relatively wet period, salt concentrations in the upper soil horizons might become low enough for dispersion to occur and it is possible that these badlands experience most rapid development during the wet conditions which occur every four or five years.

Badlands produced by piping in wadi sediments
(type 3 badlands)

In many ways badland development is more closely related to soil properties in the third area than elsewhere. Recently Crouch (1976) listed conditions which predispose a landscape to tunnel erosion. Two, a) a seasonally highly variable rainfall combined with high summer temperatures, and b) reduction or detrimental change in the vegetation cover, apply throughout the Beni Boufrah region. The other factors: 1) soil subject to cracking, 2) a relatively impermeable layer in the soil profile, and 3) a hydraulic gradient within a dispersible soil layer, are all present in this third badland area.

Soil conditions prevailing in this area are summarised in figure 4. Piping occurs in the deepest of the four sedimentary horizons, not in the very impermeable horizon (horizon 3), but beneath it. It is thought significant that piping is a phenomenon in dry regions having soils of a 'duplex' character, with a pronounced textural break between a relatively coarse surface horizon and an underlying fine-textured subsurface horizon. Although such soils (planosols, and some solonetzes and luvisols; FAO-UNESCO, 1974) are not present in the field area, a similar textural sequence of layers occurs in the area of pipe erosion. Elsewhere soils are gravelly and show relatively little textural differentiation.

Excluding horizon 1, a former Ap horizon, the profile can be summarised:

2) a highly dispersive but permeable sand horizon, 30 cm thick, with very low aggregate stability and a low volumetric shrinkage percentage (non-saline/sodic horizon).
3) a 40 cm thick impermeable layer containing about 30% clay. The volumetric shrinkage percentage, the SARp value and the water soluble salts are all very high (saline-sodic horizon); aggregate stability is relatively high and the Dispersion Index low.
4) a finely layered deposit, highly dispersive, containing fewer water soluble salts and having a higher

permeability and much lower aggregate stability than the overlying horizon. The volumetric shrinkage percentage, although high, is much lower than that of the overlying horizon (non-saline/sodic horizon).

The soil properties described could be related to piping in the following way. During summer the soil will dry to below the shrinkage limit in all horizons. Soil moisture contents in spring and autumn were also usually below this value. Differential shrinkage will cause large cracks to develop in layer 3 and to a lesser extent in layers 1 and 4 (figure 4). The first rains of autumn and winter would be insufficient in volume to saturate the entire soil mass to the depth of piping. However the wetting front would be able to penetrate through the permeable sandy horizon 2 to horizon 3, especially as this low lying area receives water from the catchment area upslope which is partly used as a school playground. The cracks in horizon 3 might allow water to pass through it without the main body of the soil mass swelling or becoming moistened. The moistening of the upper few cm of this horizon would produce swelling and reduce infiltration to extremely low values restricting percolation to larger cracks. Water concentrated in these would produce dispersion in horizon 4 provided that the electrolyte concentration remained low. The permeability of horizon 4 is sufficiently high to allow reasonable rates of flow in this layer, and is probably much higher in the individual sand layers than the laboratory k_s value for undisturbed soil cores indicates (0.25 cm/hr$^-$).

Obviously if water containing dispersed clay is to drain from layer 4 a hydraulic gradient is required. This is provided by a gully located at the downslope boundary of the badlands where water is concentrated by the Beni Boufrah/ Torres road. Some of the runoff here is derived from the badland area on the old marine sediments (figure 1). The initiation of other pipes has probably also been encouraged by the small gullies originally initiated by water draining the school play area. Once horizon 4 was breached the process of piping enabled gully networks to extend upslope in all directions to produce these badlands.

Swelling, dispersion and gully erosion

The discussion above has implicitly involved two sets of soil properties, one associated with swelling and dispersion, the other with water acceptance.

Both swelling and dispersion depend upon the well-known relationship between the SAR (ESP) and the electrolyte concentration (total dissolved solids concentration) of the soil solution mentioned earlier. This is expressed in figure 5 taken from: Kamphorst and Bolt (1976). In this the regions occupied by soil samples from the three badland areas have been entered. The position of the lines indicating swelling and dispersion thresholds will, of course, vary with the clay mineralogy and organic matter content (Quirk and Schofield, 1955). It must not be thought that smectite clays (montmorillonite) are essential for swelling, for as the results presented here indicate illites and other clays will also swell in sodium rich environments.

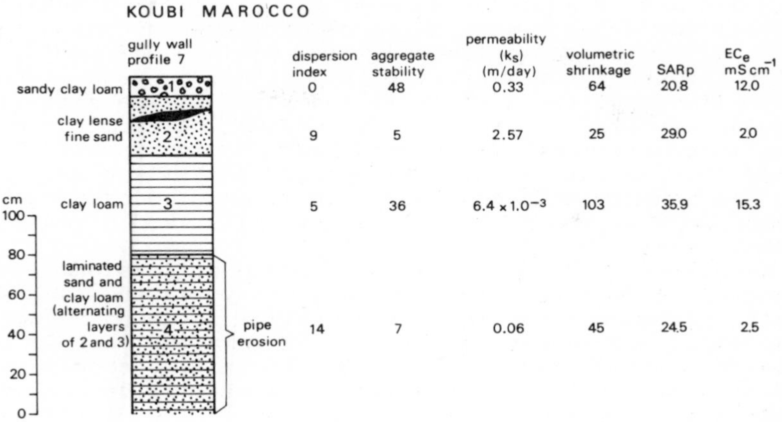

KOUBI MAROCCO

gully wall profile 7

Horizon		dispersion index	aggregate stability	permeability (k_s) (m/day)	volumetric shrinkage	SARp	EC_e $mS\ cm^{-1}$
1	sandy clay loam	0	48	0.33	64	20.8	12.0
2	clay lense, fine sand	9	5	2.57	25	29.0	2.0
3	clay loam	5	36	6.4×1.0^{-3}	103	35.9	15.3
4	laminated sand and clay loam (alternating layers of 2 and 3) – pipe erosion	14	7	0.06	45	24.5	2.5

cm: 100, 80, 60, 40, 20, 0

Figure 3.4 Soil properties of the four horizons at profile 7 (see Appendix) in the badland area affected by tunnel erosion

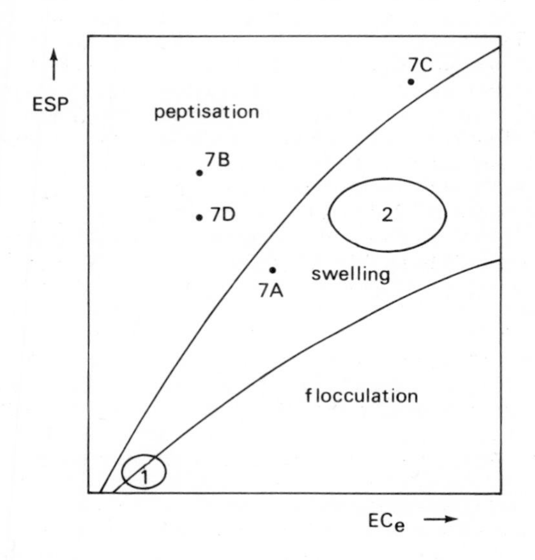

Figure 3.5 The relationship between EC_e and SAR values with respect to flocculation, swelling and peptisation. The relative positions occupied by samples from the three badland areas are shown. 1 and 2 respectively refer to samples from badland types 1 and 2, and 7 A, B, C, D to horizons 1, 2, 3 and 4 from profile 7 in the badland area affected by pipe erosion

Relationships such as those shown in figure 5 are useful in studies of gully erosion as they indicate a possible potential for erosion should the electrolyte concentration of the soil solution be lowered. The position of the dispersion threshold for a particular soil can be established empirically using the dispersion index and at the same time recording electrical conductivity, sodium, calcium and magnesium concentrations. Such measurements might provide greater insight into erosion thresholds in badland areas, which are usually only considered in terms of mechanical erosion processes (e.g. Bryan and Campbell, 1980; Imeson and Verstraten, 1981).

A weakness of the Loveday and Pyle modification of the Emerson dispersion test used in this study, is the lack of specific recommendation about the optimal ratio of volume of the solid and aqeous phases. Clearly in the case of some saline soils (high SAR values, high electrolyte concentrations), a relatively large volume of water will be required to reduce the electrolyte concentration to a level at which the soil will disperse.

Swelling and dispersion are not pre-requisites for gully erosion or badland development. In the Zin Valley badlands of the northern Negev, for example, where denudation rates are very low, Yair *et al.*, (1980) found that mixed montmorillonite and kaolinite shales swelled by less than 4% and that mechanically dispersed material flocculated immediately in water. Nevertheless, where soils do swell and become dispersed very high denudation rates can be expected. Mass wasting processes along gully sides, deep piping and low infiltration rates are usually found. Perhaps the most important point however, is that sediment entrainment and erosion cease to become simply a function of mechanical laws. Consequently relatively low flow events may be capable of evacuating large amounts of sediment from a gully system which at first sight may seem extremely large compared to the volume of water draining it (Imeson and Verstraten, 1981).

Acknowledgements

We thank Prof.Dr. P.D. Jungerius, Dr. H.van der Wusten and Drs. M.Schröder for the fruitful discussions we have had about soil erosion in northern Morocco. The fieldwork reported was undertaken in the framework of the REDRA project, supported by the Netherlands Council for Pure Scientific Research (Z.W.O.).

Laboratory analyses were undertaken by Mr. L.de Lange; the X-ray analyses were undertaken by Mr. B.de Leeuw. The illustrations were drawn by Mrs. O.M. Bergmeijer-de Vré and the manuscript prepared by Mrs. M.C.G. Keijzer-v.d.Lubbe and Mrs. G.J.M. Scholts.

4 Difference between "calanchi" and "biancane" badlands in Italy

D. Alexander

The landforms known in Italy as 'calanchi' consist of heavily dissected terrain with steep, bare slopes and channels which rapidly incise and extend headwards, but which are frequently obliterated by mass-movement debris. They are widespread in eleven of the twenty regions of Italy (figure 1) and are especially common on the Plio-Pleistocene marine clays of the Calabrian deposits. Calanchi can be divided into several types: closely-spaced rills and v-shaped channel incisions are characteristic of the calanchi proper, whereas cones and hummocks separated by flatter areas of surface wash deposits are associated with 'biancane'. Topography composed of both elements exists in Italy, and slope development can be either symmetrical or asymmetrical with respect to slope orientation. This paper will describe the basic characteristics of calanchi, will show that they are geomorphologically distinct from the classic badlands of the western United States and will account for the difference between calanchi of the rill-and-gully form and biancane.

Distinctiveness of the calanchi

The calanchi are not, of course, an exclusively Italian phenomenon, and the term could be used in any context to describe the particular kind of erosional landform that results from the accelerated erosion of cohesive sediments, including clays, silt-clays, clays interbedded with sands, soft shales and mudstones. Calanchi evolve rapidly, responding with great sensitivity to many factors, including variations in micro-climate, the pattern of tension cracks and bedding planes in the underlying sediment, and the principal orientation of the drainage network. They show badland characteristics but tend towards a greater internal disorder than classic badlands (Carson, 1958).

Many calanchi begin when shallow landslides break up the weathered regolith on a slope, steepening the overall gradient and exposing greater areas of sediment to rapid weathering and future instability. On clay slopes the regolith generally consists of sediment which has become friable after penetration by rainwater, frost and plant roots; and its susceptibility to mass-movement is a function of the depth to 'bedrock' (usually overconsolidated or desiccated

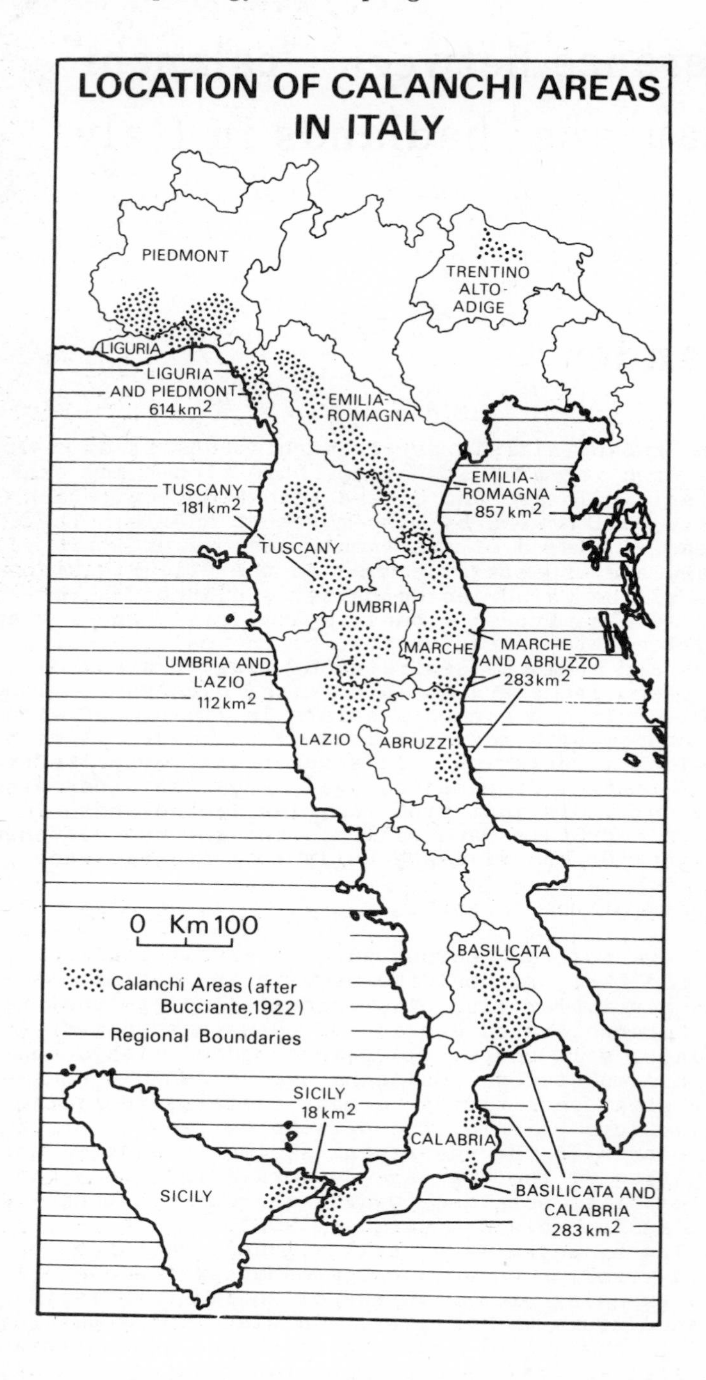

Figure 4.1 Location of calanchi areas in Italy

Plate 4.1 Calanchi at Mesola della Zazzera in the Salandrella Valley, Basilicata

clay) and of random factors related to the spatial distribution of pore water pressures during periods of intense precipitation. Fully established rill-and-gully systems in the calanchi (plate 1) tend to maintain an uneasy balance between creating small lengths of channel and obliterating them with mass-movement debris emanating from the sideslopes. In the most extreme cases, this makes analysis of stream networks and dynamic equilibrium behaviour inappropriate and thus demands that many of the tenets of classical badland geomorphology be abandoned. For example, the principle of the 'constant of channel maintenance', that each length of channel is maintained through successive phases of rapid erosion by a proportional area of 'catchment' (Schumm, 1956), rarely applies, as mass-movement and the development of pipes continually alter the length, orientation and accessible tributary area of channels.

In the case of biancane (plate 2), the dendritic network is replaced by two indistinct reticulated networks, one on the surface and one sub-surface in the form of piping and solutional cracks. Biancane are therefore closely allied to pseudokarst landforms identified in the U.S.A. (Mears, 1963; Quinlan, 1974) and volcano-karst landforms such as those found in Göreme, Turkey (Andolfato and Zucchi, 1971). The principal difference is that as a result of forming in cohesive sediments biancane are particularly susceptible to mass-movements, especially mudflows, and to dispersion of sediment particles during erosion by sheetwash.

Plate 4.2 Biancane near Aliano in the Agri catchment, Basilicata

Morphology and erosion rates

Morphological surveys and measurements of erosion rates have been carried out on calanchi and biancane at study sites in Basilicata, southern Italy (figure 2) since 1978. The aims of this work are to determine the characteristic morphology of the calanchi, to understand the processes currently acting upon them and to determine the differences in morphology and erosion rates between rilled calanchi and biancane cones.

Table 1 summarizes the results of 62 accurate surveys of selected examples of biancane, mudflows and calanchi rill systems, which were made with an Abney level, using a 0.5 m sampling interval. It can be seen that there is a difference in size, slope angle and relative relief between biancane and calanchi: the latter, however, are similar to mudflow chutes, reflecting the importance of viscous flows in creating rill morphology. But although biancane are smaller and of gentler relief than calanchi rills and mudflow chutes, they do not necessarily erode at a slower rate.

Erosion rates were measured on calanchi and biancane at La Rondinella in the Basento Valley and on biancane at Stazione Craco in the Salandrella Valley, Basilicata (see

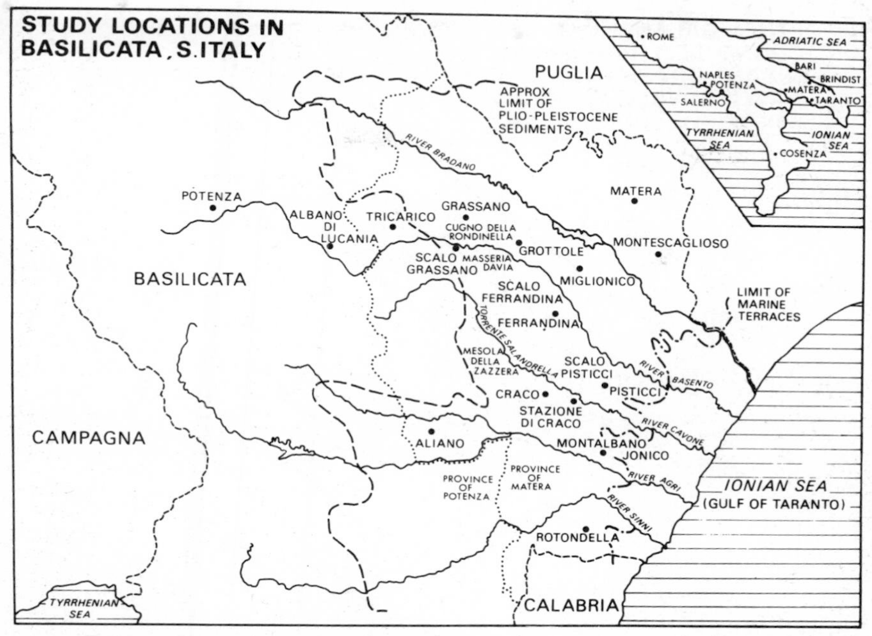

Figure 4.2 Study locations in Basilicata, southern Italy

Table 4.1 Morphological characteristics of calanchi and biancane

Landform		Mean slope angle	Max. slope angle	Min. slope angle	Mean slope length m	Mean slope height m	Mean rel. relief	No. of samples
biancana	slope	38.89	54.5	23.2	8.69	5.41	0.816	12
"	backslope	34.17	52.2	13.3	9.76	5.62	0.691	9
"	basal fan	8.92	17.0	4.0	2.74	0.43	0.160	11
calanco	inter-rill	41.05	60.7	14.5	22.00	14.43	0.873	5
"	rill	40.63	46.7	22.2	28.74	18.73	0.862	8
"	basal fan	17.71	23.1	15.3	7.40	2.16	0.321	7
mudflow	chute	40.94	53.3	22.1	23.57	13.83	0.896	6
"	basal fan	16.69	40.1	2.8	6.00	1.71	0.301	4
								N = 62

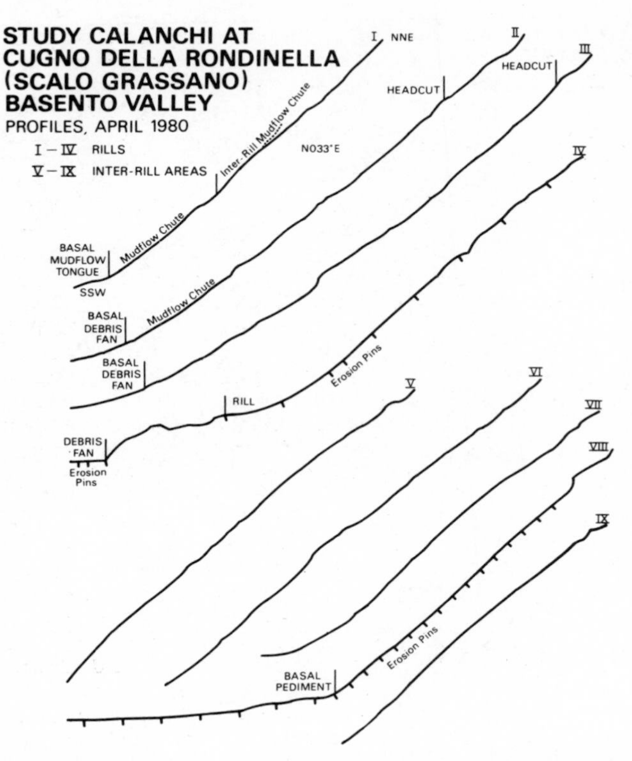

Figure 4.3 Study calanchi at Cugno della Rondinella, Basilicata

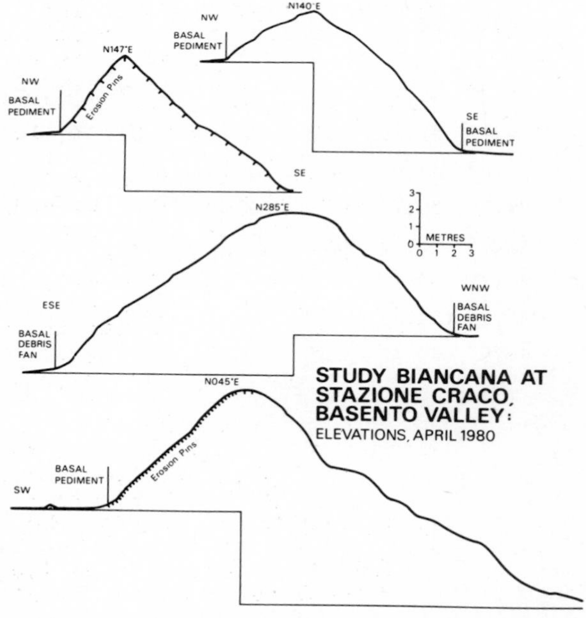

Figure 4.4 Study biancana at Stazione Craco, Basilicata

figures 2 - 4) over the period 1978-80. Rates of surface lowering were obtained at six-monthly intervals using 150 mm erosion pins, 1.0 m erosion stakes and Abney level surveys from a fixed datum point on four rill slopes, four inter-rill slopes, one biancana and two basal slopes at La Rondinella, and on three biancana slopes at Stazione Craco. Pins and stakes were inserted during October 1978 in lines of up to 45 pins spaced at intervals of 0.25, 0.5 or 1.0 m. Clearly the use of erosion pins involves some disruption of the slope, but there was little indication that the ground surface close to the pins began to differ from its surrounds as a result of inserting the pins, and it is probable that the cohesiveness of the sediment prevented surface break-up. The length of time over which measurements have been made is obviously very short and until a longer series of data has been obtained the measurements will only serve to illustrate the magnitude of contemporary erosion and its distribution across and among slopes, rather than mean rates over time. Meteorological data for the period are unfortunately not yet available.

Erosion rate data are summarized in table 2. The measurements suggest that erosion rates may be greater on biancane than on calanchi but that the erosion is not a stable process in time or with respect to individual slopes or landforms, being highly variable from point to point and with variations in the dominant erosional process. Erosion is in general randomly distributed on slopes and down rills, and thus characteristic slope angles and morphologies tend to be maintained over time. But during heavy erosion on interfluves the material eroded is often temporarily stored in rill channels when small viscous mass-movements cease to flow. Movement of material downslope tends to be sporadic and slopes undergoing net erosion may have several zones where sediments temporarily accumulate. Cones and inter-rill areas uniformly degrade, whereas basal pediments and wash-slopes invariably aggrade at a rate which is proportional to their characteristic slope. Rills undergo both aggradation and degradation with an eventual bias towards the latter. In calanchi rills variations at a point are often one or two orders of magnitude greater than the overall mean and the tops of rills erode slower than the sideslopes. The tops of biancane also erode slower than their sideslopes, making them more leptokurtic in shape as time progresses. Variation in the rate of erosion on biancane slopes with different orientations is qualified by the length and steepness of the slopes, which sometimes outweighs the south-westerly bias in the intensity of solar insolation.

With mean annual rates of surface lowering of the order of 20 - 30 mm/yr these calanchi and biancane are eroding much more rapidly than a rilled hillslope of 18° at La Rondinella, on which Rendell (personal communication) observed 10.5 mm/yr of degradation in rills and 3.5 mm/yr on inter-rill areas over the period 1974-77. All such rates are considerably higher than the 0.6 and 0.85 mm/yr observed by the Servizio Idrografico Italiano (1953) for the whole of the Bradano and Sinni basins, respectively, in Basilicata.

Measurements of the rate of headcut advance indicate that some headcuts in the silt-clays of Cugno Davia (location

Table 4.2 Erosion rates of calanchi and biancane at study sites in Basilicata, 1978-80 in mm of surface lowering

RONDINELLA	Annually			Overall
	10.78-10.79	4.79-4.80	10.79-11.80	10.78-11.80
degrading rills	-13.375	+ 0.125	- 5.25	-18.625
degrading inter-rill areas	- 9.72	- 9.19	-13.57	-23.29
aggrading basal pediments*	+12.33	+18.00	+26.00	+38.33
aggrading basal wash-slopes**	+ 5.86	+ 6.43	+ 9.00	+14.86
degrading biancane cones	-22.84	-31.84	-39.67	-62.50

* mean angle 6.67^{o}

** mean angle 4.5^{o}

STAZIONE CRACO BIANCANA	Annually			Overall
	10.78-10.79	4.79-4.80	10.79-11.80	10.78-11.80
N.E. slopes	-17.00	-41.92	-56.75	-73-75
S.W. slopes	-28.23	-28.12	-16.74	-43.97
S.E. slopes	-	-21.21	-20.51	-13.64***
Tops	-12.00	-17.50	-12.83	-24.83
OVERALL MEAN	-19.08	-27.19	-26.71	

*** 4.79-11.80

shown in figure 2) are currently advancing into the shallow regolith of the Basento Valley sideslope at about 1.5 m/yr, whereas those in the scaly clays and mudstones at Albano Lucano are advancing at more than 2.2 m/yr. There is no evidence on the ground - for example, widespread fresh mudflows or channel incision - to indicate that this is a response to exceptionally intense rainfall events.

Genitive processes differentiating calanchi and biancane

The calanchi pose many important and diverse problems in engineering, agriculture and environmental management. Given that the two landforms have differences in morphology and erosion rates, one of the most fundamental geomorphological problems is to account for the dichotomy between calanchi and biancane in terms of material properties, relative relief and piping, so that more effective strategies for ameliorating the detrimental effect of calanchi erosion on the human environment can be developed.

Calanchi and biancane are often found on the same and differentiation can be extremely subtle. Some bia are the end product of calanchi erosion and emerge when deep-furrowed rilling converges from two parallel gullies until the intervening interfluve is partially destroyed. Others emerge directly from the erosion of a planar slope, in isolation from the processes giving rise to calanchi. Without long-term observations it is difficult to attribute either origin to particular examples of biancane. Penetrometer tests on biancane in Basilicata indicate that they are almost all erosional rather than depositional (cf. Mears, 1963), and those that occur in overconsolidated sediments are composed of materials that are more than 90 percent unweathered, indicating rapid loss of surface material. Thus biancane are residual landforms - effectively piles of undisturbed sediments that remain untouched by advanced calanchi erosion or the development of a highly ramified or reticulated network of subsurface pipes.

There is a difference in both the scale and nature of processes operating on calanchi and biancane that must be at least partially a function of properties of the underlying material. Table 3 gives summary statistics for 130 landslides measured at Scalo Grassano in the middle Basento Valley in Basilicata in a zone of calanchi but not biancane. These are the kind of slides that give rise to incipient calanchi and, although blockfalls, mudflows and sheetwash become increasingly important as calanchi develop, slides such as those measured remain important throughout. The predominance of shallow slides highlights the role of the weathered layer and the relatively low mean depth of slides (0.67 m, normal to the slope) reflects the thinness of the surface cover of softened or weathered sediments on these slopes. However, the weathered layer tends to be even thinner, perhaps half the depth, on biancane, and the predominant processes are slopewash and hydration to liquid limit of material softened by cycles of desiccation and rehydration (Lulli and Ronchetti, 1973).

Such processes must be considered with respect to the properties of materials on which they act, but research has not always been successful in differentiating the material properties of calanchi and biancane.

Material properties differentiating calanchi and biancane

Cori and Vittorini (1974) and Vittorini (1977) analysed sediment samples from the Era, Orcia, Piombo and Basento Valleys in Italy and found that calanchi contain more clay and less sand than biancane, whilst intermediate landforms appear to have an intermediate mix of sediment sizes. Re-examination of Vittorini's data using a test of means and standardizing his definition of sediment size ranges calls the significance of his conclusions into question, and so the present author analysed fifty samples of his own, derived from calanchi, biancane and intermediate forms mostly occurring in the 'Argille Subappennine' marine silt-clays of the Calabrian succession (Boenzi *et al*, 1971). These sediments, in places more than 500 m thick, are remarkably uniform, and calanchi, biancane and intermediate forms tend to occur in the 'silty-

Table 4.3 Dimensions of landslides at Cugno Davia in the Basento Valley, Basilicata

	$\bar{X}$	s	max.	min.
Length (l) m	13.08	7.78	51.4	2.3
Width (w) m	5.15	3.65	27.5	1.4
Depth (d) m	0.81	0.56	4.58	0.18
Orthogonal depth (t) m	0.67	0.49	3.88	0.14
Slope degrees	34.39	4.97	58.0	25.0
Area (π.l.w)	280.52	502.71	3872.0	14.9
Depth/length (d/l)	0.0718	0.0420	0.1586	0.0175
Orth. depth/length (t/l)	0.0583	0.0298	0.1935	0.0112
Orth. depth/width (t/w)	0.1429	0.0640	0.3761	0.0460
			N =	130

Broad classification of landslides	
Shallow earthflow	70.8%
Shallow rotational slip	13.1%
Earthflow/debris slide	7.7%
Non-circular rotational slip	4.6%
Compound shallow slip	2.3%
Complex slip	1.5%

(after Skempton and Hutchinson, 1969)

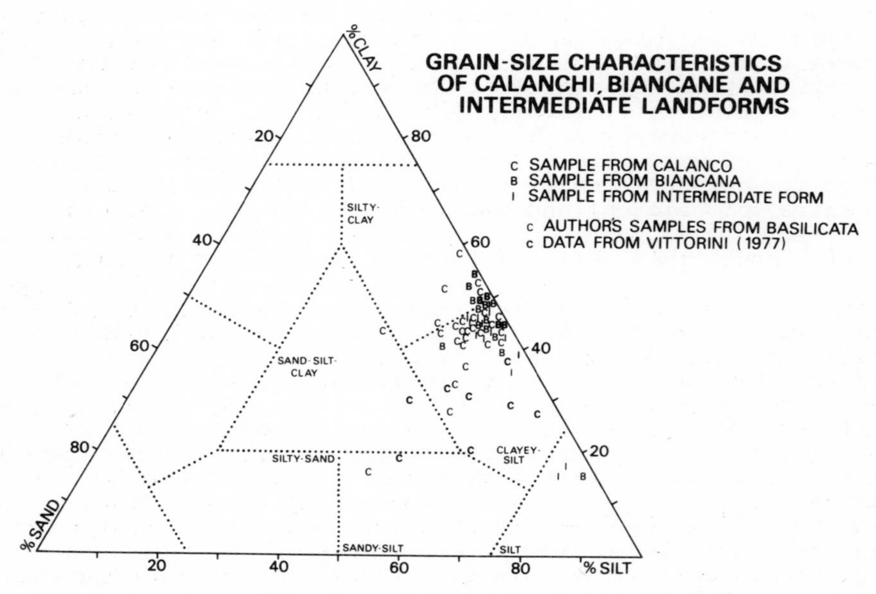

Figure 4.5 Grain-size characteristics of calanchi, biancane and intermediate landforms

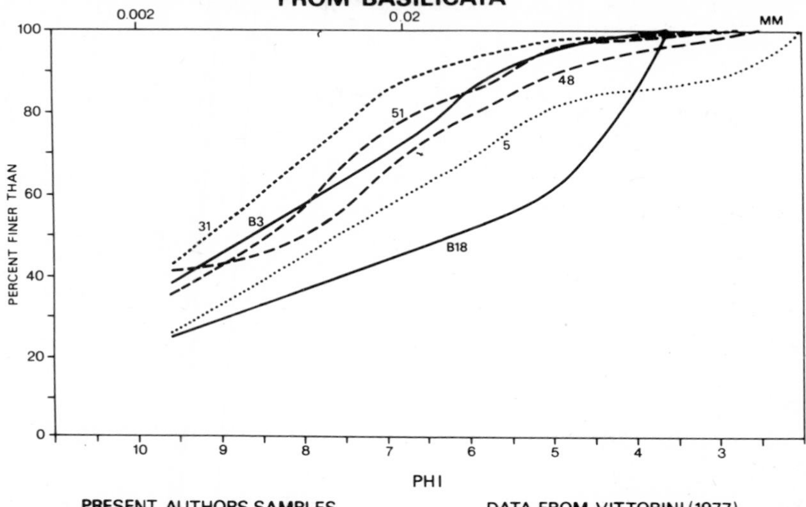

Figure 4.6 Grain size curves for some calanchi and biancane samples from Basilicata

clay' and clayey-silt' divisions of the triangular sediment-size graph with no apparent demarcation between them (figure 5). Figure 6 gives sediment-size curves for four of the samples, compared with curves for two of Vittorini's samples, and shows that the gap between curves for calanchi and biancane is narrow enough to be of questionable significance. A t-test of standard deviations, however, showed that sand content is significantly greater (0.001% level) in samples derived from calanchi than in those collected on biancane.

Using a non-parametric analysis of variance on the present author's data, a highly significant difference was observed in the clay content of samples as arranged by study site (table 4), but the difference in clay contents according to landform type at each study site was not significant. Unfortunately there are difficulties of obtaining a representative sample: areas where sediments are proportionately rich in clays could have a high proportion of biancane relative to calanchi as a result of other factors, so that collecting only a small number of samples would allow a large *statistical* regularity to go undetected. Indeed, the mean grain sizes of samples from each location (table 4) appear to vary with the type of landform which predominates there, although such variations are too complex to be easily interpretable.

Given such heterogeneity in sediment sizes it is necessary to look for variations in the properties pertaining to sediments of a given size, in particular the cohesive,

Table 4.4 Variations in sediment size with study site in Basilicata

Location	Landform types	n	%Clay $\overline{X}$	%Silt $\overline{X}$	%Sand $\overline{X}$
Masseria Davia	Symmetrical calanchi	8	40.57	49.72	9.71
Trecancelli	Symmetrical/asymmetrical calanchi	4	40.51	50.00	9.49
Albano Lucano	Calanchi in variegated mudstones	4	49.31	40.27	10.42
Rotondella	Asymmetrical calanchi	2	51.44	47.02	1.55
Rondinella	Intermediate forms	12	40.84	66.85	11.88
Torrente Carlillo/ Scalo Ferrandina	Calanchi/biancane	3	43.57	46.65	9.77
Mesola della Zazzera	Calanchi/biancane	3	41.74	54.08	4.18
Scalo Pistici	Mainly biancane	21	40.53	55.68	3.80
Aliano	Biancane/intermediate forms	6	21.74	54.58	23.78

clay-size, in order to find an explanation of the dichotomy between calanchi and biancane. The cation exchange capacity (CEC) of sediments is one indication of the nature of materials present and their possible behaviour during erosion by rainwater. Kelley (1964) reviewed studies of CEC and the behaviour of semi-arid soils and noted that the clay fraction is likely to be the principal source of exchangeable cations. CEC values obtained from the Basilicata calanchi samples indicate that the clay is an impure montmorillonite. This is supported by X-ray diffraction studies (Rendell, 1975; Vittorini, 1977) and by a differential thermograph for a sample from the Pisticci area (Kayser, 1964). Average CEC remains fairly constant at 36 - 52 meq/100 gm for the three landform categories (which is predictable if the source of montmorillonite is fairly uniform (Kelley, 1964, p. 82)) and all study sites except Mesola della Zazzera, which is particularly notable for unusually complex intermingling of calanchi and biancane, where it hardly varies from a mean of 85.44 meq/100 gm.

The presence of more than a few percent of exchangeable sodium among the cations will tend to make the sediment dispersive in water (Rolfe *et al*, 1960; Mitchell, 1976). A mean exchangeable sodium percentage (ESP) of 36.07 percent for the samples indicates that dispersion will occur rapidly and coupled with the high swelling capacity of montmorillonite, ensures that almost all the samples fall into the 'highly erodible' category.

It has been suggested above that there is usually a difference between calanchi and biancane in the depth of the weathered layer and in overall rates of slope erosion. As CEC is relatively constant among the Basilicata samples, recognizable patterns of variation in ESP may help to in-

dicate how clay 'activity rates' are related to landform.
The following table shows that ESP differs greatly between calanchi and intermediate forms on the one hand and biancane on the other:

	n	Mean	Standard deviation	Minimum	Maximum
Overall	64	36.07	20.58		
Calanchi	28	29.50	14.92	8.18	95.25
Intermediate forms	16	29.06	13.11	4.95	48.70
Biancane	20	50.92	24.78	5.54	67.33

In addition, with a mean of 52.55, ESP is significantly higher at Scalo Pisticci, in a zone predominantly occupied by biancane (see figure 7), than in the zones that are largely filled with calanchi. This corroborates the high rates of erosion observed and accounts for the small depth of softened material on the surface of biancane occurring in overconsolidated sediments.

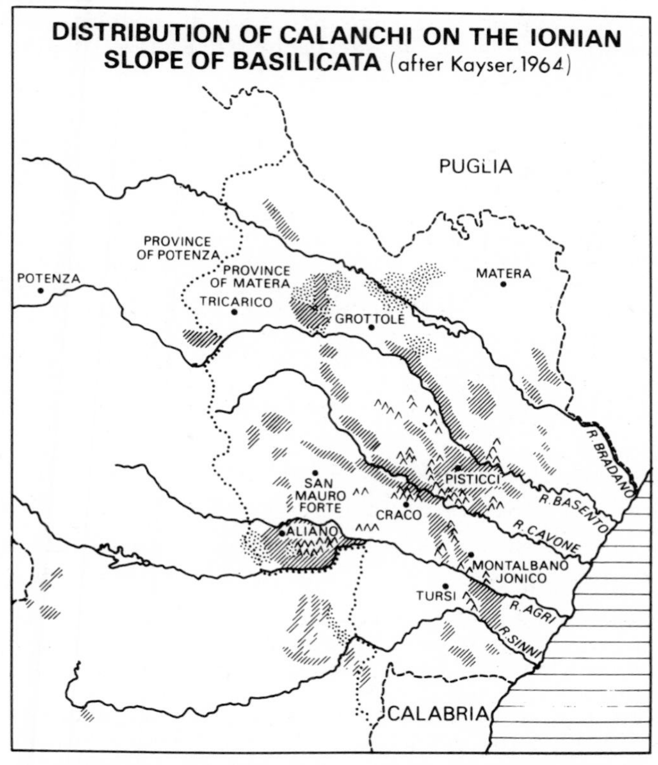

Figure 4.7 Distribution of calanchi on the Ionian slope of Basilicata (after Kayser, 1964):

cross-hatching: unvegetated calanchi
stippling: vegetated calanchi
inverted 'v's: biancane

Difference between calanchi and biancane: material properties in relation to other factors

An ESP of 12 percent is sufficient to produce the deflocculation and dispersion that are characteristic of landscapes subject to piping (Stocking, 1977; Masannat, 1980). Almost all samples from Basilicata have an ESP in excess of 12 percent and all the study sites manifest widespread piping, but there is a difference between piping on calanchi and biancane: vertical pipes are much more common in calanchi landscapes, whereas piping in biancane terrain is usually more gently inclined, having approximately the same relative relief as the basal fans (see table 1). It is most likely that the degree of relative relief at the initial stage of accelerated erosion is the main factor influencing the slope of the pipe network. Vertical slopes which are created when streams undercut a hillslope will be especially susceptible to the formation of vertical and steeply inclined pipes, especially if vertical fissuring occurs in the newly-exposed sediments.

High percentages of exchangeable sodium encourage the rapid enlargement of pipes, as sediment is easily dispersed and transported away by percolating rainwater (Sherard *et al*, 1972). Recurrent mass-movements and the presence of steeply-inclined pipes tend to make calanchi systems self-maintaining, so that where relative relief is sufficient slopes maintain a high limiting angle and the calanchi rills and gullies tend not to degrade into biancane.

Where overall relative relief is low, the enlargement of gently inclined, reticulated pipe networks may cause sections of the landscape to subside and form low-angle miniature pediment-like features, leaving biancana cones as residuals. Material is carried from the surface of cones by dispersion and suspension during slopewash, but as most dispersion occurs in clay-size particles, sediments of larger diameter tend to be deposited on the gentler slopes of the 'pediments' which thus have a low ESP, as the following example shows:

	%clay	%silt	%sand	Slope	CEC	ESP
Biancana surface	45.28	51.32	3.40	46°	32.80	57.41
Wash slope at base	15.73	82.04	2.23	3°	36.10	5.54

(Depth of samples 0 - 3 cm; orientation N318^{o}E (NW)).

For such a system to be self-maintaining (so that the biancane do not quickly disappear) the underlying sediments must be highly erosive, and an ESP of at least 30 percent is usually necessary.

Lack of an explanation for the close correspondence between sodium contents of samples obtained from calanchi and intermediate forms is perplexing, but may be related to two factors: firstly the complexity of the landforms means that many aspects of morphology must be grouped within the broader landform categories (for example, all three types have eroding slopes and depositional basal fans); and secondly, the sediments may be subject to 'demixing' between sodium- and calcium-rich areas (Graf and Arora, 1972, p. 106), which

imparts a potentially random element to the sodium content of individual samples. Some samples show a progression from calanchi, through intermediate forms, to biancane in terms of exchangeable sodium; and the sodium absorption ratio (SAR), relating sodium to total dissolved salts, indicates that this relationship may hold for diverse concentrations of salts:

Sample	Total dissolved salts Ca+Mg+Na+K meq/litre	ESP	SAR meq/litre
1 Trecancelli calanchi	2.94	21.29	5.56
5 Scalo Grassano calanchi	5.10	20.14	3.97
24 Aliano intermediate form	1.97	38.01	10.36
25 Aliano intermediate form	7.84	46.86	10.50
21 Carlillo biancana	8.10	66.57	13.65

However, further analysis is required before such results can be considered representative of a general rule.

In addition to material properties, climatic factors may be crucially important to the rate and nature of calanchi development. Investigation of climatic influences on the calanchi, notably humidity (Vittorini, 1974), temperature (Rapetti and Vittorini, 1975), prevailing wind (Rapetti and Vittorini, 1972), desiccation (Lulli and Ronchetti, 1973) and wetting and drying cycles (Panicucci, 1972; Sfalanga and Rizzo, 1974), all confirm that southerly slope orientation can crucially affect the slope angle by exposing the underlying cohesive sediments to increased weathering and erosion. Table 5 gives some climatic parameters derived from the daily rainfall record for the period 1951-68 at four locations in the Basento Valley in Basilicata, of which Grassano and Ferrandina are, broadly speaking, zones in which the growth of biancane is much more restricted than the growth of calanchi, whereas Pisticci and Bernalda are predominantly zones of biancane. It can be seen that relative to the other two locations Pisticci and Bernalda experience long dry spells and short, intense wet spells (precipitation intensities of 90 mm/hr have been recorded in this area (Alexander, 1977)). Such weather patterns increase the intensity of desiccation cracking and penetration by rainwater during rehydration and thus both slope erosion and piping may be more intense there (Vittorini, 1979).

Conclusion

The variety of landforms known as 'calanchi' are distinct from badlands because their morphology is derived specifically from the behaviour of cohesive sediments. The calanchi are at their most characteristic in homogeneous, overconsolidated silt-clays or clayey-silts that have recently been uncovered and very rapidly incised. Calanchi erosion involves a relationship between mass-movement and the incision of channels, and morphological variations are complicated by the presence of pipes, which may be vertical or inclined, and which provide an alternative to surface drainage. Cohesive sediments with a high percentage of exchangeable

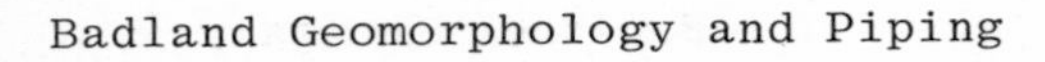

Table 4.5 Climatic parameters derived from daily rainfall data, 1951-68, for locations in the Middle Basento Valley

	Grassano	Ferrandina	Pisticci	Bernalda
Mean no. of dry days per annum	267.0	266.4	288.9	290.3
Mean length of dry spell (days)	5	5	6	6
Max. length of dry spell (days)	65	53	68	92
Max. rainfall of wet spell (mm)	212.7	232.3	398.1	232.3
Mean total rainfall of wet spell (mm)	13.32	11.03	15.95	14.98
Mean daily rainfall of wet spell (mm)	6.02	5.16	7.43	7.46
Mean length of wet spell (days)	1	1	1	1
Max. length of wet spell (days)	14	10	10	10
Max. rainfall in one day 1951-68 (mm)	137.0	139.0	314.6	155.0
Mean recurrence interval of storms over 50 mm/day (days)	429.4	636.4	225.5	265.7
Mean antecedent precipitation of storms over 50 mm/day (mm)	17.05	31.84	19.83	25.87
Mean no. of wet days immed-iately prior to 50 mm+ day	1.53	1.55	1.42	1.50
Mean no. of dry days prior to wet spell containing 50 mm+ day	2.87	4.18	6.46	4.71

sodium are most susceptible to the formation of biancane, especially if they have a low overall relief and occur in zones where the climate is sufficiently intense to produce substantial wetting and drying cycles.

It is clear that calanchi in Italy have a wide variety of causes, among which no particular process is exclusively dominant. Although the first scientific studies of the calanchi began 100 years ago (Bombicci, 1881), the behaviour of these landforms is still very poorly understood. There is now a special need for detailed studies of the inter-relationships among various cation-exchange systems which influence the erodibility of sediments in which calanchi form, and of the role of mass-movements and climate in creating typical calanchi morphology.

Calanchi erosion is frequently both spectacular and destructive. It occupies at least 2500 km^2 of Italy and threatens settlements, buildings and lines of communication throughout the country (Almagià, 1907, 1910), but the main effect is to destroy farmland by eroding productive soils, steepening slopes beyond the limits of cultivation and breaking up cultivable land into a series of cones or knife-edged ridges. Various strategies of amelioration have been adopted, including soil conditioning (Chisci, 1978), re-planting of bare slopes (Scategni, 1971) and bulldozing eroded slopes into a profile suitable for recultivation (Calzecchi-Onesti, 1957). All have yielded short-term results, but it is too early to be able to observe the long-term effects. Reasoning deductively, calanchi and biancane should require diverse strategies of amelioration: different surface profiles and surface and sub-surface drainage networks have to be prevented from re-emerging. It is therefore crucial to understand the fundamental difference between them in order to apply appropriate strategies.

Acknowledgements

The author is grateful to Dr H.M. Rendell for the landslide data reproduced in table 3, and to the Natural Environment Research Council (U.K.) for financial support.

5 Spatial variations in infiltration, runoff and erosion on hillslopes in semi-arid Spain

H. Scoging

A number of papers since Betson (1964) and Hewlett and Hibbert (1967) on the processes of runoff generation in humid regions have suggested that the non-uniformity of response is due to spatial variations in soil moisture conditions. The most effective areas in terms of contribution to storm runoff are those zones marginal to the channel network where the soil is near saturation. This concept of partial area contribution has not found much favour in arid or semi-arid areas where rainfall is scarce, evaporation high, and soil cover, where it exists, is shallow, patchy and liable to crusting. Under such conditions it would seem more probable that spatial variations in the hillslope response are due rather to differences in surface properties.

Various different surface characteristics have been used to explain patterns of runoff response: Yair (1974) noted that such spatial variations in a first order basin are controlled by two topographic variables, slope length and slope angle, which are responsible for variations in other surface properties, notably soil depth. In the more extreme conditions of a scree slope, lateral variations in scree mantle properties had a decisive effect on runoff generation (Yair and Lavee, 1974). In their study of Dinosaur Provincial Park Badlands Bryan *et al.* (1978) reported considerable non-uniformity in runoff production, despite insignificant differences in antecedent moisture patterns, due to variations in the composition and structure of surface material. Bork and Rohdenburg (1979) attribute surface runoff patterns to topographic elements, and more importantly to the volume of coarse pores in the surface layer.

The classical Hortonian model of overland flow (Horton, 1945) assumes that runoff at a distance X from the divide is equal to a simple difference between rainfall intensity and infiltration capacity.

Eq. 1 $$q_x = X(p - i)$$ where q = runoff, p = rainfall intensity, i = infiltration capacity.

In discussions concerning the predominance of throughflow on vegetated slopes with thick soil covers (Whipkey, 1965), the applicability of the Hortonian model has been questioned (e.g. Kirkby and Chorley, 1967). Variations in the areal distribution of soil moisture, throughflow and surface runoff were thought to be a function of the lateral anisotrophy of subsoil and slope microtopography. Numerous authors suggested that the Hortonian model would be most appropriate in arid and semi-arid areas with little soil development and low infiltration capacity. However, recent research is now casting doubt on that assumption. Yair and Klein (1973) showed no clear relationship between slope angle and runoff, and an inverse relationship between slope angle and erosion due to the textural variations of slopes. De Ploey *et al.* (1976) argued that observed discrepancies between the final infiltration rate and rainfall intensity cannot be explained by a simple difference, as assumed in the generation of Hortonian runoff, when the former is greater than the latter. Considerable variation in infiltration characteristics over hillslopes is reported by Sharma *et al.* (1980). The question is made more complex on unprotected surfaces by the effect of raindrop impact which can destroy surface soil aggregates and form a continuous crust with a lower infiltration capacity. Morin and Benyamini (1977) argue that crust development on bare soils is more important in controlling water absorption than antecedent soil moisture.

Mein and Larson (1973) developed an infiltration model for situations where rainfall intensity is less than the infiltration capacity of the soil. There is a period after the onset of rainfall, for any given intensity, during which no runoff is produced because the surface is unsaturated, and infiltration is flux controlled. Once surface saturation occurs runoff ensues, and infiltration begins to fall systematically due to profile control. For a given set of antecedent conditions, the lower the rainfall intensity the longer is the time to reach saturation, or the time to runoff. The resulting family of curves (figure 1) has been called the infiltration envelope (Smith, 1972). This model has been developed for semi-arid Spain (Scoging and Thornes, 1979), where very short times to runoff were shown to reflect a limited available soil storage volume rather than profile controlled infiltration as assumed in Hortonian flow.

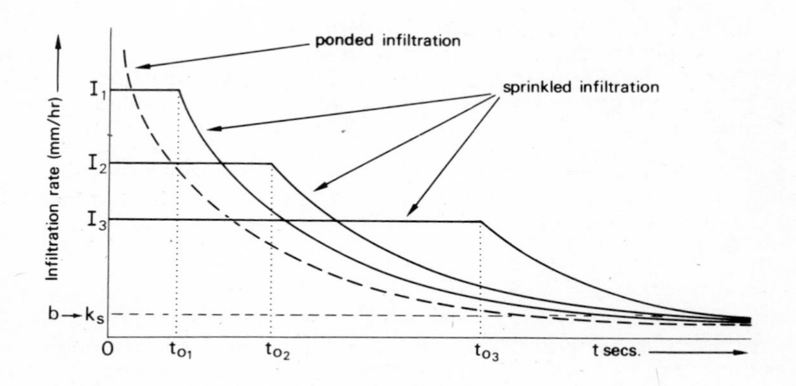

Figure 5.1 The infiltration envelope (Smith, 1972)

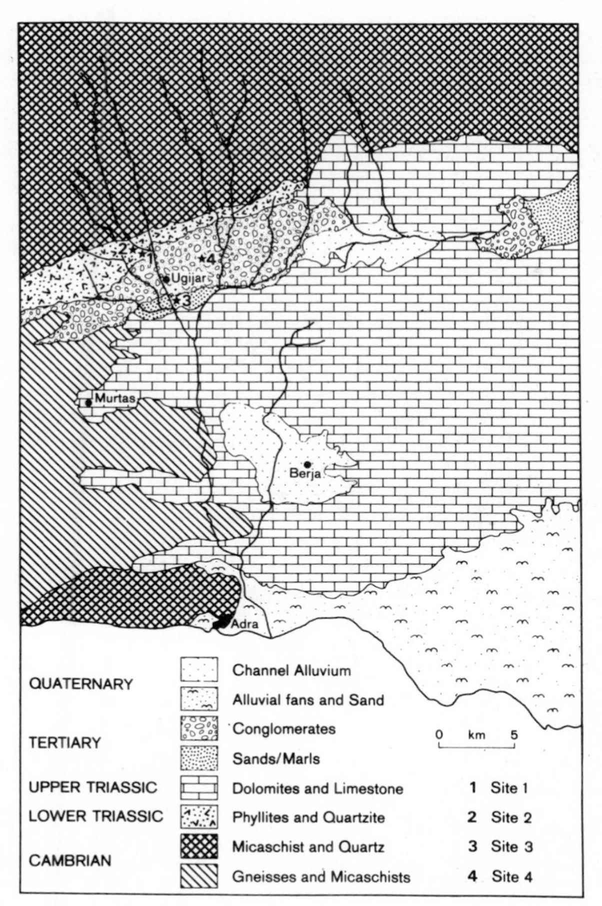

Figure 5.2 Ugijar regional geology

The Study Area

In southeast Spain a belt of severely eroded lands, largely coincident with the Betic-Penebetic Cordillera system, extends from Valencia to Malaga. Fieldwork was conducted in the central part of this badland region, near the town of Ugijar. Thornes (1976) provides a detailed account of the area, which essentially comprises a Tertiary and Quaternary deposit-filled central depression, bounded to the north by the highlands of the Sierra Nevada, and to the south by the folded nappes of the Alpujarride system (figure 2).

The area has a Mediterranean climate with aridity increasing from the Sierras toward both east and south. A hot, dry summer from May to October is followed by a cool, moist winter season, with depressional rainfall peaks in December and April. Daily and seasonal rainfall characteristics are described by Scoging (1976) and Elias (1963).

Analysis of 143 storm traces, 1973-75, indicate that over 62% of maximum 10 minute rainfall intensities are less than 0.75 mm, 57% of all storms produce less than 1.5 mm rainfall, and that 76% of storms are less than half an hour long, of which half are only 10 minutes in duration.

Soils are poorly developed, often little more than weathered regolith, and are judged by ICONA as highly susceptible to erosion. Aggravating the association of poor soils with steep slopes subject to intense rainfall is the comparative absence of any protective vegetation cover. Where it exists it takes the form of a patchy, degraded *mattoral*.

Four experimental sites were selected for a consideration of soil erosion processes and patterns, and site characteristics are presented in table 1. Although on very different lithologies the dominant process on each site was considered to be slope wash (despite localised rilling on part of site 3). Evidence for this included the absence of incipient or developed rills and gullies, surface wash patterns (plate 1), wash deposits at major breaks of slope, and rapid infilling of soil pits after storms.

Table 5.1 Experimental site characteristics

	Site 1	Site 2	Site 3	Site 4
Dimensions W.L. metres	6 x 30	6 x 36	variable x 36	6 x 60
Mean slope°	20	35	20	15
Parent material	micaschist	micaschist	marl	Tertiary sands
Vegetation type	thorny mattoral	shrubs	grasses	stubble, weeds
Vegetation %	20	35	5	10
Soil classification	silt-loam	sandy-loam	silt	loamy-sand
Mean particle size Phi.	4.15	2.44	5.24	2.08
% soil coarse fraction by weight	12.53	40.32	10.34	41.47
Coarse-fine ratio	0.55	4.20	0.37	2.37

Field Experiments

Four types of field experiment were carried out during 1975 and 1976.

(i) ponded infiltration tests
(ii) sprinkled infiltration and runoff experiments
(iii) storm runoff and sediment loss
(iv) soil pin erosion patterns.

Plate 5.1 Surface wash patterns on site 3

The first three involve the hydrological response of the hillslope to water input, while the last provides some indication, albeit for a short period, of the spatial and temporal patterns of soil loss and deposition.

Ponded Infiltration Tests

A single ring 15 cm unbuffered infiltrometer was used following the techniques described by Hills (1970) which are reported satisfactory over the range 30-500 mm hr^{-1}. Fifty-nine tests were performed with 13-18 on each site arranged systematically across and downslope. Excavations after each test indicated that lateral seepage was not a problem, and also provided information on wetting depths. In an experiment designed to evaluate the effects of variations in ponding head on final infiltration rates no statistical relationships were found.

Two empirical curves were fitted to all infiltration series -

Eq. 2 $i = A.t^{B}$ (Kostiakov (1932) equation)

Eq. 3 $i = A + B/t$ (an equation similar to the Philip (1957) equation)

where i = infiltration rate

t = time elapsed since ponding

A, B = parameters.

In every case the explained variation associated with equation 3 was greater than that for equation 2, and in all but 5 cases, where the significance level was 95%, the level reached 99%.

Sprinkler Experiments

These small plot (1 m^2) experiments involved a portable rainfall sprinkler of the 'Warley' type constructed by Eclipse Ltd. Rates of water application were controlled by a pressure valve, and results were obtained for a delivery rate of 0.773 l m^{-1} giving an intensity of 44 mm hr^{-1}. Laboratory experiments showed a spatially uniform rate of application under still air conditions. Water was applied to 1 m long plots for approximately 10 minutes and runoff collected at the bottom of the plot at 0.25 to 0.5 minute intervals. In all cases an equilibrium discharge was reached. Total sediment lost from the plot was collected in a Gerlach (1967) trough.

With a known rate of input and a measured rate of output the difference is ascribed to infiltration losses, although this is recognised as an overestimation due to some detention storage and a small amount of vegetation interception. Some 25 paired experiments were performed providing infiltration and runoff data for soils initially dry, and repeated 15 minutes later when the soils were assumed to be near field capacity in the surface horizon. In each case the time to runoff was observed, following the procedure of Rubin and Steinhardt (1964), with the addition of a final stage for visible flow at the surface. Following the successful use of equation 3 to describe infiltration characteristics the curve was fitted to the sprinkled infiltration series with significance levels in all cases reaching 99%.

Storm Runoff

Runoff and sediment collecting devices were installed at the midpoint and base of each site. Plastic barriers enclosed a contributing area of 14.63 m^2 with a plot length of 9.88 m. With the aid of a tipping bucket rainguage, natural storm yield, runoff and sediment data were collected for six events in 1976 at both slope positions. Table 2i provides magnitude and frequency evaluations of the storms. Individual storm traces for 1973-1975 have been analysed for duration and yield characteristics (table 2ii), and the graph of relative frequency of occurrence for storm yield and duration (figure 3) suggests that observed storms are fairly representative of natural conditions, with the exception of low yield, long duration events, which are considered relatively unimportant for runoff and erosion processes.

Soil Pin Erosion Patterns

Some 240 soil pins (Haigh, 1977) were positioned systematically over all sites in October 1975 and resurveyed in May and October 1976, providing information on the spatial patterns of ground lowering for a winter and summer season respectively.

Isopleth maps (figures 4, 5) were generated from the University of London computer centre program SYMAP for soil loss patterns derived from soil pin data. The results for site 3 and site 4 are well representative of the spatial

Table 5.2i Characteristics of observed storms, Sept - Oct 1976

	1	2	3	4	5	6
Storm yield, mm	2.75	4.25	0.75	4.75	3.0	0.25
Storm length, mins	30	50	10	30	10	10
% frequency* Duration	20.3	4.1	49.5	20.3	49.5	49.5
% frequency* Yield	5.59	18.88	18.88	18.88	5.59	28.67

*calculated from all storms 1973-1975

Table 5.2ii Duration and yield frequency distributions for storms 1973-1975

Yield mm	% frequency	Duration mins	% frequency
0.1 - 0.5	28.67	10	49.5
0.6 - 1.0	18.88	20	17.9
1.1 - 1.5	9.79	30	20.3
1.6 - 2.0	8.39	40	5.7
2.1 - 2.5	4.19	50	4.1
2.6 - 3.0	5.59	60	2.4
3.1 - 3.5	2.79	(86% of all storms are < 60mins)	
3.6 - 4.0	2.79		
4.1 <	18.88		

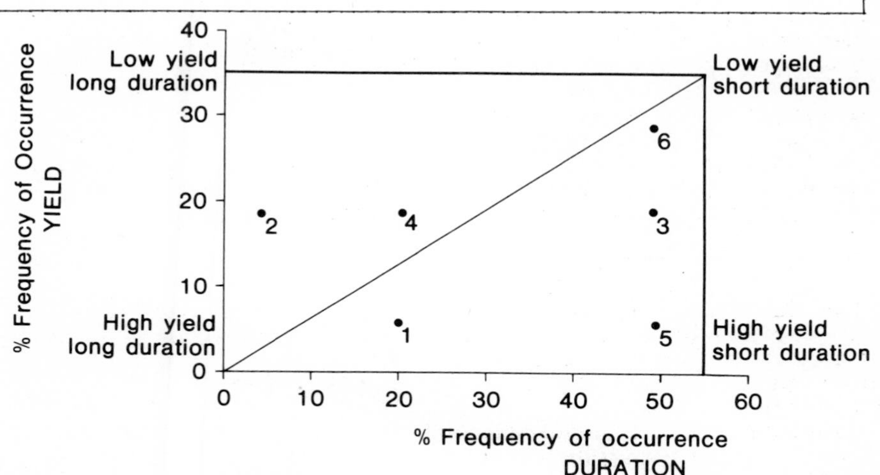

Figure 5.3 Yield and duration characteristics for six observed storms

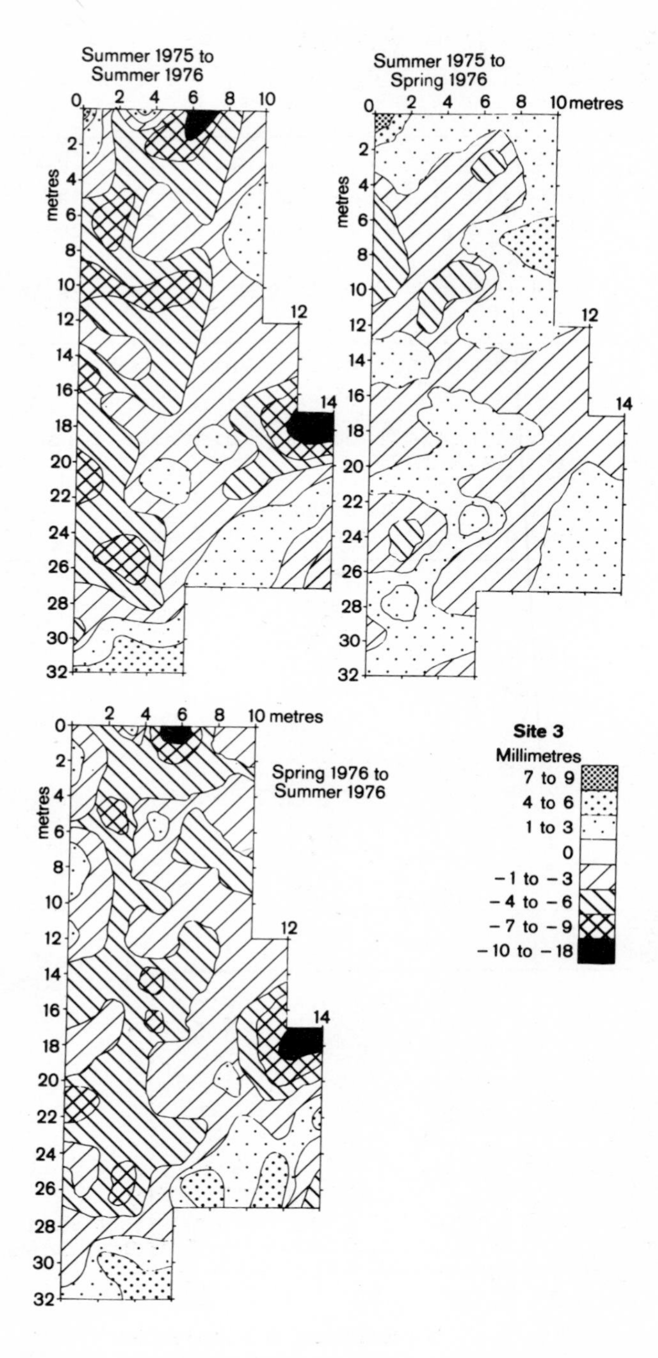

Figure 5.4 Patterns of soil erosion on site 3 (fine)

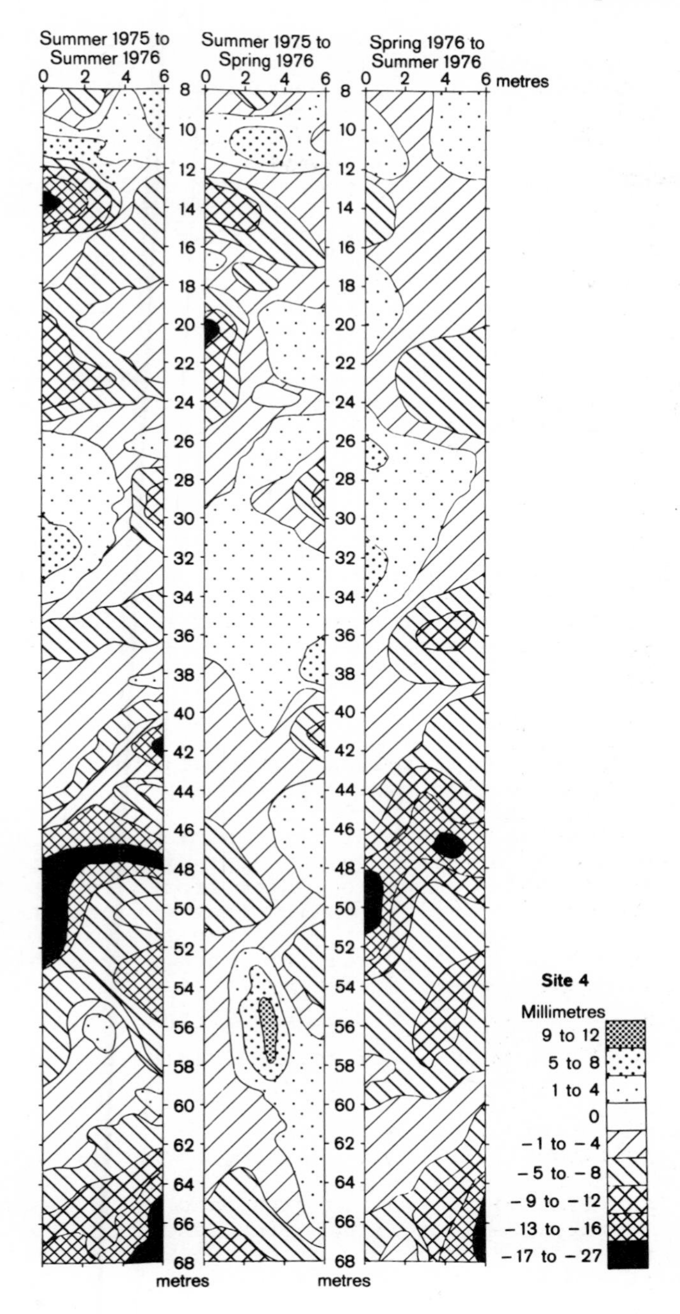

Figure 5.5 Patterns of soil erosion on site 4 (coarse)

patterns of erosion and deposition for fine and coarse grained soils respectively.

The annual pattern of erosion on site 3 (fine) shows a very small area of deposition at the top of the site derived from material eroded from a steeper slope above. The upper part of the site is dominated by soil loss (locally up to 18 mm ground lowering, but averaging 6 mm lowering). At the base of the slope, deposition, up to 6 mm is occurring.

The pattern on site 4 (coarse) is complex despite the uniformity of slope angle over this site: localised areas of high erosion (up to 27 mm ground lowering) are interspersed with patches of deposition bearing no relationship to local slope regimes. Overall however more erosion occurs towards the base of the site.

Important seasonal contrasts exist on all sites. Winter erosion is considerably less than erosion during the summer months. Summary statistics for soil pin data are presented in table 7. On all but site 1, where summer and winter annual equivalents for soil loss are approximately equal, erosion is dominated by summer activity. Summer loss on site 3 is up to 20 times winter loss. This may be explained by rare but intense summer storms of short duration which have higher kinetic energy than winter depressional rains of longer duration but lower intensities. In addition, the first rains marking the end of summer appear to flush out dry, friable soil, reducing the pre-detached supply of material for erosion by winter rainfall.

Experimental Results

Tables 3i-iii give mean site values for the fitted A and B parameters, measured wetting depths for ponded infiltration tests, and some indication of variations in parameter values with distance downslope. The best fit equation, $i = A + B/t$, is directly analogous to the Green and Ampt (1911) equation if it is assumed that the amount of water stored in the soil is a linear function of time. Assuming that A represents the final infiltration rate when $t \rightarrow \infty$, then the volume of water stored to the point where rainfall intensity equals the final infiltration rate is given by

Eq. 4 $V = pB/(p - A)$ where V = storage volume

p = rainfall intensity

A, B = parameters.

This expression converges on parameter B, directly related to soil storage capacity, when $(p - A)$ approaches p, i.e. when A is small or p is very large.

Both A, the final infiltration rate, and B, an indicator of soil storage, reflect lithological and soil characteristics such that the highest values are associated with the coarse particle sands and schists of sites 2 and 4, (with correspondingly higher variances), the lowest with the fine silts of site 3, and intermediate values with the silt-sands of site 1. This pattern is also mirrored in mean wetting depths, although the expected difference between medium and coarse sites is rather less due to the very shallow soils of site 2.

Table 5.3i Spatial variations in ponded infiltration characteristics. A coefficient, mm min^{-1}

Site 3		Site 1		Site 2		Site 4	
m downslope	A	m downslope	A	m downslope	A	m downslope	A
1	3.01	2	5.12/6.23	1	4.85/3.21	2	6.08/5.32
2	2.16/2.52	5	5.61	3	5.35	5	5.76
4	1.27/1.58	9	4.11/3.06	5	6.55/4.38	7	5.68/6.46
6	0.97	12	3.18	7	4.24	10	5.76
8	2.61/1.85	15	1.43/1.10	9	4.66/5.63	13	5.86
10	1.70	19	1.24	11	9.99	15	5.75
11	1.80	20	1.40/2.38	13	6.87/9.19	18	4.42/5.44
13	0.85	22	0.69	15	8.94	20	5.97
15	1.05	24	0.70/1.19	17	8.86/9.82	24	7.39
17	0.49					26	5.57
						29	11.87
						32	6.59/7.07
						34	7.98
						36	9.09
Summary statistics							
$\overline{X}$	1.68		2.67		6.61		6.55
σ	0.72		1.84		2.24		1.66
Regression analysis							
R^2	0.476		0.852		0.679		0.330
sig%	99.0		99.9		99.9		97.5
a	2.46		5.88		3.49		4.97
b	-0.101		-0.227		+0.346		+0.080

Spatial variations in A, B and wetting depths for ponded infiltration tests were analysed through simple linear regressions, with distance downslope as the independent variable. All regressions were significant at the 99% level with the exception of site 4, A coefficient (97.5%) and B coefficient (not significant). On coarse sites 2 and 4 the A coefficient, representing the final infiltration rate, is strongly positively related to distance downslope, while the reverse is true for the fine sites 1 and 3. This relationship is also reflected in wetting depths. With the exception of site 4 the opposite relationship holds for the B coefficient, B values increasing downslope on fine sites, and decreasing on coarse sites. High B values indicate a slower decline of the infiltration curve through time, with a lower rate of water absorption than for smaller B values.

Although the results of the ponded infiltration tests provide good relative data on the spatial patterns of the A and B parameters it is recognised that the final infiltration rates are rather high, and certainly higher than any natural rainfall intensities if a simple Hortonian difference

Table 5.3ii Spatial variations in ponded infiltration characteristics. B coefficient, mm

Site 3		Site 1		Site 2		Site 4	
m downslope	B	m downslope	B	m downslope	B	m downslope	B
1	2.22	2	0.74/0.81	1	3.27/3.79	2	3.24/4.62
2	1.05/0.94	5	1.32	3	2.92	5	6.04
4	1.06/1.58	9	1.59/1.75	5	2.63/2.91	7	1.34/4.74
6	1.92	12	1.91	7	2.96	10	2.86
8	2.09/2.03	15	1.81/1.92	9	7.96/2.48	13	1.15
10	2.35	19	1.44	11	2.20	15	5.42
11	2.21	20	2.30/2.51	13	2.71/2.71	18	3.39/3.44
13	2.13	22	2.02	15	1.71	20	3.34
15	2.97	24	3.33/2.58	17	1.06/0.48	24	1.89
17	2.70					26	2.49
						29	22.11
						32	2.05/6.35
						34	1.66
						36	10.75
Summary statistics							
$\overline{X}$	1.94		1.86		2.84		4.78
σ	0.59		0.66		1.64		4.78
Regression analysis							
R^2	0.628		0.744		0.730		0.090
sig%	99.0		99.9		99.9		NS
a	1.205		0.777		3.650		2.300
b	+0.095		+0.076		-0.133		+0.130

(rain-infiltration rates) is assumed for runoff generation. Some indication of natural conditions is provided by the results of the sprinkler experiments. Table 4 gives mean A and B parameters for the sprinkled infiltration series, and observed values for the time to runoff, T_0. Mean A and B values for paired experiments (soil initially dry, and near field capacity at the surface) show the expected relationship with coarse and fine sites: low final infiltration rates, A, and low storage values, B, on fine sites, and higher A and B values on coarse sites for both dry and wet initial conditions.

It is significant to note that two equations, derived from the infiltration equation (3) when p = A, i.e. at T_0, for storage volume, V,

Eq. 4 $V = pB/(p - A)$

and for predicted time to runoff, T_0,

Eq. 5 $T_0 = B/(p - A)$ under conditions of constant rainfall intensity, both converge on a fixed positive value related

Table 5.3iii Spatial variations in ponded infiltration characteristics. Wetting depth, mm

Site 3		Site 1		Site 2		Site 4	
m downslope		m downslope		m downslope		m downslope	
1	190	2	270/255	1	210/190	2	210/205
2	195/185	3	340	3	190	5	290
4	185/175	9	280/310	5	180/205	7	230/270
6	135	12	330	7	230	10	325
8	175/130	15	295/320	9	255/290	13	240
10	120	19	270	11	310	15	470
11	105	20	210/260	13	340/355	18	430/420
13	115	22	110	15	375	20	360
15	105	24	165/110	17	260/385	24	480
17	90					26	465
						29	420
						32	430/450
						34	550
						36	480
Summary statistics							
$\overline{X}$	147		252		270		374
σ	36.96		73.10		70.20		105.89
Regression analysis							
R^2	0.852		0.435		0.723		0.750
sig%	99.9		99.0		99.9		99.9
a	199.80		342.84		169.06		221.89
b	-6.83		-6.43		+11.17		+8.27

to B, assumed to represent the storage value of a soil. Figures 6i-iii show the relationship between observed T_0, and the fitted A and B parameters and storage volume, V, for both series and indicate the very strong correlation (99% significance level) between time to runoff and soil storage.

Nomographs of T_0 and V against B values for various A coefficients are presented in figure 7 for a uniform rainfall intensity. With the evidence derived from the relationships between A and B parameters and distance downslope rather different values for T_0 and V are predicted for coarse and fine sites. At the top of a fine site, high A and low B values predict small T_0 and V values (figure 7, point F1). At the base of a fine site, lower A and higher B values would suggest higher values of T_0 and V (point F2). The limited data on times to runoff for sites 1 and 3 in table 5 provides some experimental evidence for this predicted relationship. Conversely, the combination of increasing A and decreasing B values downslope on coarse sites would lead to relatively uniform values of T_0 and V (points C1 and C2)

Table 5.4 Mean A and B parameters, and observed time to runoff, T_o for sprinkler tests

a) Soil surface initially dry

		Site 3	Site 1	Site 2	Site 4
A coefficient.	$\bar{X}$	.1660	.1513	.2743	.3594
	σ	.0172	.0294	.0399	.1091
B coefficient.	$\bar{X}$	.1901	.1550	.2557	.2073
	σ	.0796	.1130	.0848	.0809
Time to runoff.	$\bar{X}$	23.00	15.00	31.66	29.23
	σ	10.99	12.11	11.23	7.72

b) Soil surface initially wet

		Site 3	Site 1	Site 2	Site 4
A coefficient.	$\bar{X}$	.1139	.1667	.2426	.2328
	σ	.0614	.0511	.0161	.0301
B coefficient.	$\bar{X}$	.1326	.1149	.1604	.1575
	σ	.0389	.0364	.0619	.0256
Time to runoff.	$\bar{X}$	12.50	11.75	18.66	17.00
	σ	4.88	4.85	7.66	3.36

Table 5.5 Spatial variations of observed times to runoff, T_o, secs. (dry/wet initial conditions)

Site 3		Site 1		Site 2		Site 4	
m	T_o	m	T_o	m	T_o	m	T_o
1	12/11	2	7/9	5	44/27	2	34/12
2	23/9	7	(9/6)	9	23/-	7	20/14
8	22/11	12	9/9	13	29/17	13	29/20
10	35/18	19	33/19	17	22/12	20	44/17
12	36/19	24	11/10			24	24/22
13	10/7	25	20/-			32	27/17
15	23/1					36	27/17

Mean calculated for paired data only

( ) Input rate at 0.561 mm min^{-1}

all others at 0.733 mm min^{-1}

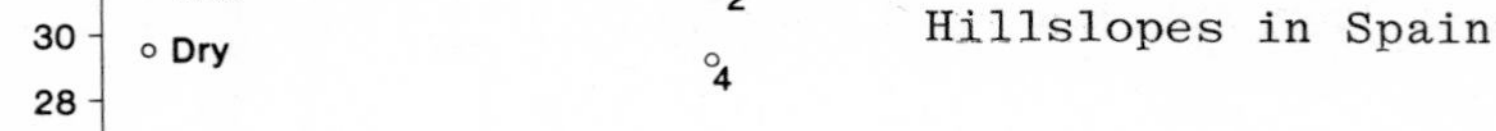

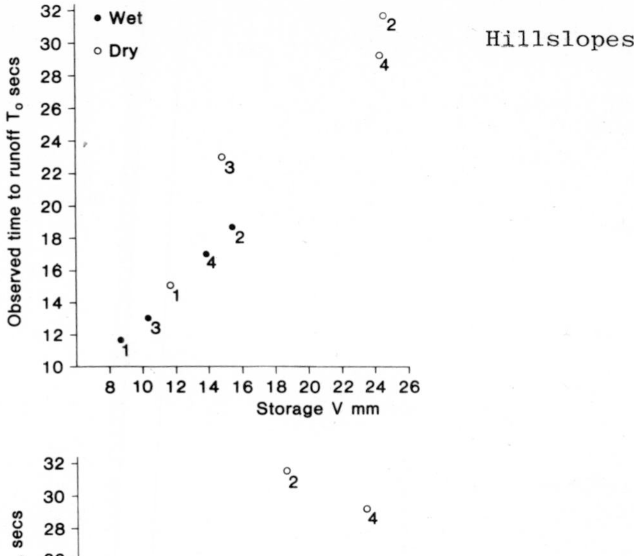

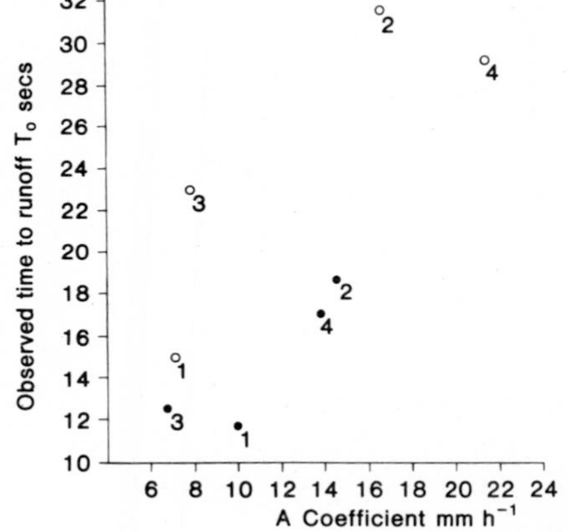

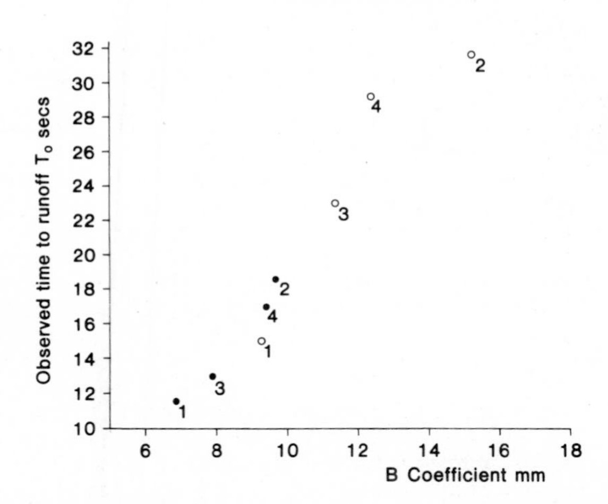

Figure 5.6 Relationships sprinkler parameters and time to runoff
i) Time to runoff and storage, V
ii) Time to runoff and A coefficient
iii) Time to runoff and B coefficient

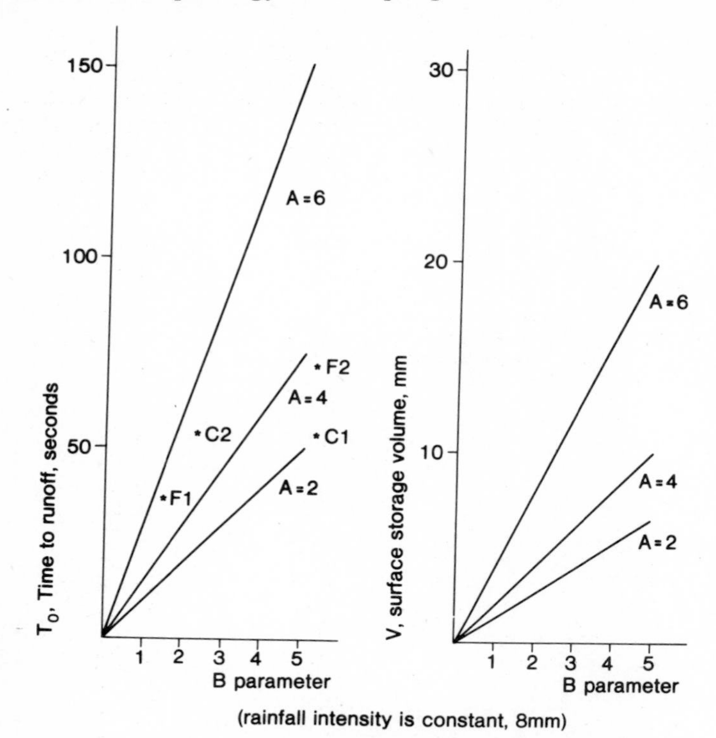

ıre 5.7 Nomographs of T_0 and V against B for various A coefficients

the whole site. Observed values of T_0 for site 2, and ·articular site 4 in table 5 confirm this uniform spatial ern.

With infiltration characteristics and runoff response s clearly dependent on the storage capacity of soils, relationship between these parameters and soil character- cs is expected. Spatial variations in particle size dis- utions are presented in table 6. A decrease in particle seness downslope is apparent for sites 1 and 3, propor- s of sand and coarse fraction decreasing from top to base lope by 12% and 9% respectively, matched by an increase he silt fraction. The implication is that fine material referentially removed from the top of the site and accum- ing towards the base. This is in accord with runoff onse rate being faster at the top of fine sites. On se sites, however, a slight increase in particle coarse- occurs downslope with increases in sand and coarse frac- of 7% and 8% (site 2) and 4% and 6% (site 4) respec- ly. One explanation is that finer material is selectively ved from the base of the slope leaving behind a coarse deposit. Low storage values at the base of coarse sites d account for greater runoff occurring at this position.

Table 5.6 Spatial patterns of particle size distributions

	metres downslope	Fine soil analysis %sand	%silt	%error	Total soil % by weight of coarse fraction >2mm
Site 1.	3	42.82	54.00	3.18	16.07
	9	36.65	59.65	3.70	11.64
	15	39.32	57.65	3.03	12.89
	21	30.99	66.99	2.02	7.30
Site 3.	2	16.66	82.98	0.36	10.15
	7	15.32	84.50	0.18	11.06
	8	15.15	84.82	0.03	11.81
	12	15.82	83.99	0.19	9.31
	14	11.15	88.82	0.03	4.69
	17	11.82	88.16	0.02	8.96
Site 2.	3	62.82	36.82	0.36	38.97
	7	62.75	37.00	0.25	40.45
	11	70.25	29.75	0	42.24
	15	69.32	30.66	0.02	46.51
Site 4.	3	71.49	28.49	0.02	38.38
	8	72.15	27.82	0.03	41.40
	13	75.16	24.83	0.01	40.39
	18	76.32	23.66	0.02	41.93
	23	74.66	24.49	0.85	42.01
	28	73.83	26.16	0.01	43.60
	33	74.99	24.99	0.02	45.93
	38	75.32	24.66	0.02	44.35

Results for infiltration and runoff response for small plot studies give some indication of the spatial variation of these characteristics, but afford little evidence for total runoff patterns. The relationship between small scale studies under artificial conditions to larger plots for natural storm events is notoriously difficult (Amerman and McGuiness, 1967), but certain trends seem common to both sets of data.

Table 8 and figures 8 and 9 show runoff (l) and sediment (g) for each of six rainfall events at the top and base of all sites. Since runoff plots are equal in size direct comparisons between runoff and sediment loss for all sites and events may be made. Runoff coefficients (percentage of rainfall which is converted to runoff, m^{-2}) and sediment concentrations (gms $l^{-1}m^{-2}$) are also calculated (Table 8 and figures 10, 11).

As expected highest runoff values occur on fine sites, lowest on coarse sites. On the former more runoff occurs at the top of the slope than at the base, while the reverse is true for coarse sites. A slight decrease in the values of runoff coefficients occurs on all sites from low to high

Table 5.7 Summary statistics for soil pin erosion

Key Kg/m² tons / acre mm lowering	Site 1	Site 2	Site 3	Site 4
Summer 1975 - Summer 1976	3.69 14.73 2.31	2.97 11.86 1.86	4.40 17.54 2.75	9.46 37.71 5.91
Oct. 1975 - May 1976	2.05 8.17 1.28	1.06 4.25 0.66	0.13 0.52 0.08	2.34 9.35 1.46
Winter annual equivalent	3.51 14.01 2.19	1.82 7.28 1.14	0.22 0.89 0.14	4.02 16.02 2.51
May 1976 - Oct. 1976	1.56 6.23 0.97	1.91 7.61 1.19	3.99 15.89 2.49	6.69 26.66 4.18
Summer annual equivalent	3.75 14.96 2.34	4.58 18.27 2.86	5.58 22.24 3.49	16.06 63.98 10.03
Area of site m²	288	216	340	378
No. of pins	48	36	85	63
Missing pins	3	0	0	3

yield storms. This suggests that the conversion of rainfall to runoff is more efficient at lower rainfall intensities than for higher yields, and may be explained by the fact that both low yield storms occurred immediately after longer storms with wetter antecedent conditions.

Overall, the highest sediment losses occurred on coarse soils but sediment production is strongly influenced by the magnitude of the storm. Large storms (> 3mm) produce greater soil loss from coarse sites than from fine sites, while the relationship is reversed for lower yield events. During large storms more sediment is removed from the base of site 4 compared with losses from the top of the site, and more from the top of both fine sites, 1 and 3. No downslope pattern of soil loss emerged for site 2 under high yield conditions, nor for any site with low rainfall.

Sediment concentrations show a lack of trend from low to high yield storms. This reflects the combination of high runoff/low sediment production for low rainfall, and high sediment losses/low runoff for higher rainfall yields. Coarse sites have greater sediment concentrations than fine sites reflecting large amounts of sediment loss relative to low runoff amounts. Inter-site variations include areas of

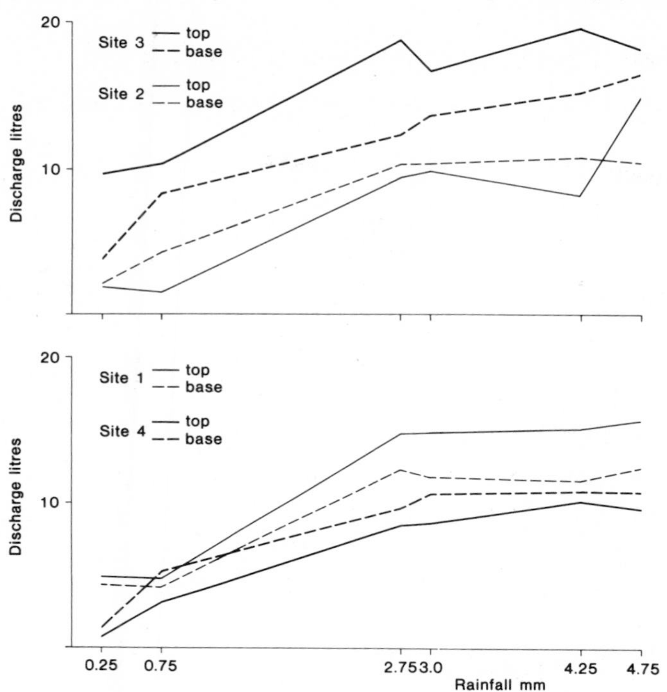

Figure 5.8 Measured runoff for six storms, top and base of site

higher sediment concentration at the top of coarse sites and at the base of fine sites due to lower runoff in proportion to sediment loads.

These results are in accord with the spatial patterns of infiltration and time to runoff values - runoff occurs more quickly and is greater on fine sites, and preferentially so at the tope of the site. Lower runoff occurs more slowly on coarse sites with higher values at their base. Furthermore these results agree with observed variations in particle sizes - maximum runoff at the top of fine sites removes fine-grained material and transports it downslope under decreasing runoff regimes where it is deposited. On coarse sites increased runoff at the base selectively removes fine material leaving behind a coarse lag deposit. These results hold only for larger storms. Under low rainfall intensity runoff is unable to remove sediment from coarse slopes.

One of the most surprising results is the relative magnitude of sediment loss on coarse sites compared with fine ones, despite small runoff values. Two factors contribute to an explanation - differences in soil erodibility, and crusting mechanisms on fine sites. The erodibility of soils, characterised by shear strength, is different on coarse and fine sites. Former soils are loose and friable, and unaffected by particle cohesion. The loose packing of the coarse fraction creates large voids from which finer

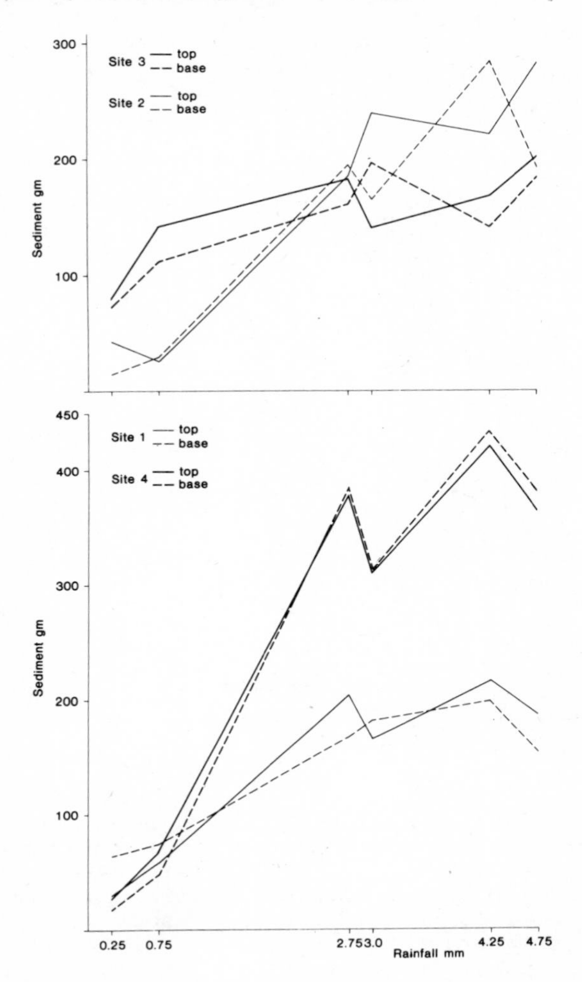

Figure 5.9 Measured sediment for six storms, top and base of site

material is relatively easily removed, thereby undermining gravels and pebbles which tend to roll downslope under relatively low runoff forces. Fine sites however have soils with stronger particle cohesion, a strength which is enhanced at the surface by the development of surface crusts which permit greater runoff. It is only when runoff is sufficient to break the crust that erosion occurs.

Conclusions and Discussion

This study provides observations on the spatial variations in hydrologic behaviour of semi-arid hillslopes of disparate character and at different scales of process operation. In an attempt to link these scales with overall soil loss

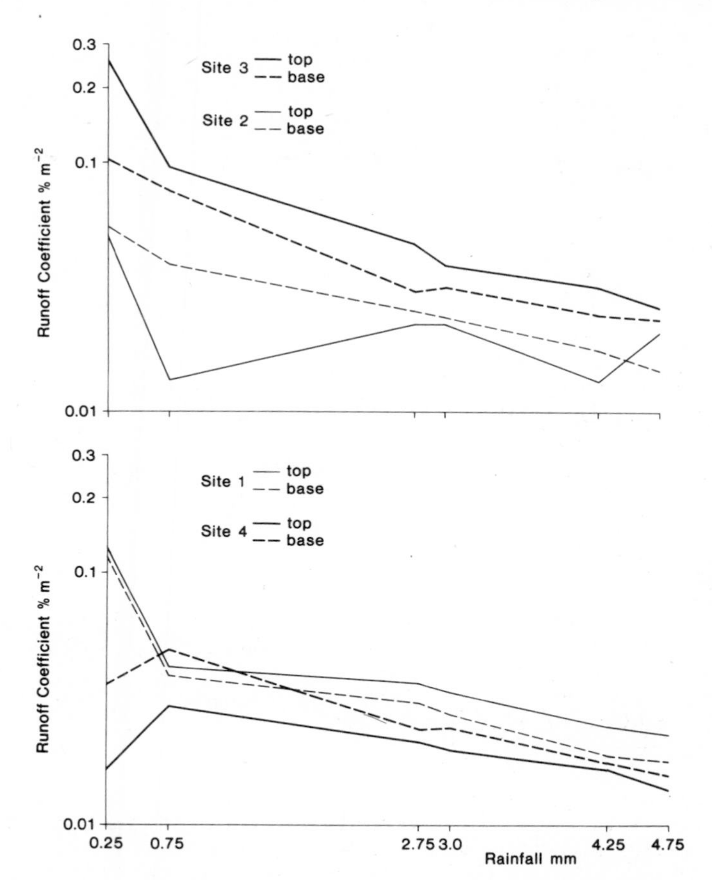

Figure 5.10 Runoff coefficients for six storms, top and base of site

patterns the following conclusions may be drawn:

1. The B parameter, reflecting soil moisture storage has been shown to be more important than the final infiltration rate in determining times to runoff and total runoff volume. It would therefore seem inappropriate to assume a simple Hortonian rainfall excess model under these circumstances.
2. A complex spatial relationship exists on the sites between storage, final infiltration rates, runoff generation and surface soil characteristics. At the top of fine sites larger runoff amounts occur more quickly in response to small storage values, B, despite high A values. Wash processes acting preferentially at this slope position remove material downslope where it is deposited under decreasing runoff regimes. Such a pattern of upslope erosion and downslope deposition predicted by the hydrological data is confirmed by soil

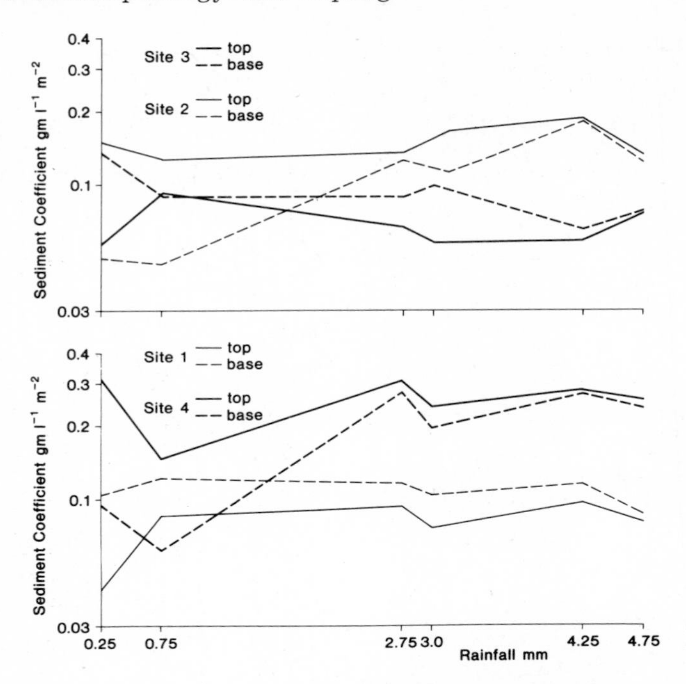

Figure 5.11 Sediment coefficients for six storms, top and base of site

loss/deposition patterns derived from soil pin data, and directly conflicts with the Hortonian notion of a belt of no erosion, and increasing discharge and erosion downslope. On coarse sites the spatial combination of infiltration parameters predicts a relatively uniform runoff response rate over the slopes, but low storage values at the base of these sites encourage higher runoff volumes. At this position silt is selectively removed, along with some coarser material, leaving the coarse fraction behind as a surface lag or armour deposit. This result agrees with observed soil loss patterns for site 4 where erosion is dominant at the base.

The concept of Hortonian overland flow derived from a simple difference in infiltration and rainfall rates, and assumed to produce a uniform runoff response, has been shown to be unsuitable for the study region, superficially because of wide variations in infiltration capacities, themselves often considerably higher than observed rainfall intensities. More fundamental is a real difference in process in denying the suitability of the Hortonian model for runoff generation, which is seen to be related rather to soil storage volumes, which in turn are associated with spatial patterns of particle size distributions.

The nature of this relationship remains unresolved although some tentative suggestions may be proposed. The difficulty revolves around determining cause and effect in the

Table 5.8 Runoff and sediment data for six storms, measured at the top (T) and base (B) of each site

q = runoff, litres
qc = runoff coefficient (%rainfall convered to runoff m^{-2})
s = sediment, gms
sc = sediment concentration (gm $l^{-1}m^{-2}$)

Storm,mm		4.75	4.25	3.0	2.75	0.75	0.25
Site 3.	Tq	18.14	19.62	16.80	18.94	10.37	9.63
	Ts	203	169	142	183	141	80
	Bq	16.40	15.11	13.65	12.37	9.39	3.72
	Bs	184	142	197	161	111	74
	Tqc	.0261	.0315	.0382	.0470	.0945	.2630
	Tsc	.764	.588	.577	.660	.929	.567
	Bqc	.0235	.0243	.0311	.0300	.0764	.1017
	Bsc	.766	.642	.986	.889	.904	1.359
Site 1.	Tq	15.67	15.11	14.89	14.72	4.69	4.71
	Ts	187	216	166	203	59	29
	Bq	12.43	11.61	11.90	12.36	4.21	4.25
	Bs	153	198	182	166	74	64
	Tqc	.0225	.0243	.0339	.0365	.0427	.1287
	Tsc	.815	.977	.762	.942	.859	.420
	Bqc	.0178	.0186	.0271	.0307	.0383	.1161
	Bsc	.841	1.165	1.045	1.165	1.201	1.029
Site 2.	Tq	14.69	8.21	9.94	9.42	1.47	1.86
	Ts	283	221	240	184	26	43
	Bq	10.39	10.73	10.27	10.31	4.24	1.96
	Bs	192	286	166	195	29	14
	Tqc	.0221	.0132	.0226	.0234	.0133	.0508
	Tsc	1.316	1.839	1.650	1.335	1.208	1.580
	Bqc	.0149	.0172	.0233	.0256	.0386	.0535
	Bsc	1.263	1.821	1.104	1.292	.467	.488
Site 4.	Tq	9.66	10.28	8.70	8.46	3.21	0.61
	Ts	364	421	308	382	68	27
	Bq	10.84	10.84	10.61	9.61	5.33	1.30
	Bs	383	433	309	385	48	18
	Tqc	.0139	.0165	.0198	.0210	.0292	.0166
	Tsc	2.575	2.799	2.419	3.086	1.447	3.025
	Bqc	.0155	.0174	.0241	.0238	.0485	.0355
	Bsc	2.415	2.730	1.990	2.735	.615	.946

relationship between storage, and soil characteristics. The resulting erosion and deposition patterns at the top and base of fine sites respectively are not only the result of hillslope hydrology but in turn will affect the hydrological response. As finer material accumulates at the base, soil storage values would be expected to decrease relative to increased storage values at the top. This in turn will produce a change in the runoff pattern which is likely to increase at the base, enabling sediment previously accumulating at this point to be removed. On coarse sites with larger runoff and erosion occurring at the base of the slope the development of lag deposits will not only increase soil storage and thereby diminish runoff and erosion at this position but will also protect the surface from all but the most potent storms. In this way runoff and erosion will be relatively higher at the top of coarse sites. The nature of these spatial feedback mechanisms in runoff and erosion processes would, in the long term, lead to major shifts in their spatial patterns, allowing sites to be eroded over their whole length.

Finally, the results obtained cast some doubt on the generally accepted assumption that high runoff and sediment yields are characteristic of badland areas. It is only at localised positions on the slope that ground lowering reaches values commonly agreed as typical of badland erosion (Wise *et al.*, this volume). The results do however indicate that runoff and erosion are considerably more variable than generally recognised, and that such variability is due primarily to patterns of surface soil characteristics within units of uniform lithology.

6 Sheetwash and rill development by surface flow

J. Savat and J. De Ploey

Baur (1952) has defined sheet erosion as 'removal of a fairly uniform layer of soil or material from the land surface by the action of rainfall and runoff.' The same author gave a workable definition of rill erosion as being 'removal of soil by running water with formation of shallow channels that can be smoothed out completely by cultivation.' When this operation becomes impossible, gullying starts. The evolution of gully walls is determined by different sets of hillslope processes, including both surface and subsurface erosion. Smith and Wischmeier (1957) have clearly pointed out that the Universal Soil Loss Equation, developed in the U.S.A., applies to slope wash which is generally a combination of sheet and rill wash.

Ephemeral rills may be continuous in time when the shallow channels are rapidly shifting across the slope, so that they may split up or regroup according to the local dynamics of the flow pattern. Of course, there is also the question of ephemeral rills when channels disappear by sediment deposition or by subsequent erosion, e.g. by wind or by splashing. Permanent rills show a continuity both in time and in space. The more or less fixed channels may deepen and transform the rill into a gully, but often a sort of self-limiting process seems to occur so that entrenchment does not exceed a few decimetres. This means that rill heads may still extend upslope, whereas the major downslope portion is more or less stabilized and functions as a sort of conveyor for sediment eroded upslope and possibly in some intermediate sections.

It is not the intention to discuss in this paper the role of subsurface flow and pipeflow in rill erosion. Attention is mainly focussed on the origin of rillflow, as a result of surface flow combined with the effect of raindrop impact. This option does not preclude the reality of subsurface erosion in rill formation and badlands development, as shown by Bryan, *et al.*, (1978).

Some attention will also be paid to sheetwash on sandy material. Ellison (1947) was among the first to discuss sheet and rill erosion processes qualitatively, taking into consideration such factors as slope soil erodibility and the transporting capacity of the flow. This paper first examines the impact of slope on rill and gully generation, and then discusses the origin and scouring capacity of rillflow in the light of thin-film hydraulics and of the results of some laboratory experiments.

FIELD DATA

A review of literature was made in order to check the reality of critical slope angles for the generation of permanent rills on farmland and on deforestated soils. Humid, temperate and tropical areas were mainly considered, as one may assume that rill generation in these belts is for the most part controlled by surface erosion. Available literature is primarily concerned with erosion on loamy and loamy-sandy soils. Some authors give complete information on a) critical slope angle for the development of permanent rills, and b) the critical slope at which rills evolve into gullies. Others provide only partial data on rill erosion, without mentioning the lower and upper threshold slopes.

Figure 1 summarizes field observation from Europe and tropical Africa. A first set of data comes from European farmlands on loess, which generally correspond to rolling landscapes. The silt content of the loamy topsoils varies between 65% and 77%, according to the author's analysis, whereas clay percentages are between 10% and 20%. In Belgium (Vanmaercke-Gottigny, 1967, 1977), in the Netherlands (Kierkels, 1971), in Alsace (Monnon, 1978) and on a T-2 terrace of the Rhine, near Basel (Schmidt, 1979) the critical slope angle for the generation of permanent rills is between 2° and 3°. In his extensive survey of the then recent soil erosion in Germany, Richter (1965) concludes that rill erosion on 6° slopes is highly probable when topsoils consist of silty loams on loess, but from the text it can be deduced that rills often generate on more gentle slopes. In Flanders, on sandy loams, rill erosion was described on a test plot with a 3° slope (Gabriels *et al.*, 1977). On all these slopes permanent rills generally entrench to a limited depth of several decimetres; vertical rill development seems often to be a self-limiting process.

In vineyards of Alsace, with a clayey loamy subsoil, permanent rills have been described on slopes of 6° and above (Saba El Ghossein, 1978). In Hungary, Pinczès (1971) reports heavy erosion on the Tokay Mountain when slopes exceed 10°. Richter concludes for Germany that gullying can start on farmland with a slope in the order of 12°. Kierkels, in Dutch Limburg, mentions a gullying threshold of 20%.

It seems that on loamy sandy material, the threshold slope for permanent rill formation may increase to 12° (Richter, 1965). Morgan (1977, 1979) mentions rill erosion in Mid-Bedfordshire, on sandy soils, for slopes varying between 6° and 11°. This author explains how saturation overland flow generates rills near the bottom of the slopes. Here, baseflow, not Hortonian flow, is the cause of rills. This is also reported in the Missouri Valley region, on (sandy) loessial watersheds where saturation overland flow is activated by a water table perched on impervious glacial till (Saxton *et al.*, 1971). Of course, one may expect more baseflow in sandy mantles with a higher infiltration capacity.

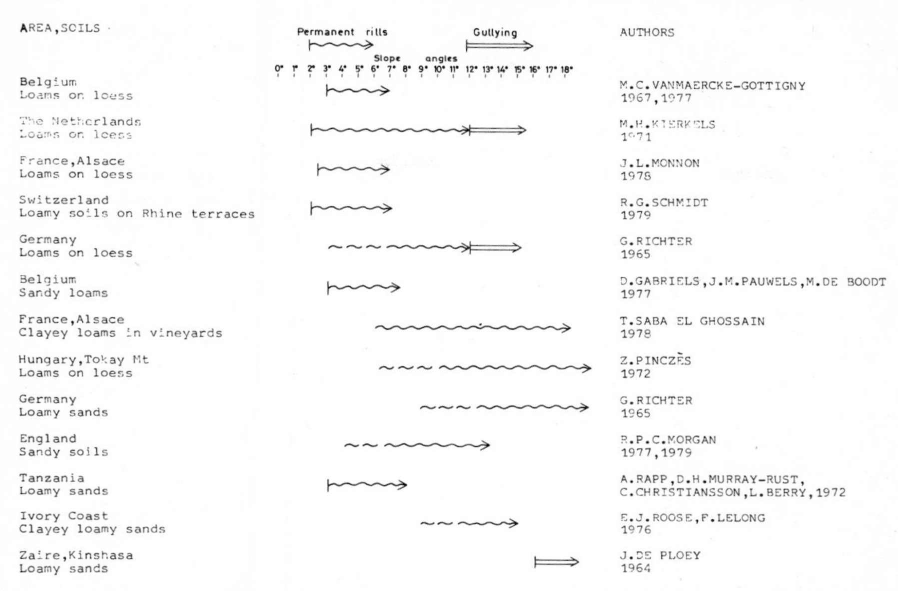

Figure 6.1 Critical slope angles for the onset of rill and gully erosion: a review of literature

In Tanzania, in the Ikowa catchment, rill erosion and gullying have been described on upper pediment 3° - 10° slopes (Rapp *et al.*, 1972), on loamy sandy and grass-covered soils. In a similar savanna belt in Shaba, Zaire, with clayey loamy subsoils, J. and S. Alexandre (1964) observed rill erosion on slopes at 3° and above, and gullying on land steeper than 15°. Rill wash is mentioned by Roose and Lelong (1976) in sandy ferralitic areas of the Ivory Coast, on slopes of 11°. Geomorphic mapping in the neighbourhood of Kinshasa showed incipient gullying on valley heads with a minimum slope of 16° (De Ploey, 1964).

Data from tropical Africa support the conclusions which can be drawn from field observations in Europe. Two main thresholds are characteristic for loamy soils with surface wash: permanent rills often start on slopes of 2° to 3°, on farmland and in areas with open vegetation; gullies often generate on land with a minimum slope of 12°-16°. General field observations confirm that there is no tendency toward gullying on slopes of between 3° and 12°. Thus, gullying is extremely rare in the hilly regions of loamy Belgium, where the steepest slopes are commonly 5° to 8°. Field data have been scrutinized from Lower Zaire and from the state of Sao Paulo, Brazil, where soil erosion is locally very active. The main impression is that shallow, permanent rills can start on low slopes, whereas gullying, incised by surface flow, needs at least moderately steep slopes. The conclusion is that the proposed threshold slopes may have a specific meaning, with reference to the hydraulic regime and the transporting capacity of runoff, as compared with the erodibility of topsoils.

There may be an upward shift of the threshold values for permanent rill erosion on sandy material, but only if rill generation results from surface flow. Pipeflow and saturation overland flow may provoke rillflow, even gullying, on gentle slopes.

Rills can start after a very short flow within the divide belt. So Schmidt (1979) reports rills within the first metre of a Rhine terrace field, near Basel. We made similar observations on loamy lands in Belgium, confirmed by 1m flume experiments (De Ploey, 1980). Finally, to be mentioned are the observations of Yair (1972) in Israeli badlands, where longitudinal rill erosion is very active on rather narrow divide belts.

The evolution of the longitudinal profile of some permanent, growing rills has recently been measured on loamy land near Leuven (Dewever, 1980; Massy, 1980). Erosion pin measurements showed complex evolution, with temporary accumulation and ablation occurring in different sections of the rill. Thus accumulation is not only restricted to a foot-slope area. Manifestly the transporting capacity of the rillflow, during rainstorms and afterflow, is often exceeded so that deposition frequently marks short stretches. Under pluvial runoff, with raindrop impact, loess can be deposited when the load concentration exceeds a certain critical value. With laminar to turbulent rill flow, on a 1-2 percent slope, it was found that the load concentration must be above 125 g/l before sedimentation starts (Mücher and De Ploey, 1977).

PROCESSES

Sheetwash on sand material

The term 'sandy material' is not only reserved for textural sands but also for silts and other sediments of very low plasticity. Topsoil liquefaction is a prerequisite for the development of multiple shallow, temporary rills, which occur in sheetwash (De Ploey, 1971).

The topsoil can be temporarily or permanently water saturated because of: a) its low permeability and infiltration capacity, b) high rainfall intensities, c) the presence of quite impervious horizons in the subsoil. Raindrop impact on a water saturated topsoil caused increasing positive pore water pressures p_w and liquefaction in non-cohesive material where the shear strength s becomes zero.

$$s = (p - p_w) \tan \phi = 0, \text{ for } p = p_w \qquad (1)$$

in which

p = normal stress per unit area, and

ϕ = angle of internal friction.

Rillflow exerts a tractive force F_t on the bed and sides which can be decomposed into τ and λ (figure 2), being respectively the drag and lift force per unit area. Liquefied by raindrop impact, the material of the rill banks is very erodible because p_w is positive and because this pressure must be added to the lift pressure vector λ, which is also directed upwards.

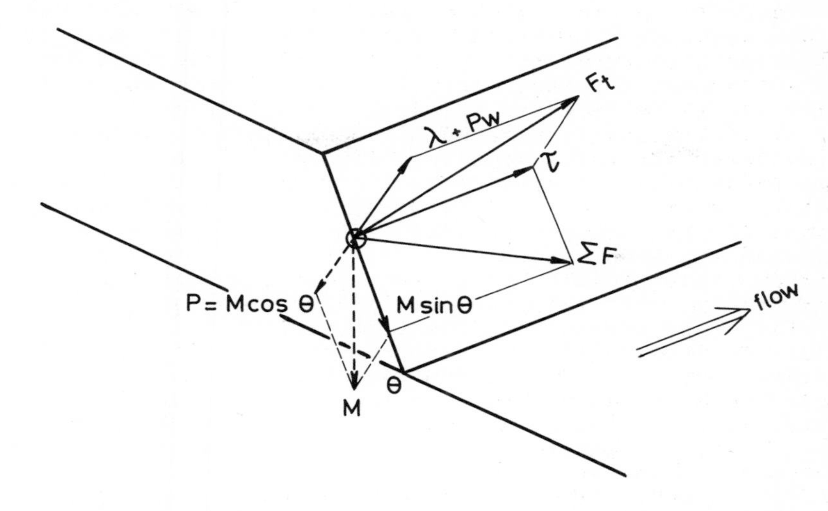

Figure 6.2 Diagram of forces acting on a particle resting on a bank (modified after Lane, 1955). M = submerged weight of particle; Θ = inclination of the bank; ΣF = total force acting upon the particle

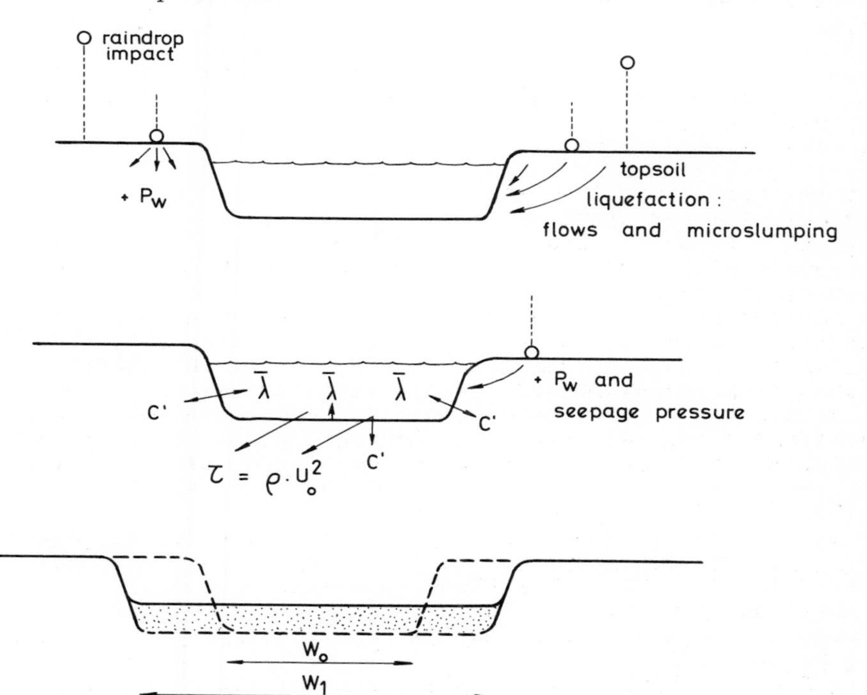

Figure 6.3 Mechanics of sheetwash in sandy material

This, of course, supposes that lift forces in the wet section are indeed directed upwards, which is the case for a flow turbulent down to the grain level (Davies and Samad, 1978). Indeed, for that case, Einstein (1950) reports the lift to equal 0.178 times the drag. The interrills are under splash erosion and from these divides material flows toward the rills as sand or silt flows (figure 3). These flows, combined with direct erosion by the rill-flow usually cause an enlargement of the rill, from W_1 to W_2. Hence, the wet section may become more shallow. This, in turn, provokes a diminishing shear velocity u_o and a reduction of the drag $C_D \rho u_o^2$. This promotes the deposition of sediment, in accordance with (2), and the possible filling up or splitting of the rill. The asymmetrical distribution of tractive forces and rates of deposition and erosion causes rapid lateral shifting of the rills. Indeed, a new steady state could be reached when the local slope of the rill has increased (Parker, 1979), whereby its bottom can be raised to the general level of the terrain. All these dynamic conditions contribute to sheetwash.

Sheetwash is less intensive when there is no full topsoil liquefaction. An unsaturated topsoil will generally be more or less cohesive, with the apparent cohesion c' determined by capillary forces. A weak cohesion may also result in the topsoil from the presence of small amounts of organic matter. Cohesive forces are in this case opposed to lift forces (see below and figure 3). Reduced liquefaction also means less frequent flows at the rill banks. The combined effect of all this is a diminishing lateral mobility of the temporary rills. Therefore, in loamy soils with c' values in the order of several thousands of N/m^2, permanent rills are the rule.

General Principles of Erosion

Transport capacity and morphology of river flow

Many transportation formulae for the computation of bedload as well as for the computation of suspended load, exist for fully turbulent river flow. In contrast, only one formula exists for sheetflow, i.e. that proposed by Yalin (Yalin, 1977; Foster and Meyer, 1972). This could also be used to determine the competence of this sheetflow. However, even this formula has some limitations. It is based on an initial movement of a grain caused by a momentary excess lift force which does not always exist (Davies and Samad, 1978). Furthermore, the formula should only be applied to cohesionless material.

In shallow laboratory rills, appreciable transport often occurs without suspension of sediment, indicating that drag alone is responsible for bed-load movement.

At present, a formula that predicts the load concentration and the competence of shallow supercritical flows in rills, especially when their regime is laminar ($Re < 600$), does not exist. Furthermore, the effect of impacting raindrops on both parameters has not received sufficient attention. Probably, when water depth exceeds a critical value of a few cm, shear at the bottom of the rill decreases, especially on gentle slopes. However, for depths under this limit, it increases (Smerdon, 1964, Glass and Smerdon, 1967, Machemehl, 1968, Yong Nam Yoon and Wenzel, 1971, Shen and Rug-Ming Li, 1973, Savat, 1977).

From several formulae predicting the intensity of bed-load transport q_s in rivers (Kalinske, 1947, Einstein, 1950, Meyer-Peter and Müller, 1948) one common statement can be derived (Rottner, 1959):

$$q_s \sim u_o^3 \cdot F\left(\rho / \rho_s' ,\ u_o^2/gD \right) \tag{2}$$

in which

q_s = rate of bed-load transport per unit width

ρ = specific mass of water

ρ_s' = submerged specific mass of sediment

u_o = shear velocity

D = diameter of grain

F, = denotes function.

The last part of the proportionality, u_o^2/gD is the square of the Froude Number relative to grain size. If the above formula (2) is to be generalized for all flow, and not only fully rough turbulent flow, the functional relationship should also include Re_o or the Reynolds Number relative to grain size (Inglis, 1968).

$$Re_o = u_o D/\nu \tag{3}$$

in which ν denotes the kinematic viscosity of the fluid (m^2/s).

It is well known that Re_o and Fr_o determine the manner in which grains are transported in rivers or, in other words, the bed-form of that river. The experiments of Simons and Albertson (1961) showed that the bed-form can be plotted on a graph of u_o/u_{ss} versus Re_o. Since the settling velocity u_{ss} of a grain is proportional to the square root of its diameter when the fall is a turbulent Newtonian one or when $D > 180$ microns, the dimensionless ratio u_o/u_{ss} could indeed be written as Fr_o.

The link between the hydraulic geometry of a channel and a general transportation formula has still to be made, despite attempts by Parker (1978, 1979). Some theoretical or empirical relationships between the geometry of the channel and: 1) the cohesion and the angle of internal friction of its banks, 2) the median grain size of the material constituting the bed, and 3) the intensity of (bed-load) transport, are, however, already established.

First, for cohesionless banks and for channels 'in regime', Ruh-Ming Li *et al.*, (1976) derived from a forces diagram (figure 2) on a particle resting on a bank, the shape of the cross sectional profile (figure 4):

$$\frac{y}{R_m} = \frac{1}{1-r}\left[\cos\left(\frac{x}{R} \tan \emptyset \sqrt{\frac{1-r}{1+r}} \right) - r \right] \tag{4}$$

y/R_m = dimensionless depth

x/R_m = dimensionless width

r = $b \tan \phi$

b = ratio of lift force λ to drag force τ

ϕ = angle of repose of a grain in a bank

R_m = maximum depth, reached in the middle where $x = 0$.

When $x = W/2$, $y = 0$ (figure 4), and therefore:

$$\frac{W}{R_m} = \frac{2}{\tan \phi} \sqrt{\frac{1+r}{1-r}} \cos^{-1}(r) \qquad (4')$$

For example, with $b = 0.178$, and with $\phi = 22^{\circ}$, $W/R_m = 7.97$. Should $(\lambda + p_w)$ increase (figure 2), the channel would become wider (see: sheetwash on sandy material). When the internal friction of the bank increases, the channel tends to become deeper and narrower.

Blench (1969) also found, empirically, that the width/depth ratio of a river increases with decreasing cohesion. of its banks. One notices the similarity with the discussion above (sheetwash on sandy material).
Blench's set of formulae can be combined to:

$$\frac{W}{R} = \sqrt[6]{\frac{Q\ F_b}{F_s}} \qquad (5)$$

in which

Q = discharge

F_b = bed factor = $1.9 \sqrt{D_{50}}\ (1 + 0.12\ C)$

F_s = side factor, which is a function of the cohesion of the banks. When c' is very great (clay), its value equals 0.3, and drops for cohesionless banks (sandy or sand-silt mixtures) to 0.1.

D_{50} = median grain size, in mm.

C = sediment load concentration, expressed in p.p.m. (this latter value must not exceed 20).

Therefore, the width to depth ratio increases with the median grain size of the material constituting the bed, and with the intensity of bed-load transportation. It decreases with F_s which is proportional to c'.

Parker (1979) used transportation theory to show that 30 percent increase in gravel load with no change in discharge implies a 25 percent decrease in centre depth, a 40 percent increase in width, and that the change will be matched by a 32 percent increase of the slope S.

The Giri and Lua rivers in Zaire (Savat, 1975) meander through forests, but are braided with a high W/R ratio through cohesionless marshy banks.

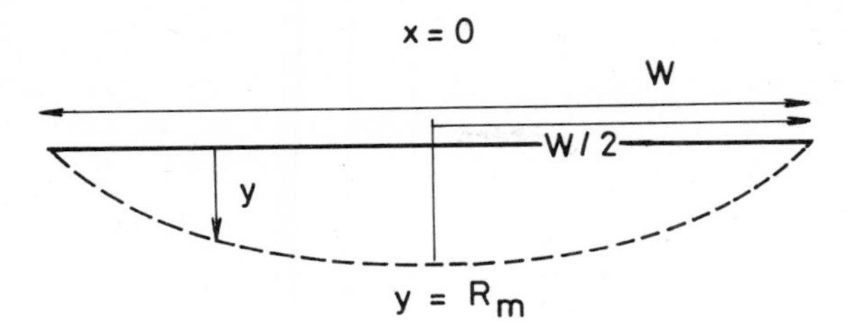

Figure 6.4 Shape of the cross-sectional profile of a river, derived from the diagram of forces (see: fig. 2)

Difference between river- and sheetflow

A small alluvial river is characterized by a Re_o Number of about twenty, at flood stage:

$$Re_o = 0.0002 \text{ m} \cdot u_o / 1.3\ 10^{-6} \text{ m}^2/\text{s} = 20 \tag{6}$$

from (3).
Hence, its shear velocity u_o equals 0.13 m/s.
From the approximate definition of the shear velocity:

$$u_o = \sqrt{g\ R\ S} \tag{7}$$

or with typical values

$$0.13 = \sqrt{9.81\ \ 1.0\ \ s} \tag{7'}$$

a slope of only 1/500 is obtained.
Mean water velocity $\bar{u}$ can be calculated with Manning's formula with an 'average' roughness coefficient of 0.025, and in metric units:

$$\bar{u} = \frac{1}{n} R^{2/3} S^{1/2} \tag{8}$$

or

$$\bar{u} = \frac{1}{0.025}\ 1\ \frac{1}{22.5} = 1.5 \text{ m/s.}$$

The Froude Number $Fr = \frac{\bar{u}}{\sqrt{g\ R}}$ only equals 0.5, even for an alluvial river in flood. Froude Numbers are thus almost always below unity, and therefore river flow is subcritical.

In contrast, figure 5 shows that Froude Numbers, according to laboratory measurements (Savat, 1977) are large in thin sheet flows resulting from steep slopes, and that they are *independent* of the unit discharge q (Q/W) when flow becomes turbulent (Re > 600). This finding merely reflects the fact that flow on a steep slope is turbulent down to grain level, and that the friction force opposed to the movement of water is nearly proportional to the square of its velocity.

Hence,

$$\bar{u} \sim \sqrt{R\ S} \tag{9}$$

$$\bar{U} / \sqrt{R} \sim Fr \sim \sqrt{S} \tag{10}$$

which means that the Froude Number is indeed solely dependent upon S.

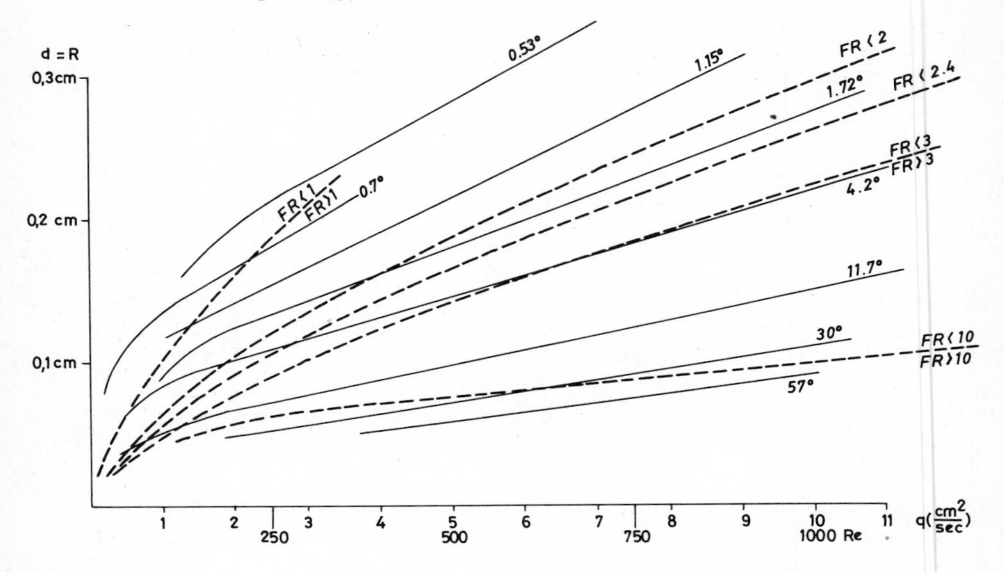

Figure 6.5 Slope angles and Froude Numbers as a function of unit-discharge or Reynolds Number (water temp. = 20.7°C) and mean water depth for shallow wide flows (mean water depth d = R)

Ishihara *et al.*, (1953) observed that wave-trains appear at the surface of a thin sheet flow with suitable depth when $Fr > 0.577$. Measurements with stereophotography (Janssens, 1979) showed that, at Froude Numbers around 2, the instantaneous depth below the crest of a wave can equal three times the mean water depth. Horton (1938) already stated that the 'force of erosion' can become five times 'the force' should the waves not appear.

Karki *et al.*, (1972) found experimentally that extensive flow separation occurs in supercritical flow when $R/D > 1.25$, where D represents the height of an obstacle (sill). The drag coefficient on the sill, defined as the ratio of the net horizontal force to the product of the projected area in the direction of flow and the dynamic pressure, is related to the Froude Number

$$C_D = 1.5\,(Fr)^{-4/3}\left[-1 + \sqrt{1 + 4\,(Fr^{2/3} + 0.5\,Fr^{-4/3} - 1.5)}\right] \quad (11)$$

which means that $C_D = 0$, at $Fr = 1.0$ and that it increases to a maximum value of 0.324 at $Fr = 3.0$. This is worthy of attention because lift seems to be absent from laminar flows (Southard, 1970) or even directed downwards at Re_o values below 5.0 (Davies and Samad, 1978) with a smooth turbulent regime. Hence the drag coefficient should be observed closely, even if there is only a faint resemblance between a sill and a grain. Flow separation implies that the downstream eddy is much greater than the upstream eddy. Photographs by Phelps (1975) clearly showed that not only the

Froude Number, but also Re_o are related to this separation. With laminar shearflow past sandgrains he found separation of the boundary layer at $Re_o = 56$, which value was exceeded during the experiments of Karki *et al.*, (1972).

Laboratory experiments with loess

Several experiments concerning the onset of rills were conducted, under simulated rainfall on a 4m long, well drained flume. Savat (1976) found that with laminar flow ($q = 2.5$; $Re = 220$) standing waves appear at a slope of 4.4^o. Subsequent steepening of the flume results in shallow rills at a slope of 6.4^o with afterflow. Under simulated rainfall these critical angles diminish to 3.4 and 4.4^o respectively.

From figure 5 it can be deduced that the critical Froude Numbers at which rills originated were about 2.8 (no rain) and 2.3 (disturbed flow). Later experiments with afterflow (Savat, 1979) discharges of $q = 1.18$ and $q = 4.0$ showed that rills start on very wet, almost cohesionless loess when Fr only equals 1.2. This value at least doubles when the loess apparent cohesion c' is in the order of 5000 N/m^2.

It is therefore concluded as a first approximation that rills originate at critical Fr Numbers of between 2.4 and 3.0 (very cohesive loess). It is read from figure 5 that these critical values are obtained at slope angles varying between 2° and 4°, and that these angles do not vary much with the unit-discharge q. This finding corresponds quite well with the critical onset of temporary rills, which many authors (figure 1) report as occurring between 2° and 3°.

From these preliminary experiments it could also be concluded that the *competence* of flow was mainly a function of its Fr number, even when the load concentration was not predicted by it alone as expected from (2). Coarser populations were exported from the flume when the Froude Number was greater; at the critical values reported, a 'mean' population was eroded away.

Laboratory experiments are thus in full agreement with field observations. If q has no great effect upon D_{50} of the eroded material, and if Fr is independent of q, it becomes obvious that critical slope angles are identical in tropical and temperate areas. Indeed, varying slope lengths and varying rainfall intensities affect on the magnitude of the unit-discharge but not the Froude Number. This also explains why rills can develop near divides.

If it is true, as for river flow, that Re and Fr numbers determine the bed-form below the flow, undulations must be created by supercritical flow. If the Froude Number exceeds about $1.2 + 0.0003\ c'$ (with the apparent cohesion expressed in N/m^2) these bed-undulations will deepen, thus creating rills even if the original surface had been perfectly plane and smooth. This general law does not invalidate the fact that an irregular surface leads sometimes more easily to water concentration and to a faster initiation of rills.

Steps originate at regular intervals in the flume plate 1), because the water in a whirlpool downstream of a step has a greatly reduced velocity (Savat, 1976). From that point on, the flow accelerates again, according to (Savat, 1977):

Plate 6.1 Flow acceleration downslope from vortices resulting from steps in loess. Arrow points toward trace of floating silver disc. Exposure time = 0.25 s. ($S = 0.06$; $Q = 8.5\ cm^3/s$; $\nu = 0.013\ cm^2/s$)

Plate 6.2 Channel fill deposits (shown by arrow) in lower portion of rill, following the splitting of the head of the rill ($S = 0.033$; $Q = 4.9\ cm^3/s$; $\nu = 0.012\ cm^2/s$)

$$\bar{u} = \sqrt{\frac{g\,S}{f}}\,(1 - e^{-ft}) \qquad (12)$$

in which f is a friction coefficient and in which t is time, which equals zero in the whirlpool. Thus, at a critical distance, the mean velocity $\bar{u}$ will again be of a sufficient magnitude to create a new step.

Hollows, which initiate rills become longer with time, in an upslope and a downslope direction. From experiments in this laboratory it could be concluded that the downslope evolution is mainly a function of the slope S or of Fr (10), while the upslope speed is controlled, to a large extent, by the apparent cohesion c' of the loess. Sediment input into the rill is essentially determined by the retreat of the head or the heads of the rill. When the rill has been split up into several tributaries, more sediment will be available and downslope sections can again fill up (plate 2). What really happens is similar again to what Parker (1970) foresaw for a gravel river that suddenly must evacuate a greater bed-load. Discontinuities between the sediment and the water input could therefore lead to the filling of a portion of the rill in the field, where rolling aggregates constitute the bed-load.

CONCLUSIONS

In many areas, temporary rills are reported to occur on slopes that are equal to or steeper than 2° to 3°, on mainly loamy or silty soils. Laboratory experiments show the same limits, irrespective of the unit-discharge. Because the competence of the flow is mainly a function of slope and Froude Number, especially when the regime is not fully rough, it must be concluded that these two parameters are prime factors controlling the onset of temporary rills. Critical Froude Numbers for onset of rill erosion on loamy soils vary between 2.0 and 3.0. Since the unit-discharge does not influence Froude Numbers, it is no surprise that in such a great variety of climates and with such varying slope lengths, the same critical angles are reported.

As in rivers, the cross sectional profile of the rills depends largely on the apparent cohesion of its banks. Measurements *in situ* of the c' values under rainfall would greatly help the understanding of this profile and the ability of the rill to shift laterally.

The onset of gullying around 12° to 16° is still an open question. It has not yet been demonstrated that overland flow creates these.

As long as adequate transport formulae for shallow flows are lacking, formulae based on drag considerations for example, the exact behaviour of a rill section cannot be explained. This remark holds particularly for the competence and for the amount of solid discharge that can be transported through a given section.

LIST OF SYMBOLS

b	ratio of lift force to drag force
c	cohesion (force/area)
c'	apparent cohesion
C	wash-load concentration, expressed in p.p.m.
C_D	drag coefficient, $F_t/(\tau \rho u_o^2 D^2/8)$
D	diameter of grain, or height of obstacle in flow
f	Darcy-Weisbach friction factor, $8\ gRS/\bar{u}^2$
F	denotes function
F_b	Blench's bed factor, function of median grain size D_{50}
F_s	Blench's side factor, function of apparent cohesion c'
Fr	Froude Number, $\bar{u}\ /\ \sqrt{gR}$
Fr_o	Froude Number, relative to grain size, $u_{(D)}/\sqrt{gD}$
F_t	fluid dynamic force
g	gravity constant
n	Manning's roughness coefficient
p	normal stress per unit area
p_w	pore water pressure
q	unit discharge of fluid (m^2/s)
q_s	unit discharge of solids
Q	discharge
R	hydraulic radius
Re	Reynolds Number, $\bar{u}R\ /\nu$
Re_o	Grain roughness Reynolds Number, $u_{(D)}D/\ \nu$
R_m	maximum depth in middle of river section
s	shear strength (N/m^2)
S	slope
t	time
$u_{(D)}$	velocity near tip of grain, $\cong u_o$
$\bar{u}$	mean velocity
u_o	shear velocity, approximately equal to $\sqrt{gRS}$
u_{ss}	settling velocity of a grain in water
W, W_o, W_1	width of rill
x	horizontal distance from middle of cross-section, where y is maximum
y	local depth anywhere in a cross section (y=O, when x=W/2)
λ	lift force or pressure
ν	kinematic viscosity (m^2/s)
ρ	specific mass of water
ρ'_s	submerged mass density of sediment, $\rho_s - \rho$
τ	drag force, equal to ρgRS
$\emptyset$	angle of internal friction

7 Hydraulic characteristics of a badland pseudo-pediment slope system during simulated rainstorm experiments

W.K. Hodges

Pediment-like complexes (Plate 1) are pervasive in badlands and resemble large-scale features commonly developed in arid zones. They have been called 'miniature' versions of large landforms (e.g., Johnson, 1932; Bradley, 1940, Schumm, 1962) since both systems are composed of similar morphological units characterized by a relatively steep upper catchment separated by a slope junction with gently sloping pediment. Each system, regardless of scale, may develop within lithologic units of varying type and resistance. The small features, although not exact counterparts of large scale phenomena, are analogous to them. The term 'pseudo-pediment' is used here for features in badlands in order to avoid confusion.

Little is known of the hydrodynamic operation of a pseudo-pediment system apart from Schumm's (1962) work. This paper is a preliminary evaluation of the hydraulic behaviour of a badland pseudo-pediment. A primary objective is to provide a simplified methodology from existing theory which can be applied to *approximate* hydraulic parameters where direct measurement of variables (e.g., flow depth) is difficult without disrupting process continuity. The scope is restricted to runoff hydraulics.

The study was located in the badlands of Dinosaur World Heritage Park, Alberta, Canada. The region is semi-arid, and the badlands are developed within a sedimentary facies of intercalated, poorly indurated sandstones and shales. More complete regional information is given by Hodges and Bryan (this volume); a map is provided by Campbell (this volume). The study is based on runoff processes monitored during simulated rainfall experiments in the field.

PSEUDO-PEDIMENT PROCESSES

General Considerations

Early studies of pseudo-pediments were conducted by Johnson (1932) in South Dakota and by Bradley (1940) in Wyoming, the former ascribing them to stream processes and lateral cutting while the latter attributed their development to weathering, sheetflow and rainsplash. Higgins (1953) carried out similar research on features developed on tuff. Smith (1958) observed natural runoff, concluding that flow was discontinuous, generating fan shaped sheetflow issuing from rills. Deposition appeared to result from waning flow and clay streaking from a change in sediment transport between sheetwash and subdivided rillwash. Smith suggested from simple runoff experiments that the slope junction checked flow velocity. This infers that sheetflow is not an effective erosional agency and that the pseudo-pediment is a graded transport surface for sediment derived from the catchment.

Schumm (1962) questioned the role of sheetflow in erosion and deposition based on long-term erosion rates. These provided evidence that relative erosion is greatest within a catchment. Deposition was not noted along the slope transition, so Schumm considered possible hydraulic characteristics of a shale catchment and pseudo-pediment to explain basal trimming. This involved comparative assumptions about surface conditions affecting flow. Relative flow velocities were calculated from Manning's equation with estimated 'n' values; 0.02 was used for pseudo-pediment and 0.06 for shale. Results suggested that flow velocities are similar for both surfaces; decreasing roughness from shale to pseudo-pediment is compensated by decreasing slope angle at the slope junction. Schumm concluded that basal trimming and erosion of the pseudo-pediment surface are attributable to sheet wash processes; runoff energy is utilized more efficiently on the pseudo-pediment than within the catchment. The pseudo-pediment is seen to lower whereas backing slopes retreat.

Emmett (1970, 1978) concurs although he suggests that 'n' values are probably higher that Schumm anticipated, but

Plate 7.1 Pseudo-pediment slope system

the relative magnitude would seem reasonable. Furthermore, Schumm's (1962) example along with data from Emmett's (1970) experiments would, according to Emmett,

> ... serve to discredit the existence of a hydraulic jump as overland flow on natural hillslopes passes from a steep slope to a moderate or flat slope... It is unlikely that overland flow is ever supercritical and thus never offers the opportunity for a hydraulic jump.

Engelen (1973), however, suggests that Schumm's flow velocity estimates may not be justified without consideration of flow segregation; flow on shale regolith may occur within concentrated micro-channels associated with desiccation cracks whereas flow on a pseudo-pediment surface is often true sheet flow.

Field Experiments

Evaluation of the hydraulic character of a pseudo-pediment system requires an understanding of morphology, lithology and surface material behaviour. Ideally where lithologic heterogeneity exists each unit should be studied in isolation before attempting analysis of collective responses. This is, however, virtually impossible within a system without disrupting on-going processes. An alternative is to monitor processes from separate slopes with lithologic and morphologic characteristics comparable to components of the system under investigation. This approach is used in the present study.

Slope Description

Examples are taken from three of the experimental slope plots described in Hodges and Bryan (this volume); topo-geological maps of the plots were prepared at a 10cm contour interval; photographic insets of surface material characteristics before and during runoff are provided with the maps (Figures 1, 2 and 3).

Plot 5 represents the complete pseudo-pediment system (Figure 1) developed on sandstone and shale; an embayment is flanked by small alluvial fans. The pseudo-pediment is mantled by alluvial silt deposits (peri-pediment). The peripediment is slightly concave with an average slope of 4.6° which is consistent with pseudo-pediment slopes in the area (variations between 3° and 7°). The upper catchment is represented by sandstone (unit 14) and two shales (units 12 and 13). The peripediment and fans form 22 per cent of the entire system.

In order to evaluate in-plot variations in hydraulic behaviour due to lithologic and morphologic changes it is necessary to consider additional slopes. Supplemental plots provide slope characteristics comparable to pseudo-pediment catchment components. Plot 9 (Figure 2) is dominated by sandstone (unit 32) and sandy shale (unit 31) which behaves like sandstone when wet. Although the upper slope is shale, the runoff response of the system to rainfall can be compared to the backing slope of the pseudo-pediment. Units 31 and

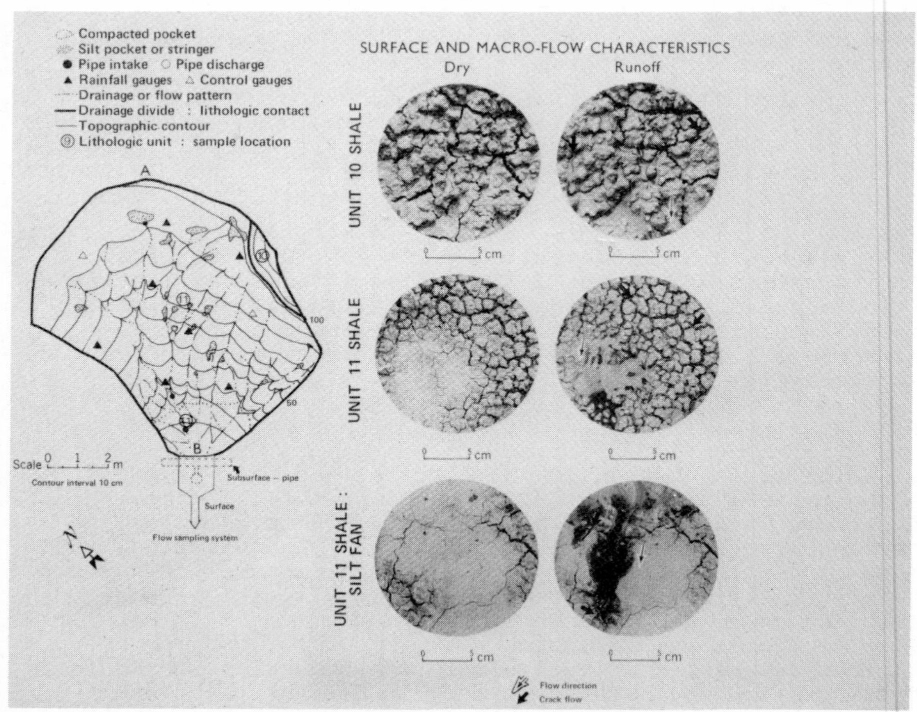

Figure 7.1 Catchment characteristics, Plot 5

32 comprise 61 per cent of the catchment with an average slope angle of 26.1^{o}. Plot 4 (Figure 3) is developed entirely in shale and is dominated by unit 11. Average slope angle is 6.5^{o}, close to the average for the shales on the pseudo-pediment system.

Surface properties of all units tested are evident in Figures 1, 2 and 3. A detailed comparison of each unit is given by Hodges and Bryan (this volume); only a brief summary is provided here. The peripediment is relatively smooth, augmented by thin laminae of silty alluvium. The mantle is effectively absent at the slope junction but increases to greater than 30cm thick at the sampling trough. Fans are particularly well developed along the steepest backing slope. The surfaces of sandstone and sandy shale units are formed on thin, friable weathering rind but may be altered by spatially variable, high density crack polygons which reflect local anisotrophism of materials. Clay-rich shales are probably the most dynamic units in badlands because of their behaviour upon hydration and dehydration. The regolith may consist of a surface and a subsurface layer overlying bedrock shard. The surface layer may be of two types, 'popcorn' (loose, puffy aggregates) or 'platelet' (thin, relatively dense plates) both of which occur in cells defined by desiccation cracks. The subsurface layer, if present, has a crust-like appearance and is denser than the surface.

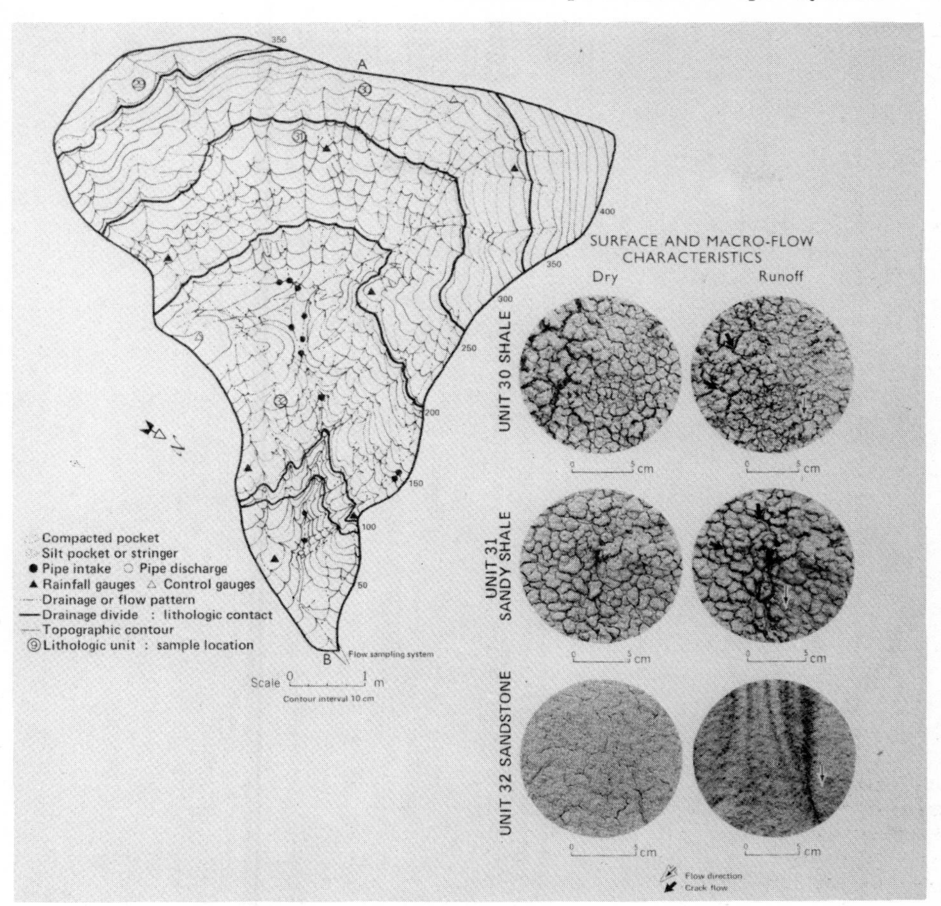

Figure 7.2 Catchment characteristics, Plot 9

Experimental procedure

Two simulated rainstorms (SST runs) were generated at each site: the first in dry antecedent conditions and the second approximately 24 hours later when regolith moisture had been raised by a known amount. Powdered dyes were used to identify incipient runoff conditions, runoff patterns and contributing areas. Flow velocities were measured by recording travel times of liquid dye across a known distance. Discharge samples were taken approximately at five minute intervals following flow initiation. Moisture content (dry weight basis) of each unit was determined from samples taken before, during and after each run. Rainfall simulator design is described in Hodges and Bryan (this volume).

Summary of experiments

Runoff discharges during runs are summarized in hydrographs (Figure 4). Runoff generation characteristics are important

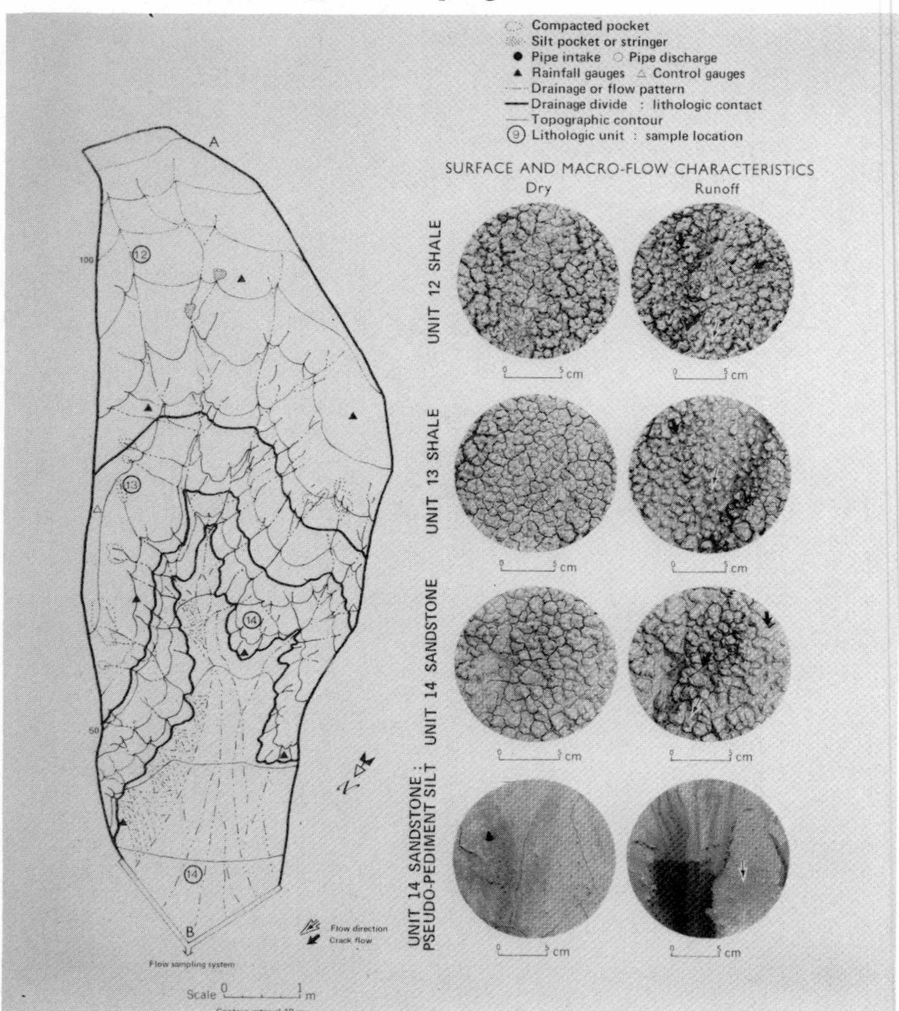

Figure 7.3 Catchment characteristics, Plot 4

as they indicate hydraulic adjustment of materials. Lag in runoff response for dry and wet antecedent moisture conditions reflects differences in material behaviour which can be used to estimate flow frequency and duration. Incipient flow on sandstone starts within 60 seconds of wetting and on sandy shale about 30 seconds later. Lag on these units may be halved during a wet run. Peripediment silts generate flow within 90 seconds during a dry run which can be reduced by one-quarter during a wet run. Shales are slowest; subtle material differences between units often result in extreme response variability (Hodges and Bryan, this volume).

Irregular discharge pulses shown by the hydrographs (Figure 4) are due primarily to wind disturbances during runs, although increases in source areas must account for

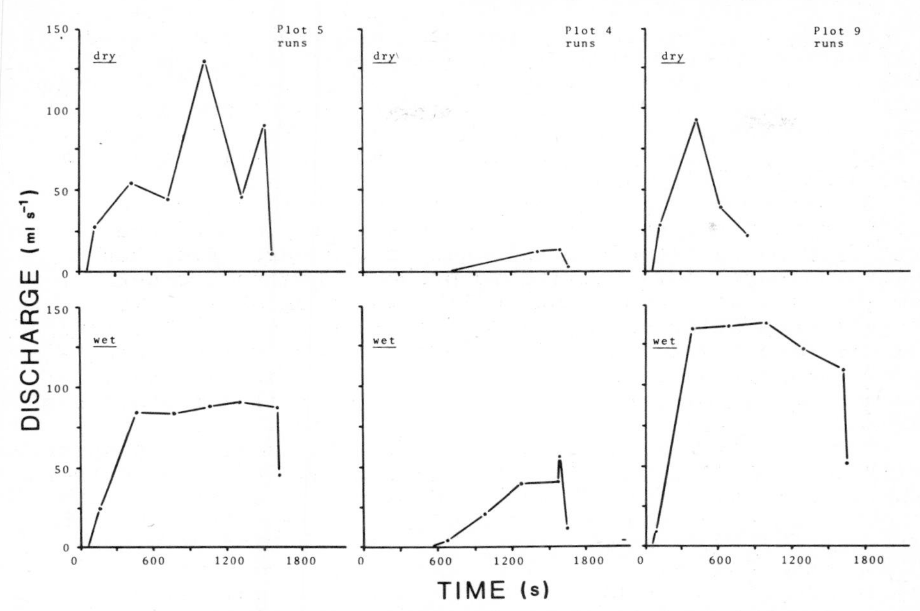

Figure 7.4 Runoff hydrographs for badland experiments

some runoff changes. For example, the increase in discharge following rainfall cessation during the wet run, plot 4, is likely attributable to an instantaneous growth in runoff supply rather than being a true 'pip' from detention storage.

Frequency, duration and magnitude of runoff events are, therefore, affected by the type and areal extent of the 'active' lithologic and morphologic unit(s). Sandstones and silt deposits (including fan and peripediment units) are by far the most active zones, and a wide range of storm intensities and durations will affect only these surfaces and materials. This would confirm that immense differences exist in the hydrological behaviour between the catchment (i.e., sandstone, shale or both) and the peripediment silts. The variations in runoff response suggest several possible response conditions between the backing slope and the pseudo-pediment surface, depending on the physical properties of the catchment. Likewise the hydraulic relationships between catchment and pseudo-pediment become extremely complex.

Observations of flow during runoff events suggest changes in flow behaviour occur in relation to surface conditions (photo insets, Figures 1, 2 and 3). Incipient runoff on shale regolith begins as a viscous slurry in rills (Bryan *et al.*, 1978), depressions, silt pockets and on silt stringers. Initial flow is thus confined to silted zones or to crack systems within compacted areas. The partial source area expands from compacted rills and pockets to popcorned interfluves. Flow depth within rills is regulated by changing crack dimensions until micro-channels are over-topped (e.g., Engelen, 1973; Haigh, 1978) and flow is controlled by rill confines alone. Lateral contributions are, however, maintained by primary crack networks adjacent to rills.

Runoff data for shales indicate that flow from silted or compacted zones during dry runs may begin after 360-750 seconds of rainfall; this may be reduced by two-thirds during wet runs. Flow from popcorned areas is delayed greatly, occurring quickest on prehydrated surfaces, the occurrence of flow here is spatially non-uniform and short-lived except under high intensity and/or long duration storms.

Sandstone surfaces generate flow which is generally spatially uniform, continuous and essentially unconfined, sheet flow which travels orthogonal to contours. Flow confinement occurs where rilling is present; however, cracks also function as micro-channels like those on shale. The net influence of cracks on sandstone or shale is to lengthen flow distance and increase flow depth whereas rills, being better integrated, tend to shorten the flow path and provide a master drainage system; the magnitude of rill effects tends to be much greater than that of crack networks.

Runoff patterns and flow behaviour on the gently sloping peripediment surface are more changeable than those on either shale or sandstone. Incipient flow is rapid, continuous and uniform, but as discharge contributions increase from upslope sources the flow is soon altered by non-channeled rivulets braided within sheet flow. The later regime persists until flow instability occurs with the inception of roll waves (Plate 2). This is followed by further instability, illustrated by chute development (rilling, Plate 3), which channelizes much of the flow in a manner similar to fan trenching. Chutes are ephemeral features; all evidence of their formation is erased by aggradation during recessional flow. A rhomboidal lattice structure (e.g., Stauffer *et al.*, 1976) of micro-rivulets is formed during waning flow as sediment deposition occurs (Plate 4).

HYDRAULIC VARIABILITY ON A PSEUDO-PEDIMENT SYSTEM

Previous related work

Although a number of studies have been completed dealing with shallow flow hydraulics (e.g., Horton *et al.*, 1934; Izzard, 1944; Emmett, 1970; Yoon and Wenzel, 1971; Savat, 1977), no attempt has been made to apply such work to badland systems. A fundamental problem is identification of conditions for flow transition. Horton *et al.*, (1934) suggested a transition between $300 \leq Re \leq 773$, while Chow (1959) felt that flow probably changes state over short distances due to surface irregularities producing mixed flow. Emmett (1970) found that laminar flow occurred when $Re < 1500$ while fully turbulent flow required $Re > 6000$. Emmett's field and laboratory experiments produced flow either laminar or transitional, agitated by raindrop impact. Savat's (1977, 1980) experiments (using $Re = 4UD/\nu s$) showed transitional flow between $1200 \leq Re \leq 2500$.

A similar controversy concerns Froude criteria. Emmett (1970, 1978) suggested that supercritical flow is unlikely in overland flow. On the contrary, Savat (1977) and Savat and De Ploey (this volume) have shown that Froude numbers do,

Plate 7.2 Development of roll waves and incipient knick scour along the peripediment mid-slope

Plate 7.3 Inception of chute (rilling) mechanism on peripediment

indeed, tend to exceed unity for flow on relatively steep slopes; Froude number is dependent upon slope alone and is independent of unit discharge when turbulence occurs.

Flow resistance is affected by friction, rainfall and surface character, including ephemeral bedforms. It is generally accepted that Manning's roughness coefficient and Darcy-Weisbach friction factor reach a minimum when $1000 \leq Re \leq 3000$ (e.g., Emmett, 1970; Savat, 1980). Rainfall influences on flow behaviour are least understood since impacting drops affect both flow and bed surfaces. Rainfall increases flow depth while retarding velocity (e.g., Smerdon, 1964). Flow resistance increases accordingly with rainfall intensity; however, experiments on slopes less than 2° (Yoon and Wenzel, 1971; Shen and Li, 1973) indicate that the friction factor tends to be independent of intensity when $Re > 2000$. Savat's (1977) studies with slopes up to 57° support depth and resistance relationships with intensity, but the effects are lessened as slopes steepen.

Another problem in slope hydraulics is flow instability, especially that associated with pulsating flow. Mayer (1959) identified two types of unstable flow, roll waves and slug flow. Several stability criteria have been reviewed earlier (Chow 1959; Hendersen, 1966; Schlichting, 1979; Karcz and Kersey, 1980) and need no elaboration.

One of the earliest geomorphic studies of roll waves was that by Horton (1938) who suggested their formation was restricted to steep slopes affected by intense rainfall. This is now known to be a condition rather than a restriction (e.g., Benjamin, 1957; Mayer 1959; Brock, 1969). Karcz and Kersey's (1980) experiments indicate flow instability prevails when $Fr = 0.5$, $Re < 200$ and $\cot S = 6/5\ Re$. Similar results were approached by Ishikhara *et al.*, (1953), and a discussion by Savat (1980) suggests supercritical laminar flow is always characterized by wave-train formation.

Determination of hydraulic parameters

Flow depth is crucial in hydraulics, yet it is very difficult to measure accurately in the field. Mathematical models of runoff systems based on applied theory derived from controlled field and laboratory experiments have been used to simplify complex catchment responses to rainfall and runoff generation. In turn, these models have been used to 'predict' events and solve for missing data.

Several models follow the kinematic approximation procedure (e.g., Henderson and Wooding, 1964; Singh, 1976; Dunne and Deitrich, 1980). Others are based primarily on continuity equations for conservation of mass and momentum. The latter approach is followed here. Horton *et al.*, (1934) developed expressions to represent laminar flow on hillslopes. Subsequent modifications by Kilinc and Richardson (1973) to determine velocity and flow depth profiles of overland flow have been applied successfully to single units. Similar applications have been made by Savat (1977, 1980) and Karcz and Kersey (1980).

Equations used to evaluate hydraulic characteristics of the pseudo-pediment system are developed below. Two assumptions are implicit: flow surfaces and flow types can be segregated, and the quadratic law of flow velocity distribution (after Jeffreys, 1925) is considered valid. Following Karcz and Kersey (1980), flow velocity can be given by

$$U = \frac{gS}{\nu} (Dy - y^2/2) \tag{1}$$

where surface velocity is

$$Us = \frac{gSD^2}{2\nu}, \text{ when } D = y, \tag{2}$$

and mean velocity is

$$\bar{U} = \frac{1}{D} \int_o^D Udy = \frac{gSD^2}{3\nu}, \tag{3}$$

since

$$Us = \frac{3}{2} \bar{U}. \tag{4}$$

Equation (3) has come to be considered the usual expression of mean velocity for laminar flow across a smooth bed (e.g., Savat, 1980); however, since sediment concentration is known to affect changes in fluid viscosity the term, ν, in equation (3) can be replaced by an estimate of the kinematic viscosity of the suspension, νs, where

$$\nu_s = (0.0157 - 0.000277T)\,(1 + C_s^{0.25}) \tag{5}$$

after Savat (1980) which is an approximation but provides values equivalent to those derived using an Einstein viscosity coefficient of 2.5 (e.g., Graf, 1971; Jopling and Forbes, 1979).

Kilinc and Richardson (1973) use equation (3) to calculate depth of flow and mean flow velocity. Following their derivation, unit discharge is given by

$$q' = q_o X = \bar{U}D \tag{6}$$

thus, mean depth of flow is

$$D = \frac{q_o X}{\bar{U}} \tag{7}$$

substituting equation (7) into equation (3) and replacing ν with ν_s will give

$$\bar{U} = \frac{gS}{3\nu_s} \left(\frac{q_o X}{\bar{U}} \right)^2 \tag{8}$$

when terms are rearranged,

$$\bar{U}^3 = \left(\frac{gS}{3\nu_s} \right) q_o^{\,2} \; X^2 \tag{9}$$

and the final relationships, after taking the cube root, are

$$\bar{U} = \left(\frac{g}{3\nu_s} \right)^{0.33} S^{0.33} \; q_o^{\;0.67} \; X^{0.67} \tag{10}$$

and

$$D = \left(\frac{3\nu s}{g} \right)^{0.33} S^{-0.33} \; q_o^{\;0.33} \; X^{0.33} \tag{11}$$

Equations (10) and (11), above, are expressions for shallow flow *without* rainfall, therefore Kilinc and Richardson suggest replacing the numerical constant, 3, with 4 to account for flow disturbances effected by raindrops. Similar changes will be suggested later.

Horton (1945) considered overland flow to be composed of two main types, sheet flow and rill flow. In this study sheet flow, crack flow and rill flow will be considered separately, and an attempt will be made to evaluate temporal changes in hydraulic parameters during runoff events.

The data base which permits estimation of hydraulic parameters for peripediment, sandstone and shale has been given in Figure 4. Measured surface slope (S) is considered a substitute for friction slope (S_f). Water temperature is given at 20°C and is based on an average for tank temperatures taken before runs. Depth of flow data was not obtained and will be computed. Reliable flow velocity data from dye traces are available for rill flow, but crack flow and sheetflow velocities will be computed using a limited number of actual measurements to define an upper limit (Umax) for the respective surfaces.

Sediment concentration data are available for each of the sample times plotted in Figure 4. Although these data are omitted here, a discussion of data range and sediment influence on kinematic viscosity are appropriate since equation (5) has been utilized in this study. The temporal variation in sediment concentration for each run was considerable in terms of sediment transport alone; however, the ranges of data collected was less significant. Ranges for the experiments are as follows (expressed in $g\ l^{-1}$):

	dry run	wet run
plot 5	19 - 59	16 - 46
plot 9	26 - 121	13 - 47
plot 4	60 - 66	16 - 27

Application of equation (5) indicates that these data effect minimal change on viscosity for the magnitude of events considered. An average of all values is probably an adequate correction factor in terms of experimental conditions.

The procedure used in calculating hydraulic parameters varies according to surface and flow type. Depth and velocity for crack flow and sheet flow can be approximated using unit discharge in the form

$$q' = \frac{q}{W} \tag{12}$$

which can be substituted for the quantity $q_o X$ in equations (10) and (11). Active flow width (W) for sheet flow and crack flow was determined from observations during runoff events. This required estimates of contributing area and measurements of crack density, especially for shale slopes. Maximum velocity for each surface and flow type was determined from isolated measurements taken during natural and simulated events. The following restrictions were effected: Umax (peripediment) ≤ 17 cm s^{-1}, Umax (sandstone) ≤ 10 cm s^{-1} and Umax (shale) ≤ 8 cm s^{-1}. Thus, to maintain a velocity criterion, it was necessary to derive surface disturbance coefficients for each surface condition similar to that suggested by Kilinc and Richardson (1973) which accounted for rainfall effects. The coefficients were 1.33 (peripediment), 3.33 (sandstone) and 10.67 (shale). The product of each and the numerical constant (3) in equations (10) and (11) become 4, 10 and 32 (above). The coefficient (4) conforms to Kilinc and Richardson's modification for rainfall on flow across a smooth surface (the peripediment in this case). The larger values are attributable to differences in micromorphological roughness between pitted sandstone and shale, thereby reflecting bed influences in addition to rainfall.

Determination of rill flow parameters was less complex than that for sheet flow or crack flow since measured velocities were available. Active flow width was based on rill width, and flow depth was determined from the relation

$$D = \frac{q}{UW} \qquad (13)$$

Additional hydraulic variables including Reynolds number, Froude number, Manning's n and Darcy-Weisbach friction factor were calculated for all flow types using standard equations.

Comparison of results

Hydraulic parameters derived for sheetflow and crackflow (Table 1) and for rillflow (Table 2) enable evaluation of hydrodynamics affected by surface and flow changes during successive rainstorms. An examination of subsequent differences between flow types is also possible.

Comparison of relative flow velocities indicates that flow confined by rills reaches velocities several times greater than those attained by sheetflow or crackflow. Flow velocities in peripediment chutes or shale rills can be three times above those for sheetflow or crackflow. Rill flow velocities in sandstone rills are greater than sheetflow by a magnitude of ten. Velocity differences between shallow flow phenomena suggest that the highest velocities occur on the gently sloping peripediment rather than an steeper sandstone; lowest velocities are associated with crackflow. These differences appear to be linked to relative efficiency dynamics and discharge variations between surface and channel. It follows that surface efficiency is affected by such factors as roughness and frictional resistance to flow.

Reynold's flow state has not been determined previously for a pseudo-pediment system. An estimate of the flow transition region can be evaluated by plotting depth of flow data as a function of Reynolds number (Figure 5). This permits direct comparison with data from Emmett (1970) and Savat (1977).

Results in Figure 5 are computed values using equations (10) and (11) for sheetflow and crackflow and equation (13) with measured velocities for rill flow. The range of Reynolds numbers are: $51 \leq Re \leq 582$ (peripediment sheetflow); $1872 < Re < 3543$ (peripediment chutes); $13 \leq Re \leq 184$ (sandstone sheetflow); $686 \leq Re \leq 12116$ (sandstone rills);

Table 7.1 Hydraulic parameters for sheetflow and crackflow

Slope	SST run	Kinematic viscosity (susp.) $cm^2\ s^{-1}$ ν_s	Calculated mean vel. $cm\ s^{-1}$ U	Calculated mean depth cm D	Reynolds number Re	Froude number Fr	Manning's number n	Darcy-Weisbach friction factor ff
Peripediment S: 0.08, W: 80	dry	0.0108	5.8	0.06	127	0.76	0.033	1.10
		0.0108	9.2	0.07	252	1.08	0.025	0.55
		0.0109	8.0	0.07	203	0.97	0.027	0.68
		0.0111	16.3	0.10	582	1.65	0.017	0.23
		0.0108	8.1	0.07	210	0.98	0.027	0.67
		0.0110	12.8	0.08	410	1.36	0.020	0.34
		0.0108	3.1	0.04	51	0.47	0.051	2.87
	wet	0.0108	5.4	0.06	114	0.72	0.035	1.23
		0.0108	12.3	0.09	387	1.35	0.020	0.35
		0.0110	12.1	0.09	378	1.32	0.021	0.37
		0.0110	12.6	0.09	400	1.36	0.020	0.34
		0.0109	12.9	0.09	417	1.39	0.020	0.33
		0.0110	12.5	0.09	396	1.35	0.020	0.35
		0.0109	8.1	0.07	208	0.98	0.027	0.67
Sandstone S: 0.44, W: 256	dry	0.0111	3.5	0.03	38	0.65	0.083	8.46
		0.0112	7.6	0.05	136	1.09	0.054	2.99
		0.0110	4.3	0.04	63	0.69	0.082	7.47
		0.0109	3.0	0.03	33	0.55	0.096	11.51
	wet	0.0108	1.7	0.02	13	0.38	0.130	23.90
		0.0109	9.9	0.05	182	1.41	0.041	1.76
		0.0109	10.0	0.05	184	1.43	0.041	1.73
		0.0110	10.0	0.05	182	1.43	0.041	1.73
		0.0110	9.2	0.05	167	1.31	0.044	2.04
		0.0110	8.6	0.05	156	1.23	0.047	2.33
		0.0110	5.8	0.04	84	0.93	0.060	4.11

Table 7.1 - continued...

Slope	SST run							
Shale S: 0.11, W: 60	dry	0.0111	2.9	0.10	105	0.29	0.113	10.54
		0.0111	3.1	0.10	112	0.31	0.106	9.23
		0.0111	1.2	0.06	26	0.16	0.194	36.95
	wet	0.0108	1.3	0.07	34	0.16	0.199	36.73
		0.0108	4.1	0.12	182	0.38	0.091	6.33
		0.0109	6.4	0.15	352	0.53	0.067	3.25
		0.0109	6.5	0.15	358	0.54	0.066	3.15
		0.0109	8.1	0.16	476	0.65	0.056	2.16
		0.0108	2.7	0.10	100	0.27	0.122	12.16

Table 7.2 Hydraulic parameters for rillflow

Slope	SST run	Measured velocity cm s^{-1}	Seconds into run	Flow depth	Reynolds number	Froude number	Manning's number	Darcy-Weisbach friction factor
		U	s	D	Re	Fr	n	ff
Peripediment S: 0.08, W: 10	dry	33.3	800	0.18	2155	2.51	0.012	0.10
		40.6	1140	0.24	3543	2.65	0.012	0.09
		28.6	1380	0.18	1872	2.15	0.014	0.14
		25.0	1505	0.36	3273	1.33	0.026	0.36
	wet	40.0	330	0.14	2036	3.41	0.009	0.06
		52.6	495	0.16	3060	4.20	0.007	0.04
		40.0	770	0.21	3055	2.79	0.011	0.08
		40.0	1080	0.22	3200	2.72	0.012	0.09
		50.0	1320	0.18	3273	3.76	0.008	0.05

Continued....

Table 7.2 - continued...

Sandstone S: 0.44, W: 4	dry	11.1	60	0.17	686	0.86	0.083	4.76
		66.7	300	0.24	5821	4.35	0.018	0.19
		66.7	450	0.31	7519	3.82	0.021	0.24
	wet	28.7	330	0.89	9288	0.97	0.098	3.73
		40.0	465	0.81	11782	1.42	0.066	1.75
		100.0	660	0.33	12000	5.56	0.014	0.11
		83.3	900	0.40	12116	4.21	0.020	0.20
		76.9	1200	0.40	11185	3.88	0.021	0.23
		83.3	1320	0.35	10602	4.50	0.018	0.17
		66.7	1620	0.35	8489	3.60	0.022	0.27
Shale S: 0.11, W:8	dry	10.0	1365	0.21	764	0.70	0.054	1.86
	wet	11.1	750	0.13	525	0.98	0.035	0.94
		15.4	1335	0.46	2576	0.72	0.059	1.72
		22.2	1590	0.44	3552	1.07	0.040	0.79
		20.0	1620	0.27	1964	1.23	0.032	0.60

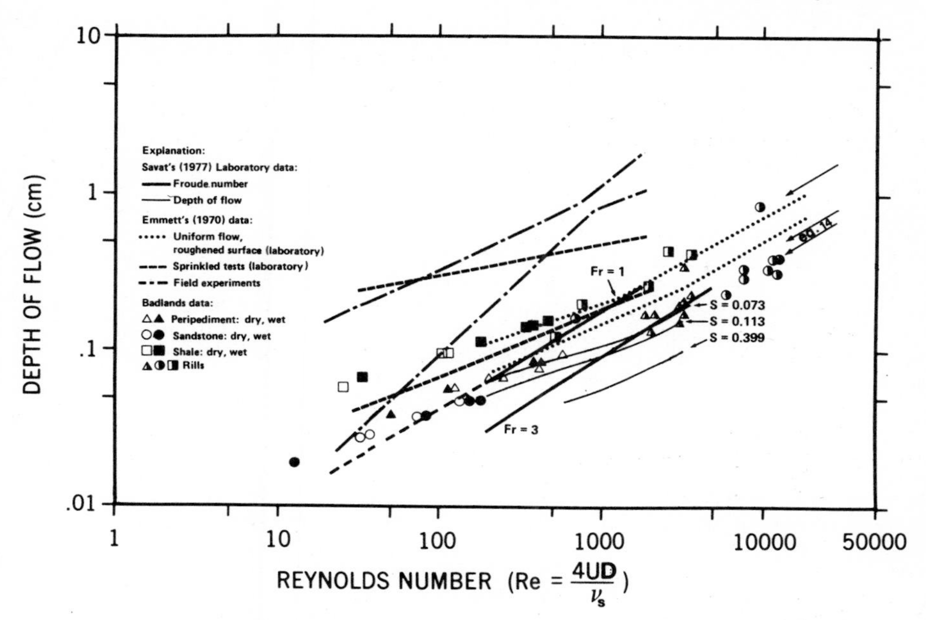

Figure 7.5 Depth of flow plotted as a function of Reynolds number

26 $\leq$ Re $\leq$ 476 (Shale crackflow); 525 $\leq$ Re $\leq$ 3552 (Shale rills). Significant scatter occurs for rillflow depths, especially for peripediment and sandstone units. Reasons are not entirely clear but probably include analysis procedure, flow pulses affected by wind disturbances and natural variability due to increased discharge and changing source areas due to material behaviour. In order to facilitate interpretation of flow transition a normalized flow depth for rills was computed from

$$D = \left(\frac{n\ \overline{U}}{s^{0.5}}\right)^{1.49} \qquad (14)$$

where Manning's n <n min for crackflow or sheetflow components were determined for line of best fit.

Figure 5 shows that sheetflow and crackflow are entirely laminar. Transitional flow occurs where 780 $\leq$ Re_c $\leq$ 1250 (approximately) which is below the range of Emmett's (1970) data but approaches the lowest limit suggested by Savat (1977, 1980) for flow on silt or sand beds. Conversely, rillflow for all surfaces is entirely *turbulent* after runoff is sustained, even on the gently sloping peripediment.

Flow depth data for sheetflow on peripediment compare well with Savat's (1977) experiments on silt beds, and values below the range of Savat's experiments appear to be consistent with extrapolation. Departures (i.e. for sandstone and shale) can be attributed to resistance or roughness differences. These data compared with Emmett's (1970) data fall within the lowermost extremes for his laboratory and field tests; however, Emmett's field test measurements may be more representative of 'braided rivulet flow' rather than of true sheetflow. This would imply that rivulet flow could also be segregated as a distinct component of overland flow.

Variations in Froude Number suggest that supercritical flow may develop in sheetflow on sandstone and on peripediment slopes except during incipient or recessive stages of runoff. Crackflow on shale slopes does not appear to reach critical stage. Rillflow on sandstone and peripediment can exceed critical state quite rapidly since discharge is confined to channels operating at much greater efficiency than adjacent surfaces. On shale slopes critical flow is inhibited by hydrological response characteristics associated with regolith and seems to occur only after continuous discharge across the surface is reached. These data agree also with Savat's results (Figure 5).

Flow instability which triggers roll waves (Plate 2) on the peripediment was nearly contemporaneous with incipient chute scour (ephemeral rilling, Plate 3). Transition from sheetflow to confined flow was instantaneous, but the hydrodynamics are difficult to isolate. Flow data suggest wave propagation during dry and wet runs had occurred before critical flow was achieved and incipient chute flow was triggered when $Fr > 2.5$. Flow depths computed on sheetflow during this period probably underestimate average depth with superimposed waves since these do not account for wave height; Savat (1980) indicates that wave height can be at least twice that of flow depth.

Flow conditions for the formation of rhomboidal structures (Plate 4) were different than those given by Karcz and Kersey (1980). The phenomenon here occurred as a result of micro-braided rivulet flow during flow recession when $Re < 50$ and $Fr < 0.5$, whereas their experiments indicate ripple structures associated with upper regime flow.

Flow resistance and bed roughness characteristics change dramatically according to surface and flow types. Calculated values for Darcy-Weisbach friction factor and Manning's n are given in Tables 1 and 2. Both ff and n reach a minimum before Re_c occurs. The pattern is most pronounced for sheetflow in the raindrop pitted sandstone surface. The effects of relatively large roughness and frictional resistance combined with flow at relatively low Re suggests it is unlikely that true sheet flow on sandstone would ever reach flow transition; any flow instability or disturbance would be due primarily to raindrops impacting flow and bed surfaces. Similar relationships are evident for shale and peripediment. As noted previously, rills tend to operate within the turbulent flow state; ff and n are reduced below minimum counterparts for shallow flow phenomena, except during incipient or recessional flow. These values are, however, what may be considered 'effective' ff and n, since their influences on flow diminish considerably with increasing flow depths.
Data for Darcy-Weisbach friction factor as a function of Re are plotted in Figure 6. As in Figure 5 a best fit relationship for turbulent rill flow can be derived using a form of Manning's equation with the Darcy-Weisbach expression.

Plate 7.4 Rhomboidal flow during waning runoff. Dark streak through photo centre is dye trace.

Since $$ff = \frac{8gDS}{U^2},\qquad (15)$$

substituting normalized flow depth (eq. 14) for depth (D) gives

$$ff = \frac{8g\left(\frac{nU}{S^{0.5}}\right)^{1.49} S}{U^2},\qquad (16)$$

which reduces to $$ff = 8gn^{1.49}\,U^{-(0.51)}\,S^{0.26}\qquad (17)$$

The theoretical relative roughness relationship ($ff = 96/Re$) for a smooth surface is also given where

$$ff = \frac{k}{Re}\qquad (18)$$

and relative roughness constant is

$$k = \frac{32gD^2S}{\nu_s U}.\qquad (19)$$

Relationships shown by the data plot indicate that under experimental conditions sheetflow across the peripediment operates at k=136 which is virtually identical to Savat's (1980) shallow flow results from experiments on löess with nearly the same surface slope. On sandstone k=333 and on shale k=1050, for laminar flow. Savat (1980) has already shown that k deviates from 96 according to bed roughness and

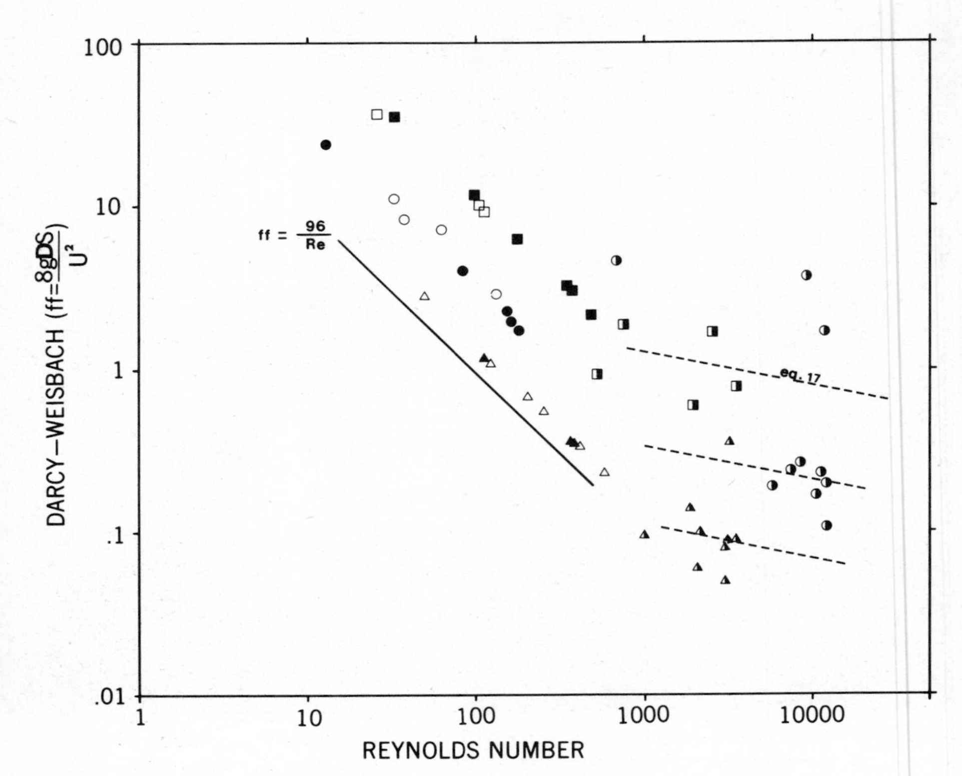

Figure 7.6 Darcy-Weisbach friction factor plotted as a function of Reynolds number. Explanation of symbols given in figure 5

slope steepness. Departures here are attributable to slope steepness and roughness both on sandstone but largely to roughness alone on shale. It will be noted, however, that the latter data do not conform with Savat's predicted values for k since surface characteristics differ from those used in his range of experiments.

During run variations in ff are evident in Figure 6. Friction factor changes occur within events and between events. The most extreme changes are associated with runoff on shale (as will be noted with Manning's n variations between wet and dry runs). Rill flow, being transitional or fully turbulent, operates outside the limits for shallow laminar flow; resistance to flow is several times lower than that for sheetflow or crackflow.

Data for Manning's n need not be illustrated, but a comparison of these values from Tables 1 and 2 reveals that the average (both runs) roughness coefficient for shale ($\bar{n}$ = 0.11) is almost double that for sandstone ($\bar{n}$ = 0.06) and nearly four times greater than peripediment (n = 0.03). The most significant changes in n for shallow flow between runs is noted for shale regolith which suggests that the greatest hydraulic smoothing occurs on a pre-wetted surface (wet run); $n_{min.}$ during the wet run was nearly half that of the dry run due to differences in material behaviour between

moist and dry states (Hodges and Bryan, this volume). Rill data patterns are nearly identical to those associated with ff variations.

Limitations

The analysis presented in the previous section is primarily theoretical, although much of the data is empirical and represents results of field experiments. Application is restricted to the conditions set. While results tend to agree, with other sources (e.g., Savat, 1977; 1980) it is possible cumulative errors have resulted from the analysis procedure. The hydraulic parameters appear to be a reasonable assessment, but the data must be considered *approximate* at best.

DISCUSSION AND CONCLUSIONS

Hydraulic characteristics of a pseudo-pediment system are determined by several factors: type and areal extent of lithologic and morphologic surfaces, frequency and magnitude of runoff events on each surface, slope steepness and slope length; and surface properties, including differences in material behaviour between surfaces. Type of surface and material behaviour response to rainfall provide limiting factors which control runoff timing and source areas and affect type and state of flow. It follows that hydraulic roughness and flow resistance are determined by rainfall influences, material behaviour, sediment size and ephemeral bedforms.

Apparent roughness and flow resistance characteristics of badland materials change during runoff events relative to those observed when materials are in a dry state. Flow resistance and hydraulic roughness are influenced by rainfall and micro-topographic roughness. On peripediments roughness can be *estimated* from surface sediment grain-size criteria, but additional variations are expected as a result of changing ephemeral bedforms as flow develops. Badland data are not available, but Moss *et al.*, (1980) found distinct suspension, saltation and bedload processes during laboratory experiments. Their shallow flow bedforms were smaller but similar to those of river systems, and the upper portion of their sequence seems to correspond with upper flow regime structures.

Roughness on sandstone is more complex than that of peripediment. Rainfall disturbs not only the flow surface but also that of the sandstone bed. Splash pits are retained within weathering rind after a storm. Ejecta rims are subdued by wind sediment transport following surface drying; however, the pits are replaced constantly during a runoff event. Pit dimensions are determined by rainfall intensity, drop size characteristics, approach angle and material behaviour. Particle size alone has little influence as a roughness element. Instead, flow retardation is effected primarily by splash-pit dynamics.

Shale regolith roughness is determined by aggregate and crack dimensions. Roughness elements change significantly

during runoff due to crack density and material behaviour, especially crack sealing and micro-channel smoothing from slaking and surface liquefaction during hydration.

Apart from surface variability it is possible to distinguish among types of flow which occur within a pseudo-pediment system. True sheet flow occurs on sandstone and peripediment. Crackflow occurs primarily on shale regolith slopes. Confined rill flow occurs on all three surfaces, but in general, permanent rills develop only within the catchment while ephemeral chutes develop within peripediment materials. The last type is rivulet flow which forms braids within sheetflow, especially on smooth, gently sloping unrilled surfaces like the peripediment. Two sub-types of rivulets can be distinguished. Large braids hydrodynamically similar to confined rill flow are unconfined and can be triggered by undulations in surface micro-topography which affect sheetflow convergence. The other sub-type characteristic of the peripediment occurs as rhomboidal flow and forms a lattice structure of micro-rivulets.

As runoff is generated on each unit, the relative effects of flow resistance diminish with efficiency of each unit increasing accordingly. The rate and degree of efficiency transformation occur most rapidly in rills; slope surfaces are slower to respond. Depth of flow adjustments to maintain and transport up-slope runoff contributions appear to be important to regulation of flow velocity changes and in establishing quasi-equilibrium conditions between catchment discharge input and peripediment flow transport.

Manning's roughness coefficients used by Schumm (1962) appear to be too low by 0.01 for peripediment and 0.04 for shale, assuming his values represent average conditions during flow. Comparison is restricted by the possibility that surfaces in the two study areas differ; however, results here seem to be in line with Emmett's (1970, 1978) suggestions that actual values may be higher. Comparison data for sandstone are not available, but it is significant that hydraulic roughness changes occur for each surface. Magnitudes depend on surface type, material behaviour and time elapsed since a previous runoff event (e.g., shale slopes).

Schumm's (1962) analysis and Emmett's (1970, 1978) comments assume that depth of flow across shale and pseudo-pediment are nearly the same. It is suggested that resistance to flow as a major factor in maintaining equilibrium between pseudo-pediment and hillslopes requires similar relative flow velocities, thereby compensating for differences between slope steepness. Emmett (1970) suggests a decrease in roughness on pseudo-pediment would compensate for decreasing slope steepness. This appears partially correct when considering unconfined sheetflow, but the effects of confined flow routed through rills cannot be overlooked. Comparison of roughness and resistance parameters between relatively steep, silt lined rills and the peripediment surface indicate that roughness effects may actually be greatest on the peripediment. It follows that depth of flow for sheetflow and crackflow across respective surfaces are not the same, nor are flow velocities. Flow associations between surfaces depend on relative positions of lithologic and morphologic units and, in part, on relative lag in response

times of different units to rainfall and runoff generation. Under these conditions several differing situations can exist which affect flow behaviour within a pseudo-pediment system.

Whether or not jump conditions may exist depends on lithology, morphology and flow associations. Sheetflow on sandstone may reach supercritical, laminar disturbed flow much like that during sustained runoff on peripediment, but these are conditions at mid-slope. It seems possible that flow conditions at the slope junction may be subcritical during peak runoff and attain a supercritical state further down slope. This neglects rill influences. Crackflow does not develop critical flow; therefore, a hydraulic jump would not be favoured at the peripediment junction with micro-channels on a shale surface.

It is most important that the effects of runoff routed through rills on the backing slope be considered since the bulk of catchment drainage is delivered to the peripediment as confined channel flow. Rill flow from shale and sandstone units develops supercritical, transitional or turbulent flow at relatively high velocities. Thus, at the confluence on the peripediment energy must be dissipated radially as distributary flow occurs. This translation of energy during sustained runoff may dampen hydraulic jump tendencies resulting from minor sheetflow transitions. It would follow that this mechanism may be responsible for the flow instability threshold on the peripediment and basal trimming at the backing slope. Travelling waves and/or rilling may occur as discharge increases exceed the depth/slope adjustment capacity of the peripediment surface. Since these conditions are eliminated during flow recession formation of ephemeral fans at the slope junction is possible.

The development of flow instability on the peripediment is a major hydrodynamic problem. Horton (1938) suggests that the appearance of roll waves on a slope may induce an erosional effect five times greater than the same flow discharge without waves. Chute development appears to be related to flow instability. Recent experimental studies by Karcz and Kersey (1980) define flow instability and wave formation when Fr=0.5; roll waves formed on the peripediment when $0.7 \leq Fr \leq 1$ (Plate 2); however, another transition occurred coeval with knick and chute development (i.e., when $Fr > 2.5$, Plate 3), but classification of chute flow as turbulent slug flow or type of hydraulic bore cannot yet be made with certainty.

Rilling has been reported in several localities, and the conditions for chute development here agree with findings by Savat and De Ploey (this volume) who conducted experiments on löess with slopes between 2° and 4°. Moss *et al.*, (1979) indicate that raindrops can inhibit channelling, but if sediment transport by overland flow exceeds that by 'rain flow transport' (i.e., bed particle disturbance and movement) channelling may occur. Additional work by Moss *et al.*, (1980) suggest the sediment transport power of shallow flow is relatively low on slopes less than 0.01, but power increases considerably (along with flow competence) when slope becomes greater than 0.04, at which point channels may form. Healing of ephemeral chutes on the peripediment is virtually instantaneous with flow recession, as transport power, com-

petence, and flow instability are reduced. These conditions trigger a flow transition corresponding to rhomboidal flow.

The rhomboidal regime which was observed on the peripediment was different than that discussed by Karcz and Kersey (1980). They note that rhomboid patterns occur within a large range of Fr and Re values, but much of their data are related specifically to the generation of rhomboidal bedforms associated with the upper flow regime. Many of their bedforms were ripple structures, whereas those which occurred during flow recession on peripediment were of the rivulet type under extremely low Fr and Re criteria. The peripediment rhomboid flow resembles closely lattice grooves formed by backwash on beaches rather than true ripples (see Otvos, 1965; Stauffer *et al.*, 1976). Their formation remains enigmatic, but their characteristics are very similar to an anastomotic stream network. They occur when $Fr < 0.5$ and $Re < 50$; therefore flow competence and capacity are altered considerably with flow depth and velocity to permit an anastomosing network of micro-rivulets to develop among small mid-channel bars formed as coarse sediment settles. Rivulets transport silts and clays which are ultimately deposited within grooves giving the streaked appearance to the surface after runoff events terminate.

The hydrodynamic behaviour of a pseudo-pediment system is extremely complex. The fact that ephemeral rilling occurs on a peripediment under certain threshold flow conditions while transport processes dominate others indicates pseudo-pediments are compound phenomena affected by transitory hydrogeomorphological processes. Additional experimental research which considers natural and simulated runoff events over a greater range of catchment types than given in this presentation is essential.

Acknowledgements

Field work in Dinosaur World Heritage Park was made possible through the co-operation of Mr. Jim Stomp, Chief Ranger, and by permission from Alberta Provincial Parks. Research was funded by a grant to my Ph.D. supervisor, Dr. R.B. Bryan, from the Natural Sciences and Engineering Research Council, Canada. Special thanks are extended to those who offered assistance and advice: to my wife Carol who helped with field experiments and to Professors R.B. Bryan, A. Yair, J. Savat and J. De Ploey for comments on an early draft of this paper. The author assumes sole responsibility for material content.

LIST OF SYMBOLS

C_s	sediment concentration
D	flow depth
Fr	Froude number
ff	Darcy-Weisbach friction factor
g	gravity constant
k	relative roughness coefficient
n	Manning's roughness coefficient
q	runoff discharge
q_o	rainfall excess
q'	unit discharge
Re	Reynolds number
Re_c	critical Reynolds number
s	seconds
S	surface slope
S_f	friction slope
T	water temperature
U	flow velocity
U_s	surface velocity
$\bar{U}$	mean velocity
W	flow width, channel width
x,y	coordinates of flow
ν	kinematic viscosity
ν_s	kinematic viscosity of suspension

8 Experimental study of drainage networks

R.S. Parker and S.A. Schumm

Throughout the history of geomorphology, the changing form of the landscape with time has been a primary consideration. However, due to the short time available to the investigator, models of landscape evolution have depended largely on deductions based upon measurements of erosion in restricted areas of rapidly eroding badlands (Schumm, 1956), or on a series of landform measurements placed in an assumed erosional sequence (Koons, 1955; Carter and Chorley, 1961; Ruhe, 1950; Hack, 1965).

Geomorphic problems are difficult to solve by field studies alone. However, experimental studies, if properly designed, may help to answer basic questions. This study was initiated to study drainage basin evolution using an experimental approach.

To simplify a drainage system into an experimental model, one must consider the scaling ratios between the prototype and the model, the boundary conditions required in the model, and the initial conditions inherent in the experimental design. Scaling ratios present problems which have been so great they partially explain the lack of experimental studies of drainage basin evolution.

To avoid the problems of scaling ratios one may build a watershed model sufficiently large that it can be considered as a prototype (e.g., badlands). Of course, as the size of the model approaches the prototype, the advantages of the model study are reduced. Such a tradeoff appears inevitable until suitable experimental theory exists to define scaling ratios.

In the present study the size of the basin is presumed to be sufficiently large to be considered a small prototype basin and that general conclusions obtained from a study of the miniature drainage basin can be extended to larger watersheds in the field. It is assumed that the relations derived are appropriate in a spatial sense but altered in time; time is compressed by an unknown amount in relation to the field.

Experimental work frequently results in unrealistic boundary conditions. In an experimental basin rigid sidewalls form an unyielding boundary to the watershed, whereas in natural basins competition exists among adjoining basins. Erosion and competition along these interbasin divides allow

divides to shift with time. Further, the rigid boundary produces a zone of little erosion that results in the surface near the wall maintaining its altitude through time. These boundary conditions do not, however, present serious problems in the overall experimental design. The divides of internal sub-basins do react much like natural watersheds and observations of downwearing, capture, and competition among these sub-basins can be made.

Initial conditions are also significant. For example, one must start with a particular initial surface slope which may influence the resulting basin configuration. Initial conditions, therefore, are a long-term influence on basin evolution. The identification of these deterministic components may be possible by varying the range of natural conditions and examining the resulting geomorphic configuration. Further, the range of values of particular geomorphic variables through time can be identified. The identification of the range of values a variable can assume, particularly the minimum and maximum limits, are of importance in identifying the impact of man's modification of natural watersheds.

EXPERIMENTAL DESIGN

To follow the evolution of a drainage system, a large surface was needed over which the application of precipitation could be controlled. Using the knowledge gained during construction of an earlier rainfall-runoff test plot (Dickinson *et al.*, 1967; Holland, 1969), a facility was built and designated the Rainfall Erosion Facility (REF). This is a container 9.1 m wide, 15.2 m long, and nearly 1.8 m deep (Plate 1), in which baselevel is controlled at the 1.2 m wide outlet. The effective watershed area for these experiments is 115.2 m^2.

The sprinkler system utilizes 5.1 cm aluminum irrigation pipes to supply water to lines of sprinklers mounted 3.0 m apart on both sides of the REF. Each sprinkler was controlled by a solenoid valve which could be activated nearly instantaneously. Pressure regulators controlled both areal distribution and rain drop size and were set in accordance with Holland's (1969) study to replicate natural rainfall closely. The highest intensity rainfall of 66.3 mm hr^{-1} was used on this experiment.

The material used in the REF was a mixture of sand, silt, and clay with a median diameter of 0.5 mm. Approximately 28 percent of the material was silt and clay and of this portion 5 percent was clay. About 1 percent of the material was larger than 2 mm.

During each of two experiments, an attempt was made to follow the complete evolution of the drainage pattern. Each experiment started with a different set of initial conditions. To determine if changes in the initial slope and baselevel changes would affect the resulting pattern.

In the first experiment the basin surface was graded into two intersecting planes to permit the development of an integrated drainage network. The maximum slope of this surface was 0.75 percent toward the outlet. In addition, baselevel was lowered 0.22 m before precipitation was applied.

Plate 8.1 View of Rainfall Erosion Facility (REF) before the facility was enclosed in a building

Precipitation was applied at 66.3 mm hr^{-1} for 2 hours, then the drainage network was mapped. This process was repeated 5 times until 10 hours of high intensity rainfall had been delivered to the surface. Thus, five maps at the same base-level were made during the initial development of the network.

When it was determined that the network had essentially ceased to grow at that baselevel, precipitation was applied for 4 more hours. After mapping, baselevel was lowered another 0.1 m to rejuvenate the network and to continue its growth into the undissected parts of the basin. Water was then applied for 13 hours until network growth had again essentially ceased at that baselevel and the network was again mapped. This sequence of baselevel lowering, rejuvenation of the drainage pattern, and the mapping of the system at maximum growth was repeated 5 times. At the end the network had reached what appeared to be maximal development within the watershed.

For the second experiment, the surface was again graded to two intersecting planes, but the overall slope was increased to 3.2 percent. In addition, the baselevel was not lowered before precipitation was applied. With these different initial conditions, the network was mapped 3 times (2, 5, and 10 hours) as the system evolved to a maximum for the initial relief.

Baselevel was then lowered 0.05 m to rejuvenate the network and precipitation was applied for 23 hours, at which time maximal growth of the network for that baselevel was achieved, and the system was mapped. Baselevel was again lowered 0.28 m, and the network was re-mapped when maximal growth was attained.

On the last baselevel lowering, the network was mapped several times after maximal growth was attained. Further information on the experimental facility and design can be found in Parker (1977).

PREVIOUS WORK

In an early discussion of the evolution of stream networks Glock (1931) classified the development of a drainage system into 4 growth stages, each characterized by listed processes and modes of change:

1. Initiation
2. Extension (growth of network)
    a. Elongation (headward growth)
    b. Elaboration (addition of tributaries)
3. Maximum extension (attainment of complete elaboration or maximum growth of network)
4. Integration (reduction of the network)
    a. Abstraction (loss of identity suffered by a secondary stream by encroachment of a primary stream)
    b. Absorption (disappearance of a stream immediately after rainfall)
    c. Adjustment or aggression (attempt made by main stream to reach the sea by the shortest route consistent with regional slope)

Initiation was characterized by:

(1) a lack of streams over much of the surface,
(2) indefinite termination of many stream without junction with a main stream, and (3) the failure of many streams to start the active conquest of territory typical of their future development.

He characterized the stage of extension as a period of growth for the initially abbreviated drainage system, with elongation as the active process by which major streams block out the undissected drainage area. This is followed by elaboration, which gradually changes the skeletal form of the initial stream system by addition and growth of minor streams. Glock stated that the end of extension (maximum extension), was the stage when the network had grown into all the available drainage area.

Integration is marked by the reappearance of the skeletal form of the network. The processes responsible for the loss of channels are abstraction and aggression. Continued lowering of divides does not provide sufficient internal relief to maintain all channels and some are lost. In addition, continued lateral migration of major streams eliminates small tributaries. This period is marked by piracy and general shifting of individual channels. The net result is a loss of streams and, therefore, a reduction of the total channel length, which reduces drainage density.

These growth stages are analogous to those observed during the evolution of the experimental network in the REF and they therefore provide a framework for discussion of experimental results. On the initial surface the network extended to a maximum limit and then began a process of integration during which channels were lost.

There were some problems when application of Glock's classification to the experimental results was attempted. For example, the period classed as initiation could not be separated quantitatively from extension. Although elongation and elaboration were recognized, a quantitative distinction between them was difficult to obtain. Glock apparently felt that the process of elaboration followed the period of elongation, but this sequence was not apparent in the experiments. Rather, the dominance of one of these processes at a particular stage in basin evolution appears to result from different initial conditions.

The maximum extension was evident by observation, and it can also be identified by plotting a variable such as drainage density through time. By using the variable drainage density, maximum extension can be identified as a distinct point in time rather than a period of time.

In the experimental facility integration results primarily from abstraction and adjustment by aggression. Absorption, a process of integration identified by Glock, was not obvious. This does not mean that such a process is not important in the field, but rather, that it was not permitted by the experimental design. In the experiment high-intensity precipitation was applied for long periods of time, while in the field long periods without rainfall and short periods of high-intensity rainfall combine to make absorption more important. With these exceptions, the outline of evolution as proposed by Glock serves as an outline of the dominant events observed in the evolution of the experimental watershed.

Models of growth

Glock's outline follows a temporal sequence but it is different from Davis' (1909) model of landscape evolution in that it focusses on the processes of network evolution. Davis' model is oriented toward the description of the surface forms and does not attempt description of processes.

There has been little discussion in the literature of the different types of network growth, but numerous computer models which simulate stream networks have been developed. These have taken several directions. One, first proposed by Leopold and Langbein (1962) generates a stream network by growth of streams from divide to mouth. Growth begins at the sources located on the basin divide and continues downstream to establish the master stream at the outlet. This mode of growth does not correspond with the headward growth of a natural drainage basin, the only correspondance with a natural system occurring when the total drainage system has developed. Contrary to the assertion by Leopold and Langbein (1962), this model is not analogous to the one proposed by Horton (1945). Horton suggested that on a steep, newly exposed surface parallel rills would develop and that, with time, crossgrading and micropiracy among these rills would produce an integrated network. This is similar to Glock's system. For the Leopold-Langbein model to be an effective model of this process, the initial rills would have to develop over the length of the available area and the integrated network could then be produced by piracy and crossgrading.

The second model reflects headward growth (Smart and Moruzzi, 1971; Howard 1971). In this model, networks develop fully at the edge of the as yet undissected area. Channels grow headward and bifurcate, but do not rapidly occupy the available drainage area, as in the Glock-Horton model.

Therefore, two realistic growth models have been identified previously. One is the Glock-Horton model and the second is the headward-growth model which represent such differences in growth that they may form two end members of a continuum of different growth types. In the Glock-Horton model channels develop almost instantly over the surface, and then, in time, the dendritic network is formed. In the headward-growth model a 'wave of dissection' (Howard, 1971) can be envisaged at the sources of the first-order channels. As this wave progresses into the undissected basin, the fingertip channels lengthen and bifurcate leaving behind a fully developed channel system. In this portion of the network few additions or abstractions of channels occur during continued extension of the network. The significant feature of this model is that the network is fully developed as the wave of dissection passes a particular point.

DRAINAGE NETWORK EXPERIMENTS

The evolution of two experimental conditions followed the two growth models previously described.

Drainage network growth

In order to compare the results of the two experiments, the passage of time is expressed by a ratio of the volume of water applied to the volume of water finally applied at the end of experiment 2. Drainage density for both patterns is plotted against this ratio (fig. 1). Initial conditions affected the drainage density values through time considerably, but both sets of data follow the same overall trend with drainage density increasing to a maximum value. In the second experiment this is followed by a decrease in drainage density. The decrease is inferred for the first experiment, because the experiment was concluded before abstraction began.

The categories of growth described by Glock (1931) can be identified on these curves, but there is no clear division between initiation and extension (fig. 1). Maximum extension occurs when drainage density reaches a maximum, after which abstraction begins, as indicated by decreasing drainage density values. Given an undissected surface and sufficient time, the general relation between drainage density and time will show this curvilinear form.

The initial conditions alter the form of the drainage density curve with time. After 2 hours of rainfall on each

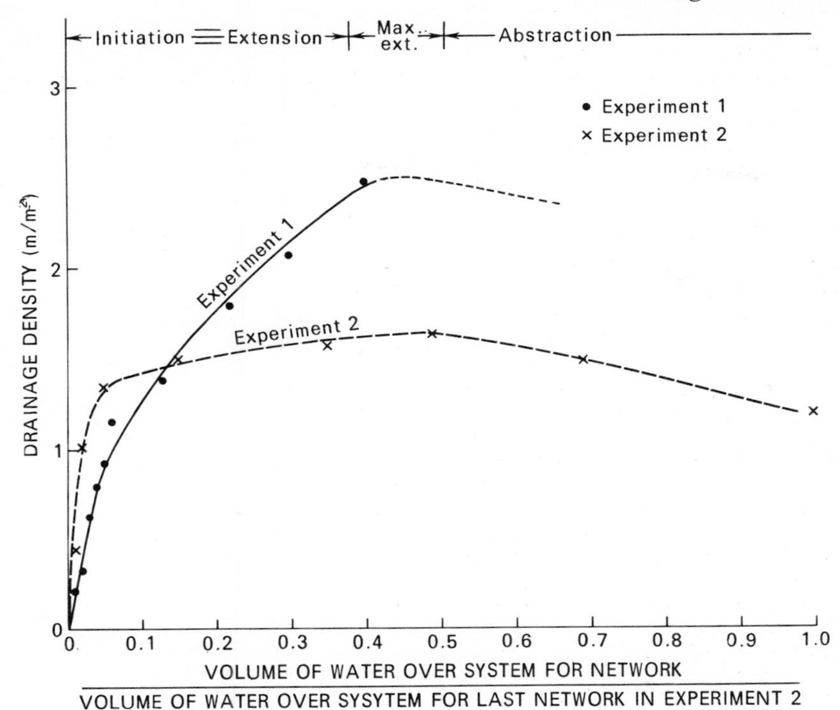

Figure 8.1 Changes in drainage density through basin evolution for experiments 1 and 2

experimental surface (first data point in each curve), the basin on the steeper initial slope without baselevel change (experiment 2) yields a higher drainage density value. This network, however, has a lower drainage density value at maximum extension. These two major differences result from different modes of growth.

After the first 2 hours of rainfall (fig. 2), several differences between the developed drainage patterns are noted. The drainage system occupies more of the available area after 2 hours when the network is formed on a steeper initial surface slope (experiment 2) with no change in baselevel. The low-order tributaries are also longer, resulting in a higher drainage density. The difference in lengths of first-order channels is apparent from the relative frequency histograms of first-order Strahler stream lengths. The histogram for the network on the lower initial slope with a baselevel change (experiment 1) is skewed to the right with a geometric mean length of 0.34 m (fig. 3). The histogram for the initial network on the steeper initial slope with no baselevel change shows a very different, rectangular distribution (fig. 4) with a poorly defined mode. Lengths range from 0.12 m to 3.05 m in an almost uniform fashion, and the geometric mean length is 0.93 m. A t-test, at the 1 percent level of significance, rejects the hypothesis that the geometric means of the first-order streams in both experiments

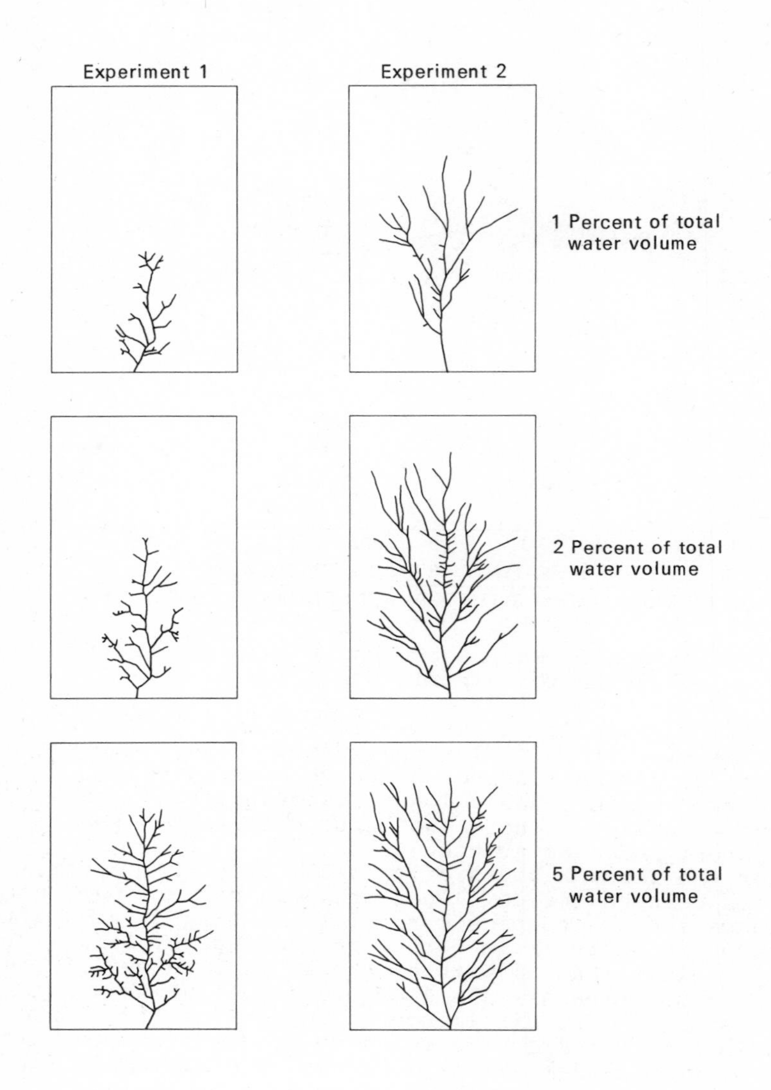

Figure 8.2 Comparison of the two experiments during initial development of the networks. The percent of total volume of water is the same time function multiplied by 100 as the ratio used on the abscissa of figure 1

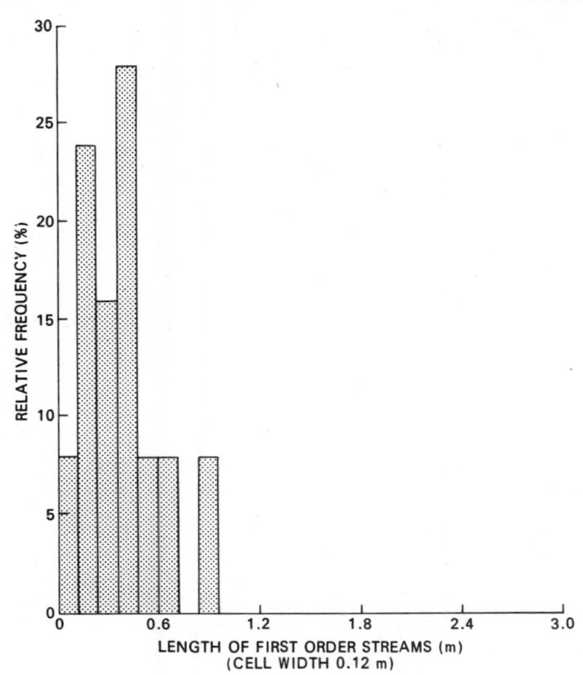

Figure 8.3 Relative frequency histogram of Strahler first-order stream lengths for first network of experiment 1

are equal. Thus, with changes in the initial conditions, the initial networks are different.

Unfortunately, the differences observed were the result of varying two components in the initial conditions - initial slope and baselevel. There is however, an additional network available from another experiment in the REF (fig. 5). The total relief of this third network was 0.88 m, lowered 0.10 m before the experiment commenced. This gave an initial slope of 12.1 percent. Due to the design of this experiment, the watershed area was reduced to 54.0 m^2, approximately half the area of the basin in the other experiments. This network, mapped after 2 hours of rainfall, had a steeper initial slope than either of the other two networks but, because the baselevel was lowered before the experiment began, it is comparable to the first network on the lower initial slope of 0.75 percent (experiment 1).

The relative frequency histogram of Strahler first-order channels for this network (fig. 6) shows again the right-skewed distribution similar to that for the network on the initial slope of 0.75 percent (fig. 3). This distribution of stream lengths had a geometric mean length of 0.41 m. A t-test was again employed and at the 1 percent level of significance, cannot reject the hypothesis that the geometric means are equal between the initial network in experiment 1 (fig. 3) and this network (fig. 6). From an examination of the network on this steep initial slope, it is noted that the net-

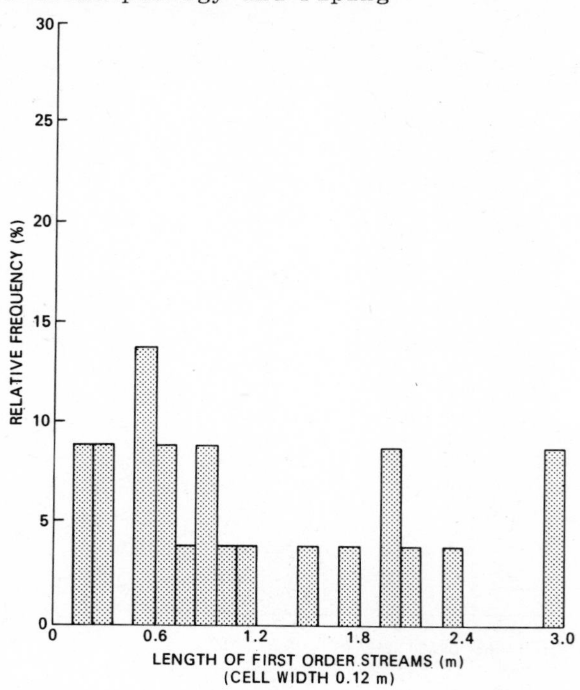

Figure 8.4 Relative frequency histogram of Strahler first-order stream lengths for first network of experiment 2

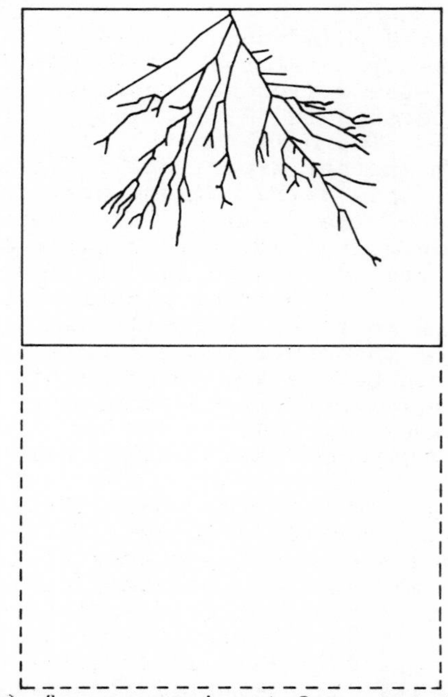

Figure 8.5 Network from experiment 3 on initial slope of 12.2 percent with a lowering of baselevel

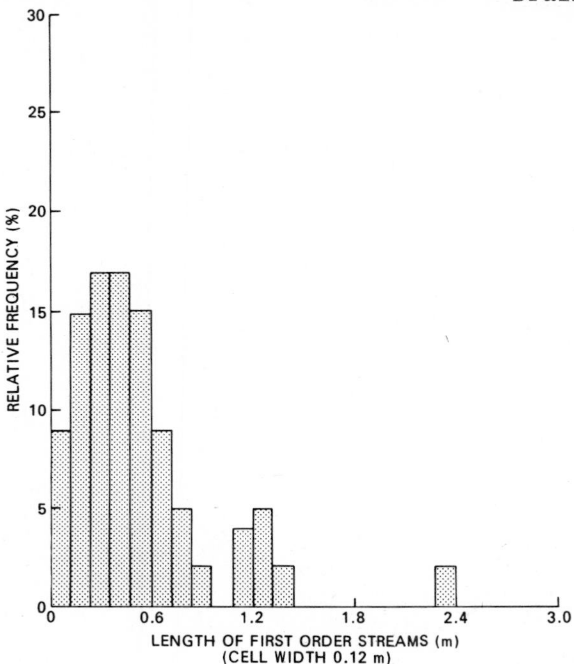

Figure 8.6 Relative frequency histogram of Strahler first-order stream lengths for network of experiment 3.

work has developed to a greater extent but the histograms for first-order streams lengths are similar.

It appears that a steepening of the initial slope produces the network more quickly in response to the added relief in the basin. Lowering the baselevel before running, however, changes the mode of growth. A lowering of baselevel produces a network that develops fully as it grows headward. This is headward growth as discussed earlier. A knickpoint develops at the outlet, where the baselevel has been lowered, and as this knickpoint migrates upstream, the channels grow and bifurcate to produce a fully developed network in which little internal growth occurs. In Glock's terminology, both elaboration and elongation progress simultaneously.

When the baselevel is not lowered before network development, long tributary channels develop, and the area between these tributaries is later filled by additional tributary growth. This process was identified by Glock as elaboration.

The network formed on the steeper initial surface (experiment 2) does not have a high drainage density value at maximum extension. At maximum extension there is a 33 percent difference between the drainage densities of the two patterns.

The relative frequency histograms of Strahler first-order stream lengths for the two experiments at maximum extension show a right-skewed distribution (figs. 7 and 8).

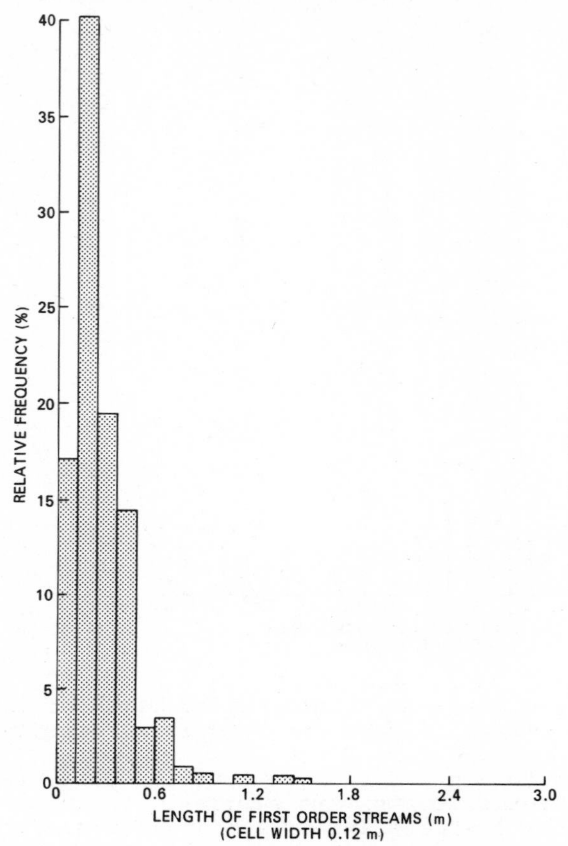

Figure 8.7 Relative frequency histogram of Strahler first-order stream lengths for network at maximum extension for experiment 1

From initiation to maximum extension the histograms of experiment 1 data show little change. On the other hand, the changes in the appearance of the histogram on the steeper initial slope are dramatic (fig. 8). Initially the histogram was rectangular, but at maximum extension the histogram of first-order stream lengths is right skewed, and it is similar to the distribution from experiment 1. Comparing the two histograms from each data set at maximum extension, the range of lengths is greater and the geometric mean length is larger for the second experiment. The geometric mean length of first-order channels at maximum extension for the first data set is 0.24 m; whereas, the geometric mean length of these channels at maximum extension on the steeper surface is 0.46 m. A t-test is significant at the 1 percent level; thus, the two means are statistically different at maximum extension.

In summary, the different initial conditions significantly influence network development. This influence is observed even at maximum extension. With baselevel lowering,

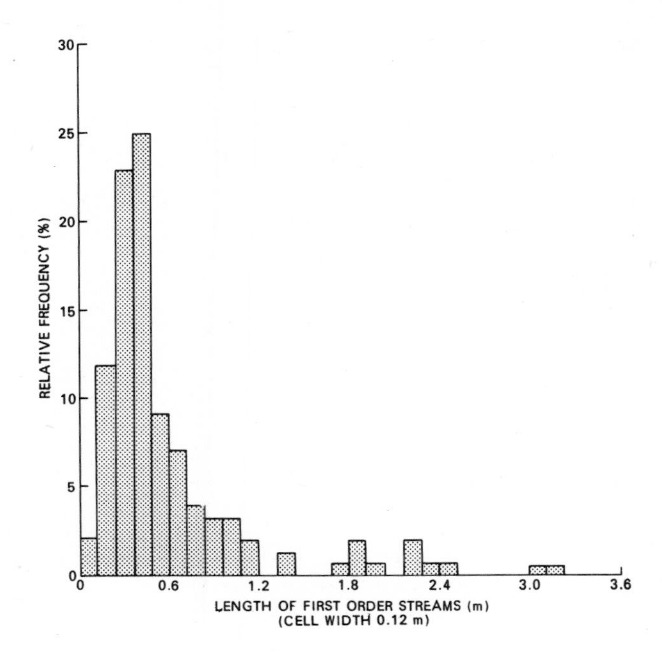

Figure 8.8 Relative frequency histogram of Strahler first-order stream lengths for network at maximum extension for experiment 2

first-order streams maintain a consistency in length during growth as shown in the histograms of length. Thus, first-order streams exhibit a regularity in the length they attain before bifurcation.

With no baselevel change, initial channels are long and develop over large portions of the watershed. Further network growth produces a number of first-order streams branching from these initial channels. Histograms of first-order stream lengths show a decrease in the mode and mean lengths through time. At maximum extension there is a substantial reduction in total channel length from the first experiment, and this is reflected in the values for drainage density.

First-order stream lengths

Changes in first-order stream lengths have been noted between initiation and maximum extension. The focus on the first-order channels is important because these channels reflect the growth of the network. For the networks already examined, the first-order stream distribution is markedly right skewed. All of the histograms except that of figure 4 show this form of distribution and, therefore, the geometric mean is used to characterize central tendency in these distributions.

To examine changes in the geometric mean length of first-order channels through extension to maximum extension, the mean lengths are plotted against relative time to maxi-

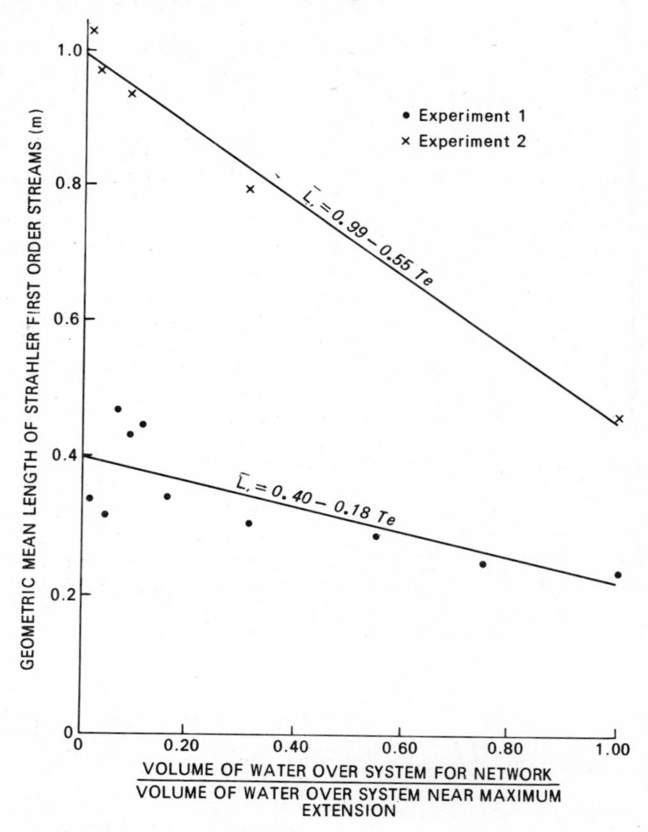

Figure 8.9 Changes in geometric mean length of Strahler first-order streams through time to maximum extension. Time is shown as a ratio with a value of 1.0 being maximum extension

mum extension (fig. 9). This time ratio is obtained by dividing the cubic feet of water applied to the system for a particular network by the amount of water applied at maximum extension. Thus, 1.0 represents the time at maximum extension.

Figure 9 shows a least-squares fit of a line to both sets of experimental data. The equation of the line for experiment 1 data is:

$$\overline{L}_1 = 0.40 - 0.18\ T_e, \tag{1}$$

where $\overline{L}$ = geometric mean length (m) of Strahler first-order streams, and T_e = ratio of time of network to time at maximum. The equation for the line through the experiment 2 data is:

$$\overline{L} = 0.99 - 0.55\ T_e. \tag{2}$$

The histograms (figs. 3, 4, 7 and 8) indicate that the distribution of the first-order stream lengths changes very little during the first experiment but that the changes are marked in the second experiment. To test this observation statistically, it is necessary to determine if the slopes of the regression equations (eqs. 1 and 2) are significantly different from zero. For equation 1, a t-test at the 1 percent level does not reject the hypothesis that the regression coefficient is equal to zero. A t-test on the regression coefficient for the second equation (eq. 2) rejects the hypothesis that the slope of the line is equal to zero at the 1 percent level.

Therefore, the geometric mean does not significantly change as the network progresses to maximum extension during experiment 1. This occurs even though there are six baselevel changes as evolution progresses.

Statistically there is a significant trend to smaller geometric mean lengths with time during experiment 2. The nearly rectangular distribution at initiation changes through time to a right-skewed distribution with a definite mode and the longer streams on the tail of the distribution are lost. Thus, patterns grow primarily by extension and bifurcation of first-order streams, but the manner of growth is significantly different during the two experiments.

SUMMARY

By plotting drainage density through time, a curvilinear relation is obtained which can be subdivided into periods of extension, maximum extension, and abstraction (fig. 1). Although this general relation appears definitive with little scatter to the data points, the relation is confounded due to changes in baselevel during the experiment. These changes in baselevel were the driving force in the experiment which allowed drainage density to increase.

If the initial conditions were constant but either the amount of clay in the material or the rainfall intensity was changed, the result would probably be a family of curves. These would not overlap but would be stacked one on top of the other.

Two very different modes of drainage network development were observed. The differences are attributed to the different initial conditions for experiments 1 and 2. On the lower initial slope of 0.75 percent with a baselevel change before the initiation of the experiment, the relative frequency histograms of Strahler first-order stream lengths are nearly equivalent at the beginning of extension and at maximum extension, and the geometric mean length of first-order streams is stable through time. The drainage density and, therefore, the total channel length also increases in a regular fashion. These observations suggest that this network is growing headward in a regular fashion as a wave of dissection progresses into the undissected area; a mode of growth characteristic of the headward-growth model.

The network on the steeper initial slope with no baselevel change produces a relative frequency histogram of Strahler first-order stream lengths that is almost rectangular for the first network. This reflects the long first-order channels that develop rapidly into the undissected area and 'block out' a large portion of the watershed. Later development includes the growth of tributaries from these long first-order channels. Again, the drainage density increases with time, but the rate of increase is slow, and the value at maximum extension is not as high for the headward growth. Although not exactly equivalent, this mode of growth suggests the Glock-Horton model of network development. Long tributaries initially develop very rapidly to produce a high value of drainage density, but then the rate of increase of drainage density slows.

These changes in initial conditions and the resultant changes in the mode of growth indicate two different types of network growth - headward growth and Glock-Horton growth. Differences in the type of growth are directly related to the initial conditions which, under natural conditions, would reflect network growth after faulting (experiment 1) or uplift and tilting (experiment 2).

In spite of all the differences noted as a result of the changes in initial conditions, the regularity in growth of the first-order channels is remarkable. For example, the geometric mean length of first-order streams remains essentially constant through time to maximum extension in the first experiment; the geometric means of first-order are equal for the initial network of experiment 1 and for the network generated on the initial slope of 12.2 percent. The histograms of these first-order stream lengths are similar, and there was a tendency for the first-order stream length histograms of the second experiment to become more nearly like the other observed histograms. These observations suggest a regularity in growth which, if better understood, could lead to better predictive capability of network development.

Acknowledgements

This research was supported by the United States Army Research Office.

9 Application of "diffusion" degradation to some aspects of drainage net development

Z.B. Begin

INTRODUCTION

The lowering of base-level is a cause of accelerated erosion in alluvial channels (Schumm and Hadley, 1957; Nordin, 1964; Livesly, 1957; Pickup, 1975). It may cause rejuvenation of a drainage network, as evidenced by the entrenchment of a drainage ditch in Iowa, which was 'responsible for much of the deep entrenchment of its tributary streams' (Daniels, 1960). This mechanism may be responsible for the development of badlands in some areas and the ability to predict the impact of such a change is of importance for land conservation.

Base-level lowering results in a headward growth of the channel network, beginning at the basin outlet and developing into the basin. Computer simulations of such a growth model were presented (Howard, 1971; Smart and Moruzzi, 1971) and an experimental study of the evolution of drainage basins was undertaken by Parker (1977). In the latter, different modes of basin development were studied, one of them being instigation of channel network due to base-level lowering.

In an experimental study of the degradation of an alluvial channel in response to base-level lowering (Begin *et al.*, 1981a), it was demonstrated that the process of degradation may be approximated analytically by a 'diffusion' model of degradation. Some corrolaries of this approach have a bearing on aspects of evolution of drainage networks, to be discussed below.

THE 'DIFFUSION' EQUATION OF DEGRADATION

In an alluvial channel with a lateral influx of sediments (bank erosion), conservation of matter is formulated by the two-dimensional equation of sediment continuity:

$$\frac{\partial y}{\partial t} = \frac{1}{\gamma_s} \frac{\partial q_s}{\partial x} + B \qquad (1)$$

where q_s is sediment discharge by weight per unit width, γ_s is the bulk weight per unit volume of sediment, y is elevation of the channel bed, x denotes distance along the channel (positive upstream, with x = 0 at the outlet), t denotes time, and B is the volume of lateral inflow of sediments, per unit time, per unit flow width, and per unit length of channel.

As noted by Gessler (1971), many sediment transport equations can be brought into the form:

$$q_s = C_1 (\tau - \tau_c)^p \quad (2)$$

where C_1 and p are empirical constants; τ and τ_c are the bottom shear stress and the critical bottom shear stress, respectively. In particular, in some sediment transport equations the power p is 3/2 and in these cases, if $\tau >> \tau_c$ (or, in other words, if the erosive ability of the flow largely exceeds the sediment resistance to erosion), eq. (2) reduces to:

$$q_s = C_1 \tau^{3/2} \quad (3)$$

The average flow velocity $\bar{u}$ is given by:

$$\bar{u} = \sqrt{(\frac{8}{f} gRS_e)} \quad (4)$$

where f is the Darcy-Weisbach friction factor; g is acceleration due to gravity; R is the hydraulic radius of the flow, and S_e is the slope of the energy line. Following Gessler (1971), it is noted that for wide channels, $\bar{u} = q_w/R$, where q_w is the water discharge per unit width, and substituting this value in equation (4) and rearranging leads to:

$$R = (q_w^2 \frac{f}{8gS_e})^{1/3} \quad (5)$$

Since: $\tau = \gamma_w RS_e$ (where γ_w is water density), the substitution of R from equation (5) yields:

$$\tau = \gamma_w (q_w^2 \frac{f}{8g})^{1/3} S_e^{2/3} = C_2 S_e^{2/3} \quad (6)$$

and substituting this value of τ in eq. (3) yields the following expression for the sediment discharge:

$$q_s = (C_1 C_2^{3/2}) S_e = \gamma_s k S_e \quad (7)$$

with

$$k = \frac{C_1}{\gamma_s} q_w \sqrt{(\gamma_w^3 f/8g)} \quad (8)$$

that is: the sediment discharge is linearly related to the energy slope, if q_w is assumed to be constant. If the energy slope S_e can be approximated by the bed slope:

$$S_e \simeq S_b = \frac{dy}{dx}$$

then substitution in eq. (7) yields:

$$q_s = \gamma_s k \frac{dy}{dx} \quad (9)$$

from which

$$\frac{\partial q_s}{\partial x} = \gamma_s k \frac{\partial^2 y}{\partial x^2} \qquad (10)$$

Substituting eq. (10) into eq. (1) yields:

$$\frac{\partial y}{\partial t} = k \frac{\partial^2 y}{\partial x^2} + B \qquad (11)$$

which is a version of the diffusion equation and k in equations (8) and (11) thus becomes a 'diffusion coefficient', with dimensions of $[L^2]/[T]$. This equation was arrived at from different lines of reasoning by Culling (1960), Ashida and Michuie (1971), deVries (1975), Wilson and Kirkby (1975), and it was solved for different initial and boundary conditions.

Assuming simple boundary conditions, at time $t < 0$ the channel profile is at equilibrium, and it is described by the linear equation

$$y_{(x,o)} = Y + bx \qquad (12)$$

where Y is a constant and b is the initial bed slope.

For channel degradation in response to base-level lowering, the initial conditions are as follows: at time $t = 0$, base level is lowered at the outlet ($x = o$) by an amount Y, to a new elevation of zero, which is maintained for all $t > o$. For the sake of simplicity it is also assumed that there is no lateral influx of sediment, and so $B = 0$. (If $B > 0$ see discussion in Begin *et al.*, 1981a).

Under the above boundary and initial conditions, the solution of equation (11) is:

$$y_{(x,t)} = Y \operatorname{erf}(x/2\sqrt{(kt)}) + bx \qquad (13)$$

Here erf is the error function, which is defined as: $\operatorname{erf} z = (2/\sqrt{\pi})\int_0^z \exp(-m^2)dm$ and its values are tabulated in many books of mathematical tables. In equation (13) z is the argument $z = x/2\sqrt{(kt)}$, where k is the diffusion coefficient of eq. (11).

Equation (13) permits the delineation of longitudinal profiles of alluvial channels at different times following base-level lowering, provided that the amount of base-level lowering Y, the initial slope b and the 'diffusion' coefficient k are all known (Fig. 1).

Figure 1 also shows the propagation of degradation of a channel to a certain amount Δy, defined as:

$$\Delta y = y_{(x,o)} - y_{(x,t)} \qquad (14)$$

where $y_{(x,o)}$ is initial channel elevation (at a certain distance x from the outlet) and $y_{(x,t)}$ is channel elevation at some later time t. We may refer to degradation as a 'disturbance' in the original surface, and calculate analytically the speed of propagation of such a disturbance. In order to permit comparison of different experiments, a non-dimensional magnitude of such a disturbance is defined relative to the amount of base-level lowering:

$$\text{disturbance magnitude} = \frac{\Delta y}{Y} \qquad (15)$$

Combining eqs. (12) to (15), and rearranging leads to the following version of eq. (13):

$$\frac{\Delta y}{Y} = 1\text{-erf } (x/2\surd(kt)) \qquad (16)$$

Within the same process of channel degradation (constant k), disturbances of different magnitudes propagate at different rates, with the smallest disturbances moving fastest upstream. For example, if the 'diffusion' coefficient k of the degradation process is equal to 1000 cm^2/minute, a disturbance of magnitude $\Delta y/Y = 0.1$ will propagate upstream according to eq. (16):

$$0.1 = 1 - \text{erf } (x/2 \surd(1000\ t))$$

from which:

$$\text{erf } (x/2 \surd 1000 \surd t) = 0.9$$

Looking up the value of the argument for which the value of the error function is 0.9 (Abramowitz and Stegun, 1964), one finds that

$$x/2\surd 1000\surd t = 1.16$$

$$x = 73.36 \surd t \qquad (17)$$

from which it is easy to calculate the different values of distance x arrived at by the said disturbance at different times t. The results of such calculations are presented in Fig. 2, for some values of $\Delta y/Y$.

Application to drainage network evolution

Parker (1977) used a 15x9 m container (the Rainfall-Erosion-Facility at Colorado State University, Fort Collins, Colorado) to study the development of drainage nets. Some of his drainage nets were developed in response to base-level lowering.

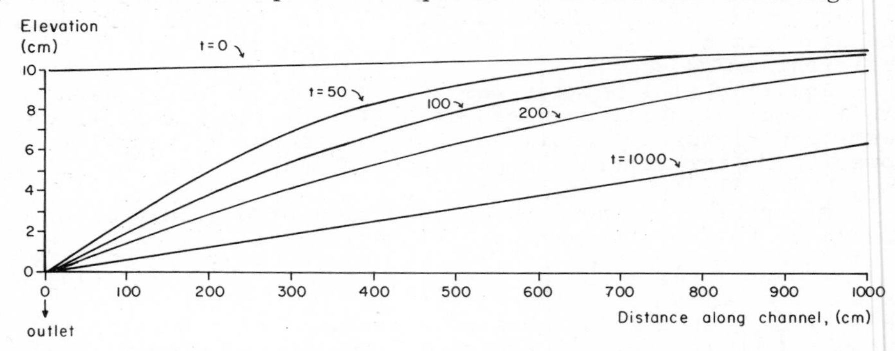

Figure 9.1 Theoretical development of longitudinal profiles of an alluvial channel in response to base-level lowering by amount Y = 10 cm. Numbers denote time in minutes from base-level lowering. The diffusion coefficient of the process is $k = 1000\ cm^2/min$, and the initial bed slope is b = 0.01

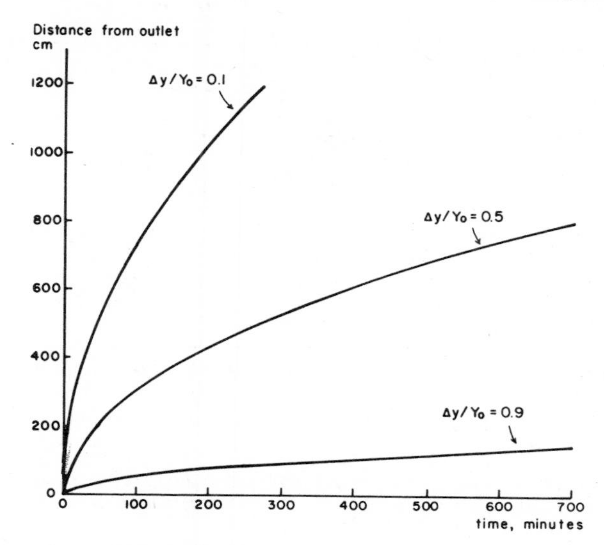

Figure 9.2 Migration curves for the propagation of disturbances of different magnitudes $\Delta y/Y$. Initial conditions as in Fig. 1. Note the slower rate of propagation of disturbances of greater magnitudes. (After Begin *et al.*, 1980)

Parker (1977, p.17) noted that for first order streams, the 'geometric mean length does not significantly change during the extension of the networks of data set 1' (instigated through base-level lowering). 'In order for this to occur, a first order channel must grow to some limit and then bifurcate, therefore being eliminated from the class of first order streams. The marked regularity in the range of lengths may reflect a length at which the probability of bifurcation is maximized' --- 'The identification of a probability distribution for channel bifurcation is not straight-forward. Such a distribution would undoubtedly vary in both time and space' (Parker, 1977, p. 17).

Indeed, if the discussion is limited to source-type first order streams (Mock, 1971) in Parker's (1977) data set 1, it seems that a vague trend exists, of increasing the length of source-type first order streams, with time. The data were obtained from five consecutive networks maps, which developed in response to lowering of base-level by 22.5 cm, and are taken from Parker's (1977) Table 4-3:

Network	Time t (hours)	Geometric mean length of S-type, 1st order streams L_m, (cm)
1	2	30.9
2	4	24.1
3	6	40.6
4	8	42.5
5	10	46.7

The least-square log-log fit to these data yields the equation:

$$L_m = 21.4t^{0.32}$$

with a correlation coefficient of 0.74 which, for the 5 data points, is not significant at the 0.05 level, but would be significant at the 0.10 level.

Using some of the ideas and relationships developed above, we may now try an analytical approach to the problem, noted by Parker, of the changing probability distribution for channel bifurcation.

First we have to address ourselves to the problem of definition of the length of a first order stream. The downstream end-point of such a stream is easily defined by its junction with the stream from which it bifurcates, but where does the tributary begin? Different operational definitions yield different results, as was pointed out in several previous works (Eyels, 1968, p. 703; Smart and Moruzzi, 1971, p. 576; Yang and Stall, 1971). The main difficulty lies in the fact that in many cases, where one walks upstream, there is a continuous passage from a clearly defined tributary to a point along its imaginary continuation where no channel is discernible. Some arbitrary decision seems to be necessary in such cases, and a certain depth of channel below the original, uneroded, surface should be used to define a point downstream of which a stream 'begins'.

In terms of the theoretical considerations presented above, this specific depth is a certain Δy (eq. 14), which will be denoted were as Δy_{tip} (Fig. 3). If the discussion is limited to headward growth of channels in response to base-level lowering, and if this process can be approximated by the 'diffusion' model presented above, then the migration upstream of Δy_{tip} can be calculated in the manner portrayed in equations 16-17. This is predicated of course on the knowledge of the amount of base-level lowering Y and the diffusion coefficient k. In other words, in response to base-level lowering by Y, a disturbance of magnitude $\Delta y/Y$ travels upstream according to eq. (16), and the distance x_{tip} travelled by it in time t is taken from eq. (16) to be:

$$x_{tip} = 2C_{tip}\sqrt{kt} \qquad (18)$$

in which C_{tip} is taken from the relationship:

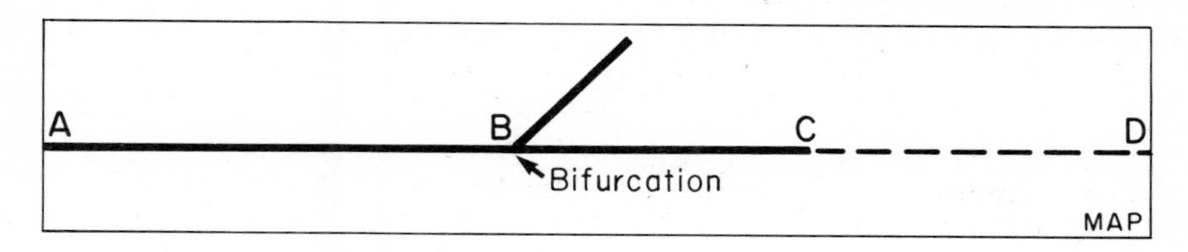

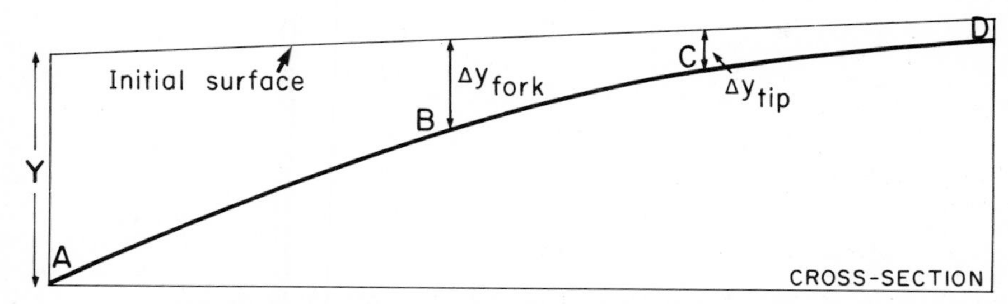

Figure 9.3 Sketch showing the definition of Δy_{tip} and Δy_{fork}

$$\operatorname{erf}(C_{tip}) = 1-\Delta y_{tip}/Y \qquad (19)$$

and assuming that Δy_{tip} and Y are given.

Now we refer again to Parker's (1977) observation that a first order channel grows to some limit and then bifurcates. This limit may be explained by taking into account the fact that in response to base-level lowering, channels not only elongate but also deepen. As the tip of the channel migrates upstream, degradation at any station continues (Fig. 1), and the channel is deepened. The deeper the channel the greater is the local relief it creates and it is reasonable to assume that the main channel has to reach a certain depth in order to act as a local base-level for a new sub-basin. When the area of this sub-basin reaches the value of the constant of channel maintenance (Schumm, 1956), the discharge contributed by this sub-basin is just enough to cut a new channel causing bifurcation. Thus, the probability that a new, source-type tributary will develop increases with increasing local relief. We suppose, then, that for a certain drainage net, the sediments and the hydraulic conditions determine a certain depth of channel which is needed to permit the development of another tributary at some point along the first channel. This specific depth will be denoted as Δy_{fork} (Fig. 3).

We now try to incorporate this assumed mechanism into the above presented 'diffusion' model of channel growth. The distance X_{fork} which at time t is travelled by a disturbance of magnitude $\Delta y_{fork}/Y$, is taken again from eq. (16) to be:

$$X_{fork} = 2C_{fork}\sqrt{kt} \qquad (20)$$

in which C_{fork} is taken from the relationship:

$$\mathrm{erf}\,(C_{fork}) = 1-\Delta y_{fork}/Y \tag{21}$$

Now we can define the length L of the source-type first order tributary. This length is simply the difference between the distance travelled at time t by the tip of the channel (Δy_{tip}) and the (smaller) distance travelled in the same time by the (greater) disturbance Δy_{fork}:

$$L = X_{tip} - X_{fork} \tag{22}$$

substituting eqs. (18) and (20) into eq. (22):

$$L = 2(C_{tip} - C_{fork})\ \sqrt{kt} \tag{23}$$

It may be noted that in all cases $C_{tip} > C_{fork}$. This stems from the obvious relationship $\Delta y_{tip} < \Delta y_{fork}$, combined with the definitions of C_{tip} and C_{fork} as given in eqs. (19) and (21).

Equation (23) shows that during the evolution of a drainage network in response to base-level lowering, the length of the source type first order tributaries is expected to increase in proportion to $t^{0.5}$. It should be noted that verification of this relationship may be difficult: If the diffusion coefficient k has a low value, one would need great spans of time to observe significant changes in L_m, taking into account the expected statistical dispersion of L at any minute, around the mean value L_m (L_m is the mean length of all source-type first order streams, mapped at a certain stage of drainage-net evolution).

The results of Parker's (1977) experiment, cited above, provide some - though weak - support for this analysis. From the least-square fit L_m is proportional to $t^{0.32}$, but this is not statistically different from the expected power of $t^{0.5}$ (it should be noted again that this statistical relationship is not significant at the 0.05 level, but only at the 0.10 level). With such limited data, no experimental proof can still be given for the above analysis, but these results show at least that such a treatment of drainage basin evolution might prove fruitfull. In such a manner, the geometrical changes of an evolving drainage net may serve as tools in an attempt to gain further insight of the processes involved in drainage net evolution.

SEDIMENT PRODUCTION OF A DRAINAGE NET IN RESPONSE TO BASE-LEVEL LOWERING

The degradation of a channel in response to base-level lowering causes increased sediment production at its outlet, which typically decreases with time (Parker, 1977; Begin *et al.*, 1981b). The following is based on Begin *et al.*, (1981b) who obtained an analytical solution for the changes in sediment production, based on eq. (13). The analysis begins with differentiation of eq. (13) with respect to distance, x:

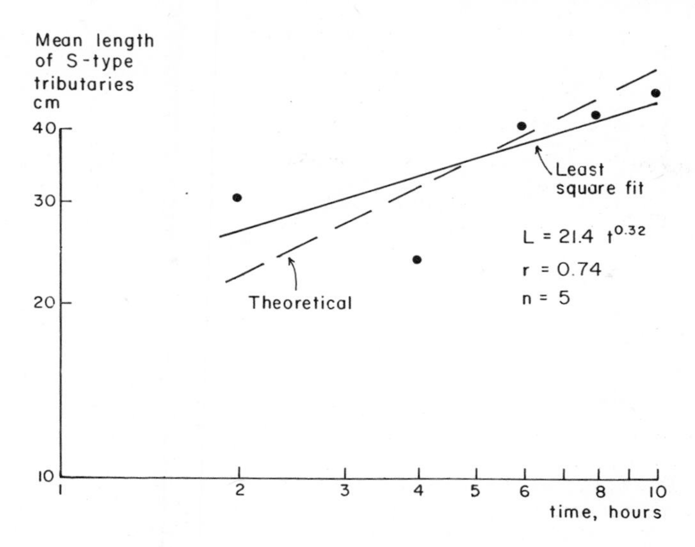

Figure 9.4 Experimental results (Parker, 1977) and theoretical line for the increase with time of the length of S-type first order tributaries

$$\frac{dy}{dx} = \frac{Y}{\sqrt{(\pi kt)}} \exp\left(-\frac{x^2}{4kt}\right) + b \tag{24}$$

and for the outlet (x = o) the changes in bed slope reduce to:

$$\frac{dy}{dx} = \frac{Y}{\sqrt{(\pi kt)}} + b \tag{25}$$

substituting eq. (25) in eq. 9, the sediment discharge per unit width of channel at its outlet is:

$$q_s = \gamma_s k \frac{dy}{dx} = \frac{\gamma_s Y \sqrt{(k/\pi)}}{\sqrt{t}} + \gamma_s kb \tag{26}$$

Under simple experimental conditions entailing one channel, sediment production in an alluvial channel in response to the lowering of its base-level was shown to follow the trend predicted by eq. (26) (Begin *et al.*, 1981b). These experimental results are very similar to the results obtained experimentally by Parker (1977) which pertain to development of drainage nets.

Extending eq. (13) to a drainage net is not trivial, because rejuvenated tributaries constitute point sources of sediments to the main channel, violating the assumption of no lateral sediment flow. However, if the tributaries are small enough, it may be assumed that the main channel disposes 'momentarily' of all the sediments produced by its tributaries, and a field study by Harvey (1977) points to the possible occurrence of such cases.

Adopting this assumption permits the examination of the effect of a rejuvenated drainage net on sediment production through the main stem, by a computer simulation based on a

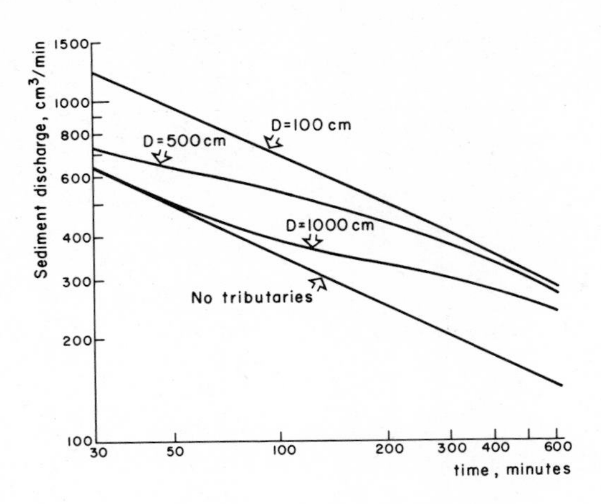

Figure 9.5 Changes with time of sediment production of an alluvial channel in response to base-level lowering, as obtained by a computer simulation (see text for values of parameters). The only parameter changed was the distance D from the outlet of main stem to the point of confluence of its two tributaries (from Begin *et al.*, 1981b)

numerical solution of eq. (11) with B = O (Program ERFUSE, Begin, 1978). To simplify matters, only two tributaries were 'added' to the main channel at one point of confluence, and the location of that confluence point was the only changing variable. Other parameters for the simulations are:

	width (cm)	initial bed slope (cm/m)	diffusion coeff. k, (cm^2/min)	amount of base-level lowering, (cm)
Main stem	20	0.005	1000	10
Tributaries	10	0.010	700	-

The results of these simulations (Fig. 5) demonstrate that even in the case of a drainage net which is topologically simple and constant, the geometrical relationships such as the location of tributaries confluence with the main channel have a marked effect on the trend and amount of sediment production of the rejuvenated net. This adds a measure of complexity (Schumm, 1975) to the response of drainage net to the simple act of base level lowering.

CONCLUSIONS

'A simple rejuvenation of a drainage basin yields a complex erosional and sedimentological response, and the behaviour of the basin itself introduces a new degree of complexity into the geomorphic process' (Schumm, 1977). 'Much of this

complexity is the result of a delayed transmission of information through the system' (Schumm, 1975). At least in some cases, this delay is ammenable to a quantitative assessment through eq. (16) which predicts the rate of upstream propagation of a disturbance of a specified magnitude. If the assumption that B = 0 is considered inapplicable, the same approach can still be taken, but the resulting equations are somewhat cumbersome (Begin *et al.*, 1981a, eq. 16). Also, if the drainage net is entrenched into heterogeneous material leading to channel armoring, appropriate corrections may be available (Begin *et al.*, 1981a, p. 58). The effect of channel armoring on knickpoint migration and sediment production in the mouth of a main stem is discussed by Begin *et al.*, (1980) and by Begin *et al.*, (1981b).

As shown above, this approach may prove useful in explaining some geometrical attributes of network growth in response to base-level lowering. The advantage of eq. (13) and its corrolaries is that it permits estimates of both the geometry of the network (stream length) and the time factor.

Referring to their drainage net simulation model, Smart and Moruzzi (1971, p. 583) noted that '... the most promising possibility for improvement is the replacement of our rather arbitrary selection of probability factors with a procedure more closely related to the processes involved'.

If the processes can be quantified through an approach such as outlined above, it may allow the linkage between topological similarities of drainage patterns and geometrical attributes of natural and simulated drainage nets. From this respect, badland areas may prove to be essential in providing natural experimental sites, in which the processes of erosion are fast enough to permit observation of network growth within a reasonable time span.

10 Sediment and solute yield from Mancos Shale hillslopes, Colorado and Utah

J. Laronne

More than a century ago Gilbert (1877) published his monumental report on the Henry Mountains. The monograph is particularly relevant because debate-generating theories and examples on the formation of pediments, channel grade and hillslope form were formulated on Mancos Shale. Hillslope erosion was once more studied on Mancos Shale by Schumm (1964) who showed the effects of seasonal variations, primarily frost-heave, on processes and rates of erosion on hillslopes in semi-arid areas. Branson and Owen (1970) and Wein and West (1973) conducted additional research on Mancos Shale hillslope erosion. Pediment formation has also been recently studied in the Mancos Shale terrain (Schumm, pers. comm.).

The Colorado River has the highest sediment and solute concentrations among the major rivers in the North American continent. A high relative relief within its basin and a semi-arid climate coupled with sparse vegetation in the lowlands are major reasons for this intense erosional activity. In addition, highly erodible shales are exposed in large portions of the Colorado River Basin. Because of its thickness, usually over 1000 m, the outcrop area of Mancos Shale is overwhelmingly the largest among these erodible sedimentary rocks and their derived soils. Cretaceous exposures, of which roughly half are composed on Mancos Shale, cover one third of the drainage basin area of the Colorado River Basin above Lees Ferry, Arizona (Fig. 1).

Not only does the Mancos Shale terrain yield much sediment, but it is also the single largest contributor of salinity (i.e., solutes or the misnomer 'total dissolved solids') from diffuse sources within the Colorado River Basin. Iorns *et al.*, (1965) showed that drainage basins underlain by Mancos Shale are major contributors of sediment and solutes to the Colorado River. Later studies confined to specific areas have confirmed and amplified these findings in the Badger Wash Basin (Lusby *et al.*, 1971) and in the Price River Basin (Mundorff, 1972).

The foregoing remarks clarify why soil and water management problems are associated with Mancos Shale, and that considerable research has already been undertaken on hillslope form and processes in Mancos Shale badlands. These have

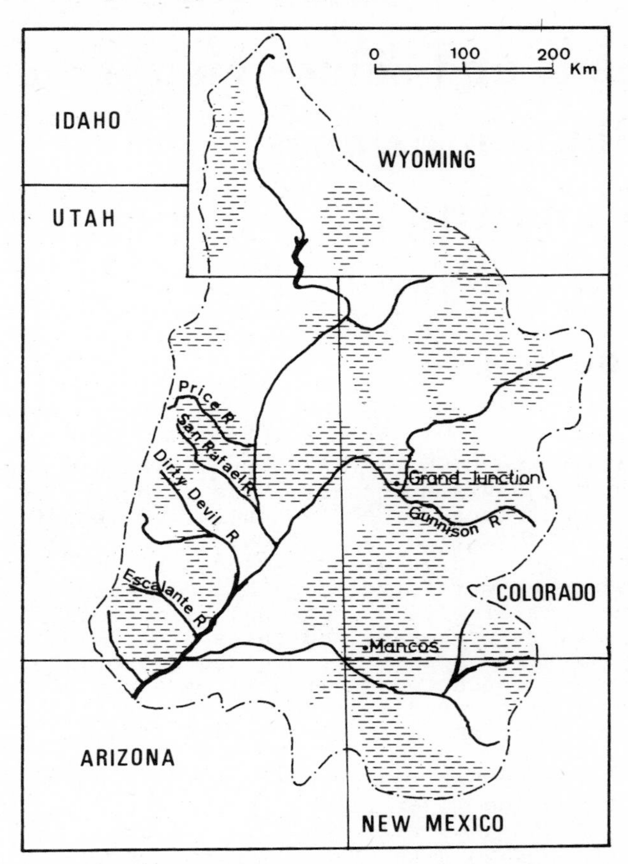

Figure 10.1 Map of the Upper Colorado River Basin showing exposures of Cretaceous rocks and location of study areas.

prompted the author to review and study the export of matter from selected Mancos Shale hillslopes.

STUDY AREA

The name Mancos was first applied in 1899 by Cross (Fisher *et al.*, 1961) to exposures of shale near the town of Mancos in southwestern Colorado (see Fig. 1). This use has since been extended to thick shales, usually in part of Colorado age and in part of Montana age, found over a large region south of the Uinta Mountains in Utah and west of the Rocky Mountains in Colorado and New Mexico. Almost all of the exposures, the largest of which is found in a belt averaging 10 km wide south of the east-west trending Book Cliffs in west central Colorado and east central Utah, are located in the High Plateaus and Canyonlands sections of the Colorado Plateau physiographic province (Fenneman, 1931).

Mancos Shale overlies the Dakota Sandstone which caps an escarpment in the Morrison Formation and typically forms hogbacks and cuestas. The Mesa Verde Group overlies the Mancos Shale. This series of sandstones with conglomeratic, shaly and calcareous members forms high escarpments capping Mancos Shale and forming buttes and mesas such as at the Mesa Verde National Park. The Mancos Shale is a shallow marine formation. It is dary gray, thinly bedded and lacking pronounced fissility when fresh. The shale abounds in veinlets of gypsum and calcite, and it is often covered with patches of 'white alkali' or salt efflorescence. It is drab gray and monotonous where weathered producing crusted flakes on the surface when dry and a friable, semi-powdery mass that is sticky and impervious when wet. It includes few thin layers of bentonite, calcareous and shaly sandstone and shaly limestone. In places it includes lenticular carbonaceous and coal-bearing shale and sandstone.

Knobel *et al.*, (1955), Swenson *et al.*, (1970) and Lusby *et al.*, (1971) have investigated and classified the soils derived from and in association with Mancos Shale. These thin gray silty shale loam soils have a pH of 8.0 and a high salinity. Montmorillonite, illite, chlorite and mica have been identified in these soils. Both fresh and somewhat weathered Mancos Shale swell considerably when wetted with a 25-50 percent volume increase in free swell tests (Schumm, 1964).

Shale members of Mancos Shale form badlands with fine to ultrafine drainage densities averaging 60 km km^{-2}. Drainage density and hillslope gradient are highest where the shales are exposed near the overlying Mesa Verde sandstones and near overlying basalt in the Gunnison River Valley. Badland hillslopes are generally less than 50 m high with an average height of 20 m and a slope of 20 percent. Mean hillslope inclination in Badger Wash sub-basins is between 14.3 and 27.8 percent (Lusby, 1979). Streams crossing the badlands and the lower Mancos Shale area have typically sinuous courses with a considerable downstream thickening of their alluvial fills. The channels are incised into the fills and have formed several terraces, more than half a dozen in some locations.

Hunt *et al.*, (1953) stated that 'as erosion advances in the badlands, the height of the hills is reduced, and as the narrow-crested divides are lowered they become smoother and broader.' The width of hillslope divides approaches a knife-like edge on hillslopes with gradients exceeding 100 percent unless these are covered by alluvial aprons. The steeper hillslopes are rilled and miniature basins are clearly delineated on all but the most gentle slopes.

The climate in the lowlands of the Upper Colorado River Basin is of a semi-arid continental type with frequent high intensity convective storms of small areal coverage. Maximum monthly precipitation occurs in July-August. In the Book Cliffs desert daily and seasonal temperatures vary widely with extremes of 42 to -41°C (Mundorff, 1972). The average potential evaporation measured with a class-A pan at the Grand Junction airport is 233 cm with a monthly average of 46.5 cm in July (Lusby, Reid and Knipe, 1971).

Mean annual precipitation is 250 mm at Price, 200 mm at Woodside in the lower Price Basin and 215 mm at Badger Wash near Grand Junction (Branson and Owen, 1970, Mundorff, 1972).

The crowns of living perennial plants cover less than 10 percent of the surface except in local areas and less than 1 percent on the steeper hillsides. Plant cover somewhat increases in early summer. Plant communities include greasewood and rabbitbrush on alluvial fills in the valley bottoms, and sagebrush on the gravel-covered pediments below piñon-juniper stands on further higher ground. Mat saltbrush and shadscale dominate the barren shale terrain that contains considerable soluble salts.

The soluble mineral content (expressed as weight per weight and herein denoted SMC) is of particular interest in studies of water quality. Laronne and Schumm (1982) have demonstrated that the SMC averages 0.7 percent in Mancos Shale surface crusts (0.1-10 cm thick) and 2 percent in underlying material, about 1 percent in Mancos Shale gully walls, 6 percent or more in channel surface crusts and 0.2-0.6 percent in terrace and channel alluvium. It should be noted, however, that the standard deviation about mean SMC values is proportional to the mean, and that a high inherent variability in SMC is characteristic of weathered Mancos Shale and derived soils (Ponce, 1975; Laronne and Schumm, 1982).

Results presented in this study are based on plots and hillslopes characteristic, in terms of slope steepness, surface micromorphology, SMC, etc., of average conditions in the shaly facies of Mancos Shale. They are, therefore, not characteristic of the entire Mancos Shale area but, rather, of average shaly hillslopes with high inherent variability in lithologic, morphologic and, therefore, hydrologic characteristics.

PLOT EXPERIMENTS

Sediment and solute concentration data from Mancos Shale hillslopes is available from two types of plot experiments. Not only was runoff generated by different means, but, unfortunately, the experiments were undertaken under different conditions. The first set of experiments was undertaken with artificial rainfall on small plots and miniature watersheds (Ponce, 1975, Ponce and Hawkins, 1978); the second set was undertaken by the author and H. W. Shen from Colorado State University on complete hillslopes of considerably higher gradients using a runoff generator without precipitation. Hence, the results presented in this study are complementary but not directly comparable.

Methodology

A Rocky Mountain infiltrometer (Dortignac, 1951) using distilled water was used by Ponce in the Price River Basin because of its portability, means of measuring precipitation, runoff and sediment as well as collecting runoff for chemical analyses; rainfall intensity could be controlled and varied (25-127 mm hr^{-1}) and the error and sampling variation of the instrument had been previously investigated. Generated raindrops fell from 3 to 6 m striking the ground at 75-94 percent of their terminal velocity. Although windscreens were used it was noticed that wind gusts affected rainfall distribution. Plot frames driven into the ground caused some soil disturbance at the frame-soil interface. Rainfall was generated on 30x77 cm plots. A larger, modified infiltrometer was used on 11.2 m^2 plots with a maximum length of 3 m and slope of approximately 10 percent. This was also used on small basins ('micro-watersheds') with a drainage basin area of 3.7 m^2, rill depth not exceeding 2.5 cm and slope of 5-15 percent. Runoff was sampled 3 min after rainfall had begun and thereafter regularly 5 times every 5 min. Ponce (1975) reported interesting results of various experiments relating rainfall to runoff and to sediment and solute concentration, but only the most relevant are described here.

A series of hillslope experiments was undertaken by the author during the summer of 1977 in the Grand Valley north of Grand Junction, Colorado. Nine sites, three of which will be discussed here, were chosen so as to be accessible by a water tank truck and to be on Mancos Shale (equivalent to the undivided Mancos Shale of Ponce, 1975). The three hillslopes selected ranged in steepness from 12 to 62 percent and were 8-30 m long (Fig. 2). The two channels were required to access variations across the slope.

Water flow of specific electrical conductance, denoted EC, of 450 μmho cm^{-1} @ 25°C and <100 mg ℓ^{-1} suspended solids was applied using a perforated 3.66 m long PVC pipe and a constant head tank delivering a flow rate of 33.1 ℓ min^{-1} quite uniformly throughout the length of the pipe. This pipe was situated rather close to the upper hillslope divide. The low-cost runoff-generator was built and used for several runoff experiments in order to arrive at a first approximation of the interaction between sediment and solute pickup mechanisms on these hillslopes. The applied water with 0.025 percent solutes could be used because of an expected much higher salinity in the generated runoff (an assumption which did not hold for the hillslope with the most gentle gradient), and because of undersaturation with respect to the major soluble minerals (Laronne, 1981).

Runoff samples were collected downslope from the pipe at locations denoted A, B and C (Fig. 2); overland flow was collected at A and channelized flow at B and C. Generated overland flow did not reach much beyond B on the mildest slope for which no data are, therefore, available for location C (see Fig. 2). Samples were also collected at two locations across the hillslope whenever runoff concentrated in more than one clearly definable channel at B_1 and B_2 and at C_1 and C_2. In addition, samples were collected where the hillslope contacted the channel proper, station D_1, and further downstream at D_2 (Fig. 2).

The leading edge of flow was sampled with minimum disturbance to the flow by constantly changing the shape of specially prepared aluminum funnels. Broad, flat funnels were used to collect overland flow and also to decrease the sampling time; narrow funnels were used in rills. The leading edge of flow was collected at each of the stations and flow was sampled again after 2,4,6,8,10,12,15,20, 30 and 45 min and thereafter at 15 min intervals. All odd-numbered

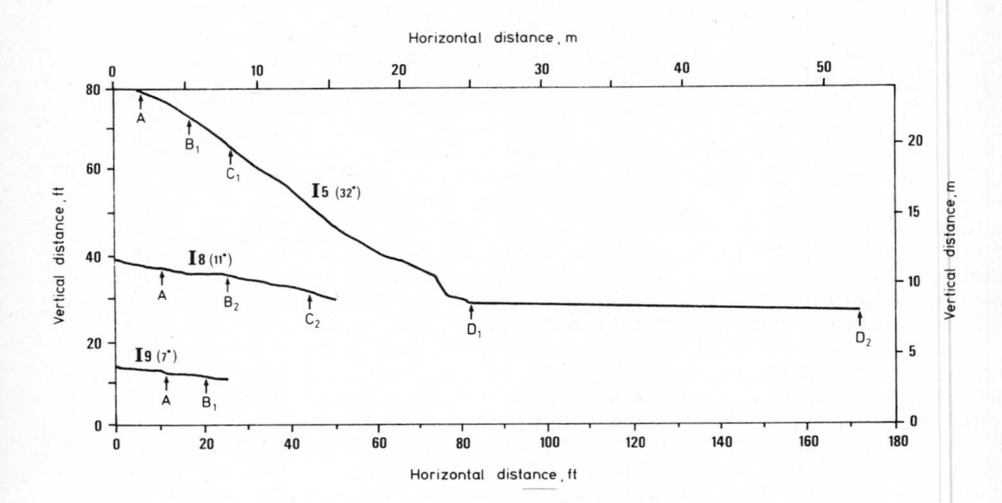

Figure 10.2 Vertically unexaggerated long profiles of the studied hillslopes showing the location of sampling stations.

samples at a particular station underwent immediate EC determination in the field (Lectro Mho Meter, Lab-Line Instruments). The precision of the EC meter was 1 percent as compared to a Beckman meter and was similar to Ponce's meter. Sediment concentration was determined by the evaporation method (Vanoni, 1975). Flow velocity ranged between 0.9 and 4.6 cm sec^{-1} and flow depth averaged 1.8 cm with a maximum of 3.8 cm in the rills. Only overland anastomosing flow was generated immediately downslope of the pipe. The leading edge of flow advanced downslope; at midslope it concentrated into one or more rills in all the studied hillslopes except in the most gentle. Anastomosing overland flow with high infiltration losses characterized the latter hillslope run, but it was present solely in broad micro depressions. Pipe flow occurred during one hillslope run not reported herein.

RESULTS

The temporal variation of rainfall, runoff, sediment and solute concentration (the latter expressed as EC) on three small plots is adapted from Ponce and Hawkins (1978) in Fig. 3. Plots 141 and 144 were located on undivided Mancos Shale and plot 82 on the Upper Blue Gate Mancos Shale member. The runoff curves are similar to those of the rainfall for all plots after initial abstractions have been met. Sediment concentration decreases with time and responds only slightly to rainfall intensity. The solute concentration remains low and constant for plot 141 (Fig. 3a), increases with time and reaches a peak when runoff rate is lowest irrespective of the

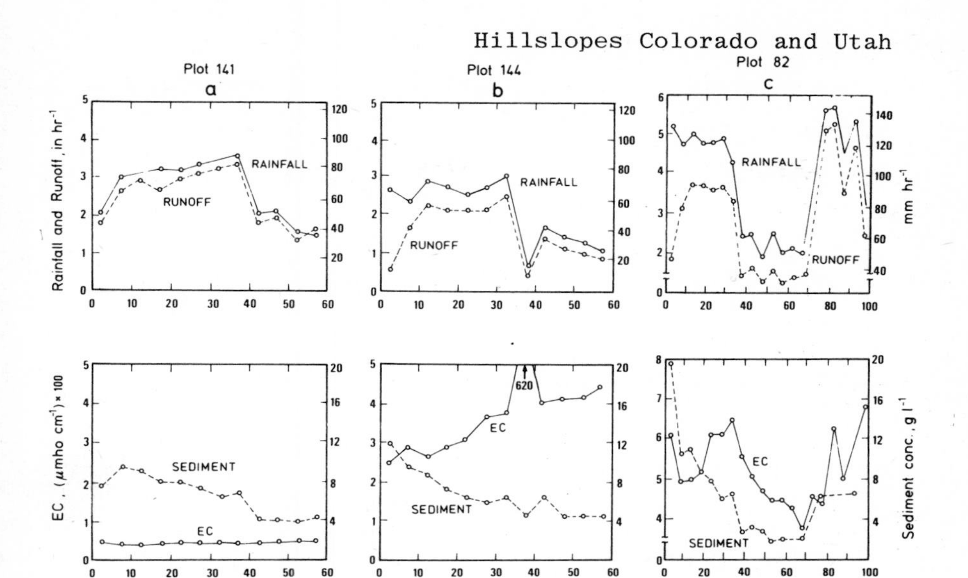

Figure 10.3 Temporal variation of applied rainfall, runoff, and sediment and solute concentrations in plot 141 (a), plot 144 (b), and plot 82 (c), after Ponce and Hawkins (1978).

continuous decrease in sediment concentration (Fig. 3b), and behaves rather erratically on plot 82 (Fig. 3c).

Fig. 4 (after Ponce, 1975) shows the variation of rainfall, runoff, sediment and solute concentration with time on micro-watersheds 1, 2, and 3 (on undivided Mancos Shale). Runoff also increases continuously with time after initial abstractions have been met and infiltration decreases although rainfall intensity varies considerably. Sediment concentration responds only slightly to runoff. Solute concentration initially decreases but thereafter increases (Fig. 4b and 4c) with a lag in Fig. 4a. The behaviour of the runoff and the corresponding solute concentration is rather variable with time. Note that the runoff/rainfall ratio is rather high though variable, which is in accordance with the high hydrologic soil complex number (CN = 93 with a standard deviation of 12) attributed to undivided Mancos Shale by Ponce (1975). Using linear regression, Ponce showed that the variability (r^2) in solute concentration accounted for by the corresponding variation in sediment concentration was as low as 11 percent but reached 69 percent with rather high standard errors. Some micro-watershed runs of Ponce not reported here have considerably higher r^2.

It is interesting to observe that solute yields, expressed in terms of solute concentration, are 2-40 times higher in the micro-watersheds than in Ponce's small plots (solute concentration expressed in mg ℓ^{-1} is roughly equal to 65 percent of the electrical conductance expressed in μmho cm^{-1}). Sediment concentration is approximately twice as high in the micro-watersheds.

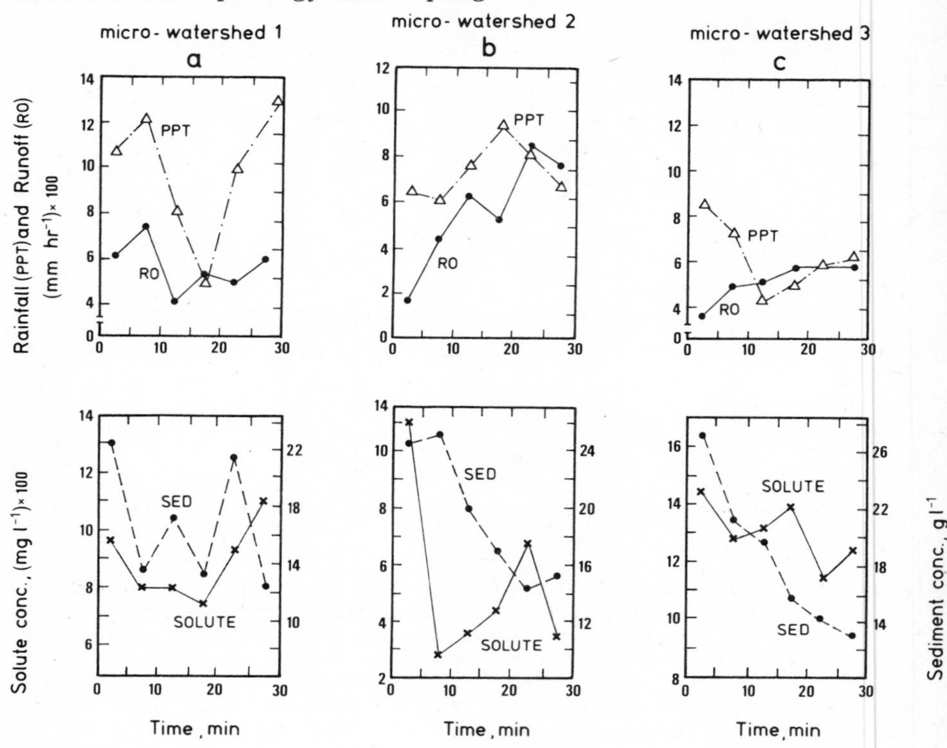

Figure 10.4 Temporal variation of applied rainfall, runoff, and sediment and solute concentrations in micro-watersheds no.1 (a), no. 2 (b), and no. 3 (c), after Ponce (1975)

Fig. 5 shows the temporal variation of solute concentration in runoff generated by the perforated pipe on hillslopes 9, 8, and 5 (Fig. 5a, 5b and 5c, respectively). The respective gradients are 12, 19, and 62 percent. The gentlest gradient corresponds to the one chosen by Ponce for his experiments. Solute concentration is low and changes only slightly with time at station A (Fig. 5a) where true overland flow is generated; it is somewhat higher at station B and decreases with time to values roughly the same as in the upper part of the hillslope. The concentration of solutes is higher at the beginning of run 8 (Fig. 5b) than at run 9 (Fig. 5a), and gradually rises to a maximum about 70 min. after the beginning of the run. It also increases downslope quite regularly. During run 5 (Fig. 5c) solute concentration is highest and it increases both downslope and downchannel very markedly. Sediment concentration changes similarly to solute concentration in run 8, peaking approximately 20 and 70 min into the run (Fig. 6a). An accordance between sediment and solute concentration also takes place during run 5 which is characterized by extremely high sediment yields (Fig. 6b). The abcissas of Fig. 6 refer to the time from the arrival of the leading edge of flow (to each station).

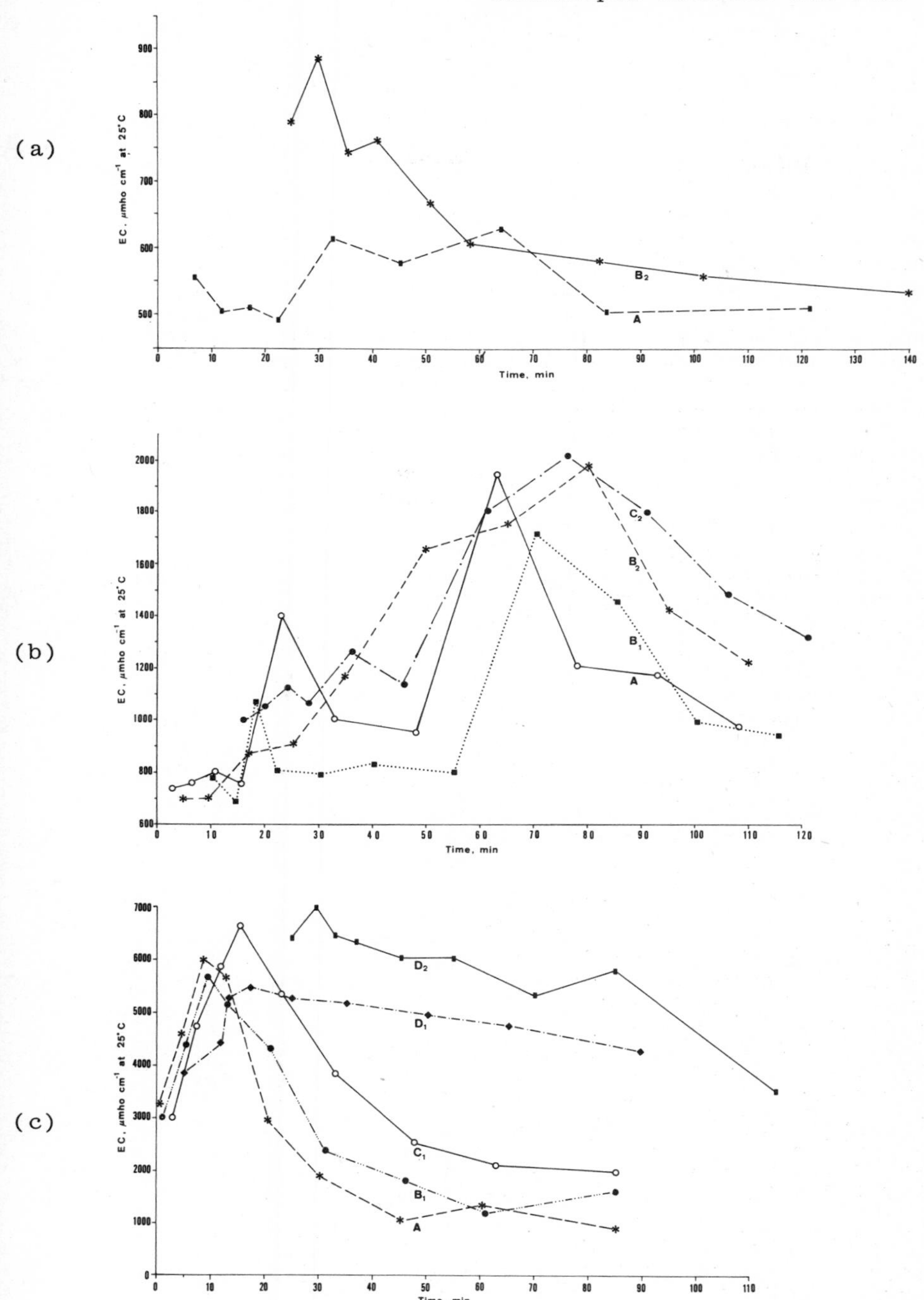

Figure 10.5 Spatial and temporal variation of runoff conductivity during run 9 (a), run 8 (b), and run 5 (c)

DISCUSSION

The results presented hitherto are not directly comparable. A Hortonian runoff-generating mechanism in Ponce's experiments implies contribution of overland flow *and transported matter* from the entire studied plot area. Runoff was generated in the author's hillslope experiments solely in the upper part of the study area, a condition equivalent to generating runoff and allowing it to flow downslope from Ponce's sprinkled area. Comparison between his results and the author's is possible, however, due to similar temporal trends in sediment and solute production. Also, the magnitude of sediment and solute concentrations are similar in the micro-watersheds and where overland flow occurred at stations A (Fig. 5a and 5b).

The initial high but thereafter continuously decreasing sediment concentration in the hillslope experiments (Fig. 6) ultimately reaches low values (4000-7000 mg ℓ^{-1}). An identical trend occurs in the micro-watershed experiments (Fig. 4) and concentrations in these are ultimately 2-3 times higher than in the small plots (Fig. 3). Such a trend with a concomittant trend in solute concentration in the small plots and in the micro-watersheds has been explained by Ponce as a weathered surface layer flushing phenomenon. This initial sediment flushing stage is more prominant and lasts longer in the hillslope experiments where flow velocity and flow depth are higher, as in the rill flow of runs 8 and 5 (see Fig. 5b, 5c and 6). Unlike overland flow, rill flow is capable of eroding the less saline surface layer quite rapidly. Sediment concentration is therefore higher overall than in the experiments conducted by Ponce. It is also higher because of the steeper and longer slope and the resulting increase in sediment concentration downslope. Solute concentration, still at under-saturation level with respect to the more soluble minerals, is also higher in the author's hillslope experiments. It should herein be stressed that the temporal increase in solute concentration in the hillslope experiments is due not only to entrenchment of rills and concomittant increase in transported, still-dissolving sediment, but it also arises due to the higher SMC of material underlying the surface layer of the rills. The concentration of solutes is similarly low in the longer but gentler slopes (micro-watersheds and station A, Fig. 5) because entrenchment does not take place in either.

Solute and sediment concentrations increase downslope (see Fig. 5 and 6). Increase in sediment concentration downslope is expected in runoff carrying fine suspended matter at undercapacity levels. Therefore, solute concentration must also increase downslope as more sediment is transported by and dissolves in the rill flow. The temporal peaks in sediment and solute concentration should, however, flatten downslope, which, indeed, they do. Water discharge decreases downslope, though at an ever decreasing rate with time into the run, because of infiltration losses. Similarly but due to principles of dissolution kinetics, rate of dissolution decreases as the solute concentration gradient (between the

(a)

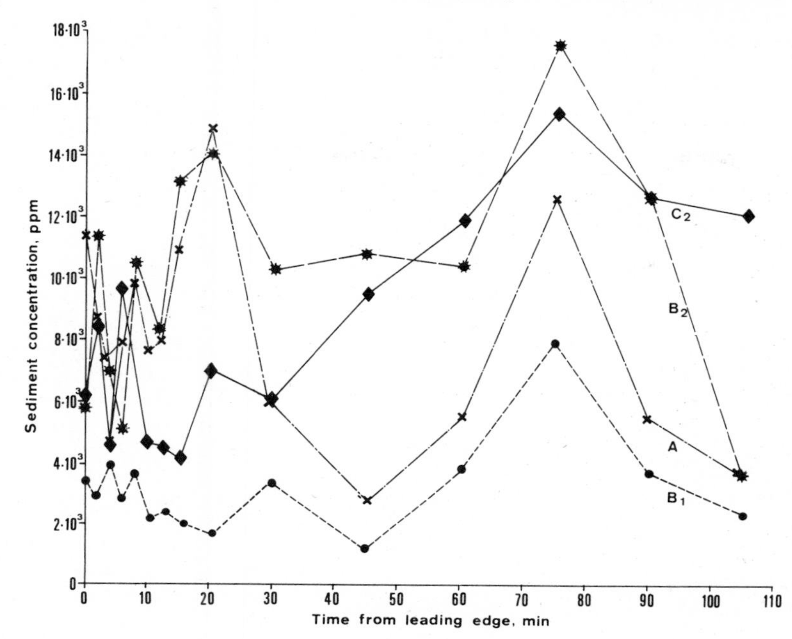

(b)

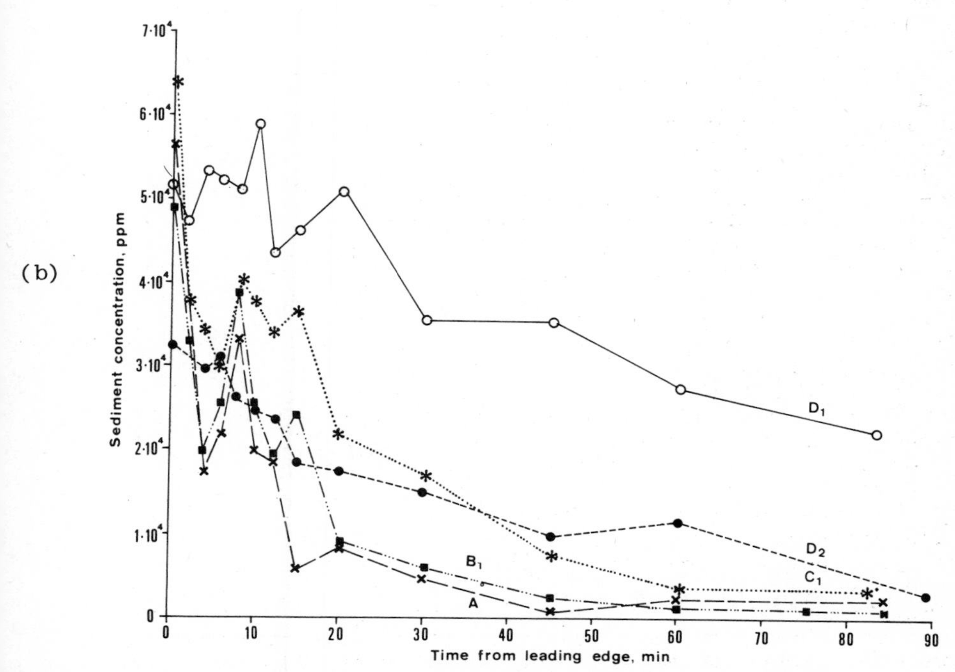

Figure 10.6 Spatial and temporal variation of sediment concentration during run 8 (a), and run 5 (b)

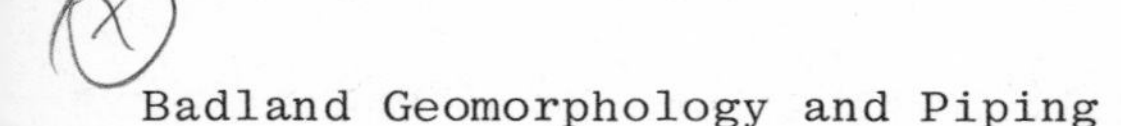

soil water and the runoff) decreases downslope. Approach to kinetic equilibrium with respect to the major dissolving minerals is thus the major cause for the downslope flattening of the solute concentration curves as well as for the shortening of the time interval necessary to arrive at solute concentration peaks with distance downslope. The causal relationship between sediment and solute yield has been briefly discussed by Shen *et al.*, (1979) and will soon be reported in detail (Laronne and Shen, 1982).

IMPLICATIONS

Sediment and solute yields are known to be closely related to runoff volumes and to runoff peaks. The results presented demonstrate that these yields are not only related to runoff volume per unit area (i.e., the power available for transport), but also to slope steepness and length on Mancos Shale terrain. This is especially noteworthy regarding solute yield from saline soils. Sheetwash is unconcentrated and therefore does not erode as deeply as rill flow. A salient characteristic of Mancos Shale hillslopes is the occurrence of a relatively leached surface layer. Solute yield derived from the partial removal of this layer by unconcentrated overland flow on gently inclined hillslopes is therefore low. Steeper hillslopes in identical stratigraphic locations within the Mancos Shale sedimentary sequence develop rills and are denuded at a much faster rate both physically and chemically.

With minor interflow and groundwater flow occurring in Mancos Shale, surface processes dominate the export of matter, particularly from the steeper hillslopes. Hadley and Lusby (1969) demonstrated that the rate of these processes, including mass wasting, is very high during high intensity storms. The operation of major sediment and solute pickup mechanisms during such infrequent high energy events on entire hillslopes is still undetermined.

A drawback of the runoff studies reported hitherto is the high rainfall and runoff intensities characterizing them. The lowest intensity reported by Ponce for a 1-hour duration, approximately equivalent to the conditions prevailing in the author's runs, is higher in magnitude than a 100-year event (Richardson, 1971, in: Ponce, 1975). Additional studies using shorter durations and lower intensities are needed to test the hypothesis that sediment and solute yield increase downslope due to rilling during less intense runoff-generating events. *Unlike the sprinkling experiment reported by Shen *et al.*, (1979), future studies should incorporate sediment and solute data from various portions within sprinkled hillslopes, and not solely from their outlet.

It has been demonstrated that in this study sediment and solute loads increase as the size of the experimental

---

* The experiments discussed in this paper are realistic for shorter durations: 5 and 10 min durations, during which most solute concentration maxima occur, have return periods slightly shorter than 1 and 0.5 yr^{-1}, respectively.

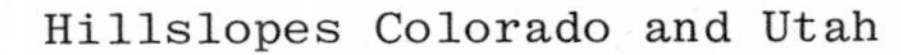

area increases from a small plot to a complete, though narrow, longitudinal hillslope strip, unlike the results presented by Yair *et al.*, (1980). This shows the importance of complementing small plot studies by examination of more extensive areas before attempting to develop complete hillslope models. The increase in solid and dissolved matter export with increase in reference area is limited; sediment delivery ratios smaller than unity and decreasing downstream attest to such a limit. Nevertheless, the high sediment delivery ratios and actual sediment yields of small watersheds (e.g., about 1000 t $km^{-2}yr^{-1}$ of exported sediment for small watersheds - 0.036-0.1 km^2 at Badger Wash, as reported by Lusby, 1979) indicate the importance of studying not only complete hillslopes but also related channel phenomena.

Acknowledgements

Sponsorship by the Office of Water Resources and Technology, U.S. Dept. of Interior, through grant no. 14-34-001-B-137 to H.W. Shen (principal investigator) and to K.K. Tanji is gratefully acknowledged. Alan, Beni, Ed, Jerry and Roger helped with fieldwork. Lazar Sharona drafted the figures.

11 Salt scalds and subsurface water: a special case of badland development in southwestern Australia

A. Conacher

This paper considers salt scalds as a special form of badland development in the non-irrigated, Western Australian wheatbelt (figure 1). Here, soluble salts are the main cause of devegetation, and the salt scalds are much less dissected than most badlands. Nevertheless, the effects of land use changes, and the relative importance of overland flow and throughflow in salt scald development, are important questions. In addition, the possible upward movements of water and salts from deep, saline aquifers is a fundamental problem.

DESCRIPTION AND DEFINITION OF SALT SCALDS

Salt scalds are post-European settlement phenomena and are distinct from the salt pans or playas which occur naturally throughout the wheatbelt. Several criteria can be used to describe and define salt scalds:

1. Salt scalds may have bare soil surfaces, be covered by dead trees, or be colonised by salt-tolerant plants.

2. Anaerobic conditions are common. Most salt scalds have perched water tables within a few centimetres of the surface, and surface seepages occur frequently on scald margins, especially in winter.

3. Surface salt concentrations are generally high, ranging between 3000 and 40 000 mg l^{-1}* in the York-Mawson area (figure 1), whereas at depths of 9 cm and greater, concentrations commonly range between 300 and 600 mg l^{-1}. These data contrast with valley-side soil salinities, which range between near-zero and 450 mg l^{-1} both at the surface and to depths of 2 m (Conacher and Murray, 1973). Detailed soil descriptions have been presented in Conacher (1975).

4. Salt scalds are highly susceptible to erosion by water and wind. On the more sloping scalds in the York-Mawson area, for example, 55% of the scalds had 20% or more of their surfaces affected by rilling (plate I) (Conacher and Murray, 1973). On the other hand the more extensive, flat, valley-floor scalds such as those in the Dalwallinu area (figure 1)

* Unless otherwise stated, salt concentration data refer to total soluble salts.

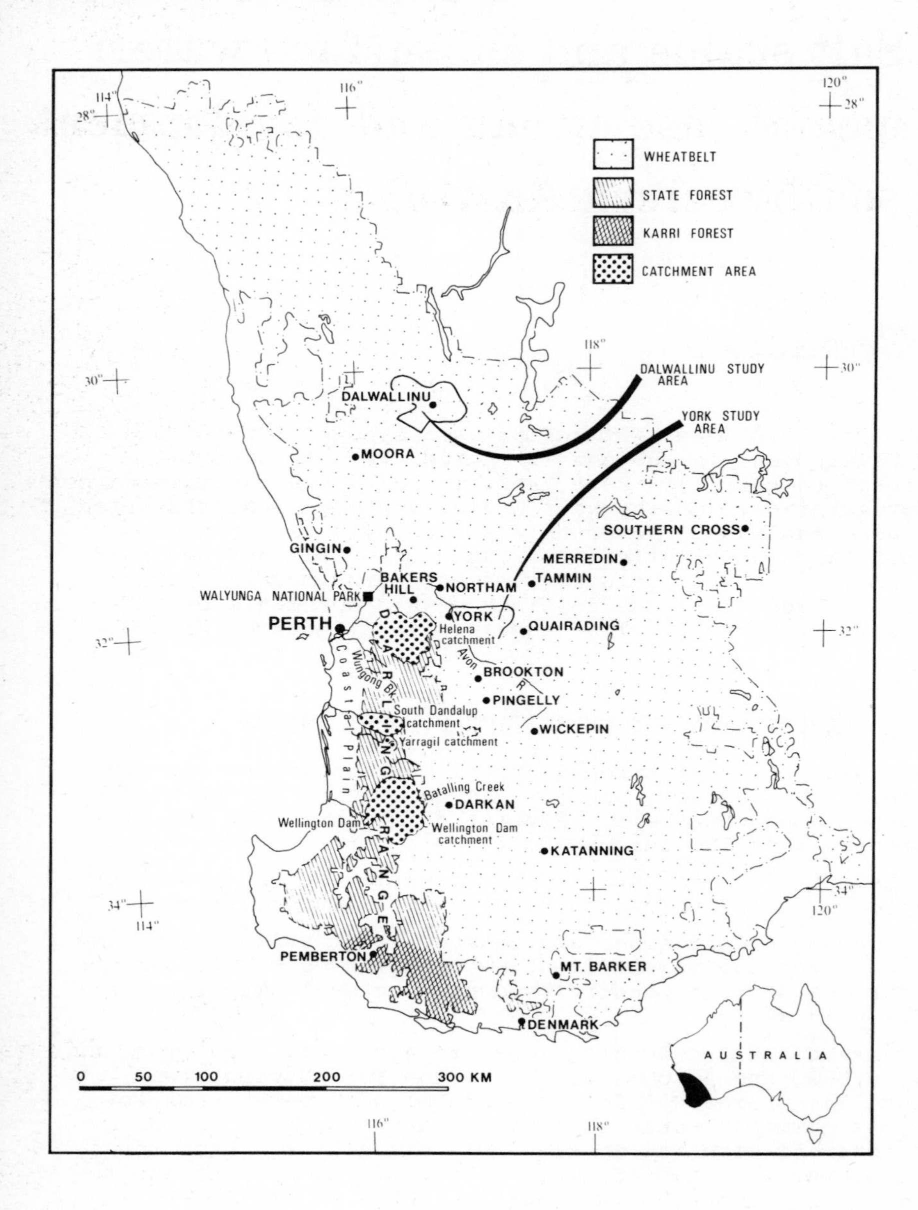

Figure 11.1 Location of places referred to in the text

Plate 11.I Rilling on a salt scald in the York-Mawson area (figure 1). Rill-heads are initiated and developed by throughflow. (All photographs by the author)

Plate 11.II Wind erosion on an extensive valley-flat salt scald near Merredin (figure 1)

are especially prone to wind erosion, particularly during summer, when salt crystallisation has loosened surface soil materials (plate II).

5. Most salt scalds are associated with ephemeral streams. In the York-Mawson area, especially where associated with higher-order stream channels, the scalds are long, narrow, winding areas adjacent to the stream, and they exhibit a number of upslope extensions. In contrast, in the much more extensive, near-flat drainage systems which characterise many parts of the wheatbelt, such as the Dalwallinu area, salt scalds may extend for tens of kilometres downvalley with widths of 2-3 km or more.

6. Salt scalds are dynamic. The extent of once agriculturally-productive land in the Western Australian wheatbelt rendered useless through salinity increased from 72 219 ha in 1955 (Burvill, 1956) to 256 314 ha in 1979 (Henschke, 1980); a net increase of 255%. Field observations give every indication that this increase is continuing.

THE SALT SCALD ENVIRONMENT

The Western Australian wheatbelt is a lateritised plateau underlain by Precambrian granites and gneisses with dolerite intrusions. The plateau has a mean altitude of 300-450 m, shallowly dissected by broad (up to 15 km), low-gradient valleys which often contain salt lakes. Maximum relative relief is only 90 m over several kilometres. Soils are deep-weathered (up to 50 m). The higher rainfall areas to the west (up to 1400 mm yr^{-1}) support a dry sclerophyll forest of jarrah (*E. marginata*), marri (*E. calophylla*) and wandoo (*E. wandoo*) on gravelly sands and lateritic duricrusts. In the drier eastern areas (300 mm + yr^{-1}), yellow, sandy soils on residuals support *Casuarina* and *Acacia* scrub alliances. Shallow, red duplex soils occur on dissections while valley fills have red or yellow duplex soils, each soil type exhibiting specific semi-arid woodland associations (Mulcahy, 1971; Pilgrim 1979; McArthur and Bettenay, 1979; soil terminology from Northcote, 1971). More than three times as much rain falls in winter (May-October) than in summer (November-April) (Southern, 1979). Wheatbelt summers are extremely hot and dry, with most centres recording more than 100 days with temperatures exceeding 30°C. Annual evaporation rates range from 1300 to 2200 mm yr^{-1} and considerably exceed mean annual rainfall.

Accumulation of salts in deep-weathered profiles

It is widely accepted that the major repositories of salts in the wheatbelt land surface are the deep-weathered soils and the groundwaters contained within them (Dimmock *et al.*, 1974) and that the major source of salts has been from rain- and dust-fall (Hingston and Galaitis, 1976) rather than weathering (Peck and Hurle, 1973). The composition of the salts generally reflects that of sea water, although there are marked variations.

The variability of the salt concentrations in the groundwaters, which are often under pressure (Smith, 1962; Bettenay *et al.*, 1964), is indicated by data from Negus (pers. comm., 1972). Groundwater salt concentrations from eight wheatbelt locations ranged between 6817 and 40 054 mg l^{-1}, with a mean of 20 375 mg l^{-1}. This variability can occur within very short distances. For example, in the Helena catchment (figure 1) to the west of the wheatbelt, Batini *et al.* (1977) found that salt concentrations of semi-confined groundwaters varied by 9 x and of perched groundwaters by 22 x within 500 metres. Within-bore variations of up to 10 x (semi-confined) and 8 x (perched) occurred; and in two bores the salt concentrations of the semi-confined aquifers exceeded those of the perched groundwaters by 12 x and 19 x.

Few data on the salinity of the deep weathered soils in the wheatbelt are available, but further west towards and in the forested areas, a considerable amount of drilling has been undertaken in recent years. Figure 2 shows the salinity profile of a reasonably representative site; but again there is considerable lateral variability, sometimes over very short distances. Batini *et al.* (1976) found that salt contents in soils in the Helena catchment varies by up to 40 x within 300 m. This was attributed to topographic situation, dominant tree species and soil type.

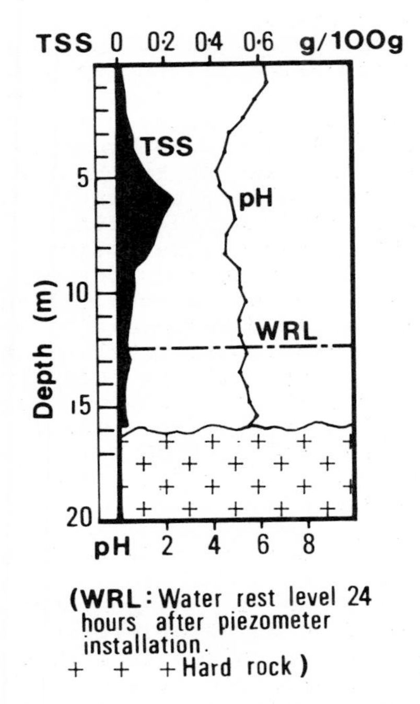

Figure 11.2 A fairly representative salt profile in a deep-weathered soil in the Helena catchment (figure 1). From Batini *et al.*, 1976, H16 on p.5

Badland Geomorphology and Piping

Herbert *et al.* (1978) noted a west to east gradient of increased salt accumulation in lateritic soil profiles which coincided with decreased rainfall and topographic changes in the Yarragil catchment (figure 1). They pointed out, however, that these variables are insufficient to explain the large variations of salt accumulations observed between *and within* sites throughout the catchment.

This distribution of salts in soils and groundwaters in uncleared areas presumably reflects an equilibrium between incoming rain- and salt-fall, outgoing water and salts in natural drainage, and outgoing water through evaporation and transpiration by the generally deep-rooted native vegetation. Peck and Hurle (1973) provide quantitative support, but point out that in undisturbed semi-arid catchments, inputs and outputs of salts may not be in equilibrium owing to climatic or botanic changes within the past 5000 - 10 000 years.

Effects of land use changes

A positive relationship between clearing natural vegetation, formation of salt scalds and the salinisation of watercourses is now clearly established (Wood, 1924; Teakle, 1938; Conacher and Murray, 1973; Peck and Hurle, 1973; Trotman, 1974; Batini and Selkirk, 1978). It is difficult to be precise about the lag time, as different parts of farms have been cleared at different times, and clearing on adjacent farms may be important. Data from Vaux and Morony (1977) indicate the range of lag times between clearing and the development of salt scalds (table I) and suggest that, in the majority of cases, at least 30 years are required after clearing before salt scald expansion ceases.

Salts must have been mobilised and removed from certain positions within the landsurface and redeposited or concentrated in others, but there is little quantitative evidence of such movement. Teakle and Burvill (1938) found that removal of native vegetation led to a reduction of salts by about 3 kg m^{-2} in the top 1 m of soil in the area contributing to a seep in the wheatbelt (Peck, 1978a). In the southwest karri forests, calculations from Valentine's data (1976) show that about 70 percent of soluble salts were leached from the near-surface horizons of the red earths (McArthur and Clifton, 1975) 2 to 7 years after clear felling.

Despite this paucity of data, it seems clear that hydrological changes are the connecting link between vegetation clearing and salinisation.

HYDROLOGICAL RESPONSES TO LAND USE CHANGES AND EFFECTS ON SOIL SALINISATION

In Western Australia certain characteristics of the native forest and woodland vegetation influence its transpiration capabilities relative to introduced exotics, especially crops and pastures. Deep root systems enable it to draw on groundwater at depth. Jarrah tap roots, for example, have been found to proliferate into a secondary fine root system between 0.6 and 0.9 m above a water table at a depth of 15 m (Kimber, 1974). In addition, exotic pastures and crops are

Table 11.I Relationship between clearing and the appearance of salt scalds on eleven wheatbelt properties. Source: Vaux and Morony, 1977, Table 6. For locations refer figure 1

Locality	Year salt scald noticed	When cleared	Lag (years)	Is scald increasing (1977)
Brookton	1957	1947	10	Yes
Brookton	1960	1927	33	Yes
Brookton	1962-67	1917	40-45	No
Brookton	1944	pre 1940	>4	?
Brookton	1948	1902-1940	8-46	?
Quairading	1932	1908	24	Yes
Quairading	1950	1937	13	No
Quairading	1915	1904	11	?
Gingin	1961	1956	5	Yes
Tammin	1945	1917	28	Yes
Wickepin	1957	1947	10	Yes

annuals which grow in winter-spring, while evergreen native forests continue to transpire through the summer and autumn (Peck, 1978a).

Peck and Hurle (1973) calculated the increased groundwater recharge following clearing of 30-70% of seven catchments in the Darling Ranges, compared with groundwater recharges in eight forested catchments. They found an increase from 2.4 and 6.5 cm^3 cm^{-2} yr^{-1} for six catchments, and 43 cm^3 $cm^{-2} yr^{-1}$ for the seventh. Expressing the present groundwater recharges from the cleared portions of these catchments as a percentage of the estimated pre-clearing recharge gives results ranging between 714 and 3529 percent (mean 2014%). In other words, following the replacement of natural vegetation with exotics, there is a considerable 'excess' of water available to mobilise the salts in the landsurface. The precise movements of this excess water and the practical implications arising from them have generated much recent research.

Basically, four mechanisms of water and salt movements have been invoked to explain the occurrence of salt scalds (Conacher, 1979a):

1. increased overland flow and surface runoff;
2. increased throughflow and development of perched water systems;
3. increased infiltration of soil-water to deep groundwater, raising water tables;
4. increased throughflow and infiltration to deeper groundwater, leading to a mixing of perched- and ground-water systems.

Figure 3 illustrates these mechanisms, which are now examined in further detail.

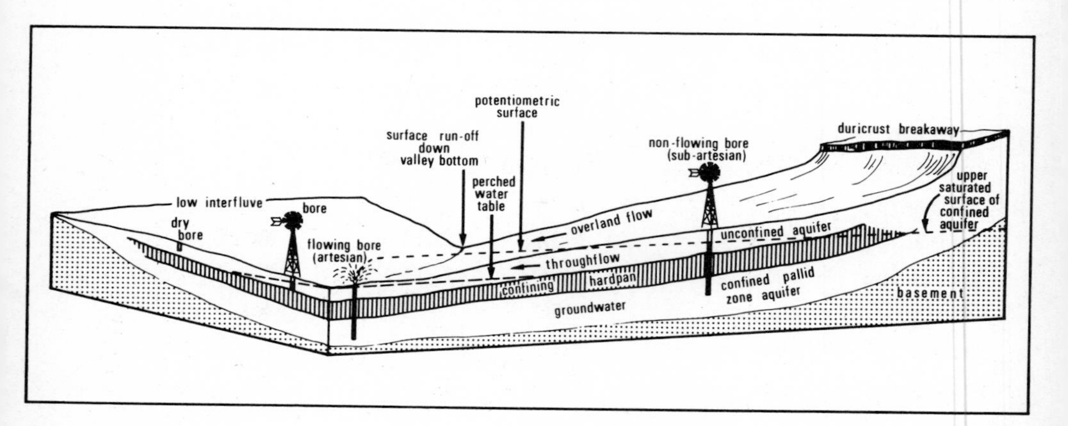

Figure 11.3 Diagrammatic representation of surface and subsurface water systems in the Western Australian wheatbelt. From Conacher, 1979b, Fig. 6

Increased overland flow and surface runoff

Bettenay *et al.* (1964) stated that clearing increased the supply of water to valley bottoms by overland flow and streamflow, leading to a higher soil-water content and a more spatially-extensive wet zone above the deep, confined aquifer. This in turn resulted in a greater upward movement of water and surface accumulation of salts in areas previously salt-free.

There is very little detailed, empirical evidence to support the overland flow part of this model. Indeed, recent work by Puvaneswaran (1981) in the Wungong catchment in the Darling Ranges (figure 1), has produced opposing data. Eight forested micro-catchments (catenas 1-4) yielded considerably more overland flow and sediment movement from seven rainfall events than the two cleared, pasture-covered, micro-catchments (catena 5: table II), despite the fact that the cleared micro-catchments have the steepest slopes. Further, area-weighted data show that stemflow is much more important than overland flow (table II), although the data also need to be weighted for the spatial distribution

Table 11.II Rainfall, stemflow, overland flow and sediment movement from seven rainfall events in June 1979 on experimental plots in the Wungong valley, Darling Ranges. Source: P. Puvaneswaran, 1981

Catena[1]	Rainfall (mm)		Stemflow (%)[2]		Experimental Plots			Overland flow (%)[4]		Sediment (gm 500 m^{-2})	
	Mean of 7 events	Range	Mean of 7 events	Range	Site[3]	Max. slope angle (o)	Area (m^2)	Mean of 7 events	Range	Mean of 7 events	Range
1	14.8	3.5-31.6	0.56	0.10-1.44	A	9	1250	0.008	0.005-0.010	0.89	0 - 2.69
					B	9	720	0.016	0.011-0.024	4.43	0 - 14.59
2	14.0	4.5-30.4	0.97	0.06 2.56	A	21	500	0.008	0 -0.016	0.16	0 - 1.15
					B	20	700	0.007	0.001-0.012	1.11	0 - 2.48
3	13.2	3.5-24.0	1.80	0.00-4.68	A	20	500	0.022	0.002-0.037	3.79	0 - 9.53
					B	20	750	0.017	0.015-0.018	0.85	0 - 4.24
4	15.5	10.8-22.0	1.02	0.02-2.58	A	35	356	0.038	0.010-0.082	10.35	0 - 17.04
					B	26	500	0.009	0.001-0.019	8.48	0 - 15.61
5	15.7	4.0-29.8	1.04	0.13-2.79	A	36	2375	0.003	0.001-0.005	0.05	0 - 0.36
					B	35	3400	0.003	0.002-0.004	0.08	0 - 0.33

(1) Catena 1 is in the headwaters of the valley with the others located sequentially downstream. Catenas 1-4 are forested; catena 5 has a ground cover of pasture grasses.
(2) Stemflow expressed as a percentage of rainfall, per tree crown area. One tree was measured on each catena.
(3) Plot A is in an upslope and plot B in a downslope position, on landsurface unit 5 (Conacher and Dalrymple, 1977), of each catena.
(4) Overland flow from each plot expressed as a percentage of rainfall.

of the trees. Obviously, this process does not occur in wholly cleared areas.

These forested areas have been disturbed by logging activities, feral pigs and rabbits and especially prescribed burning, and it is quite possible that overland flow and sediment movement have been accelerated as a result. However the effects of pre-European fires caused by lightning strikes and thousands of years of Aboriginal activities (Hallam, 1975), are not known. Nevertheless it can be concluded with some confidence that the replacement of natural forest with pasture grasses in the Wungong valley, has reduced overland flow and sediment movement on valley-side slopes.

In the wheatbelt, however, field observations show that the season, rainfall intensity and the nature of the introduced vegetation, are all important. By early autumn, many wheatbelt paddocks are quite bare, and sand-textured soil surfaces are often water-repellent (Roberts and Carbon, 1971; 1972). If the first autumn rains are intense then considerable overland flow and erosion occurs. On the other hand if the early rains are gentle, giving time for winter pastures and crops to become established and for the soil to become wetted, then, as is also shown by Puvaneswaran's data, overland flow is negligible. The varying effects of intense rainfalls on soil surfaces with different ground treatments are indicated by recent data from a severe summer storm in the major wheat-producing area of Queensland (table III). Western Australian wheatbelt farmers generally operate on a three-year rotation; that is, in any one year one-third of the property will be under pasture, one-third crop and one-third stubble. However, sheep-grazing (and hence pastures) predominate in the western portions of the 'wheatbelt' whereas wheat predominates in the lower-rainfall eastern areas. Overland flow can therefore be expected to be of relatively greater importance in the eastern wheatbelt.

Table 11.III Overland flow and soil loss from five plots in the Darling Downs, Queensland, following 92 mm of rain falling on dry soils on 5 February 1980. Source: Seymour, 1980, p.7.

Plot treatment	Overland flow (% of rainfall)	Soil loss (tonnes per ha)
Wheat stubble burnt	83	100
Stubble incorporated	41	<5
Stubble mulched	45	<2
Summer crop: sorghum	16	<2
Zero tillage	20	negligible

In general, it is concluded that increased overland flow is not a significant mechanism redistributing salts and leading to salt scald development, although it may be of relatively greater importance in the more heavily-cropped, eastern wheatbelt than in the west. The greatest portion of the 'excess' water available after clearing must therefore infiltrate into the soil.

Increased throughflow and the development of perched water systems

This mechanism was proposed and developed by Conacher (1970), Conacher and Murray (1973) and Conacher (1975). In essence, the model states that since most wheatbelt soils have sandy textures at or near the surface, with increasingly finer textures with depth, most of the water redistributed following vegetation clearance occurs as throughflow rather than overland flow or deep infiltration. Within relatively short distances downslope it is concentrated into near-linear zones. Since slope angles decline towards the base of the valley sides, throughflow, with its relatively low concentrations of dissolved salts, is checked, forming a near-stationary, perched soil-water system in the valley flats (figure 3). The spatial extent and temporal duration of these perched systems is increased after clearing. Seepages occur where relatively impermeable soil horizons are close to the surface. Capillary action and suction resulting from intense evaporation bring the soil-water to the surface and increase surface soil salinity concentrations to toxic levels, killing the vegetation. Many of the resultant salt scalds are at the heads of or adjacent to stream channels, and seepage is slowly drained downvalley. This water, characterised by high salt concentrations, contributes to the much larger valley-floor scalds downstream, whose perched-water and salt inputs are predominantly from upvalley rather than from upslope. Episodic floods may flush salts from the valley-bottom scalds but they also contribute to the perched water systems and to surface waterlogging, and to initiation and expansion of salt scalds.

This differs from the model proposed by Bettenay *et al.* (1964), as throughflow is considered to be the major cause of the 'higher soil-water content, and a more spatially-extensive wet zone (i.e. perched water system) above the deep, confined aquifer', and also the increased streamflow.

Empirical field evidence demonstrates the ubiquitous presence of seasonal throughflow zones on valley-side slopes in the wheatbelt and in the forested Darling Range, and of seasonal, perched soil-water systems beneath valley-floor salt scalds.

Valley-side throughflow was identified in the York-Mawson area primarily on the basis of morphological evidence (Conacher and Murray, 1973; Conacher 1975). This relates especially to the shape of rill-heads on salt scalds (plate I), and was confirmed by the emergence of seepages during appropriate weather conditions. It also included laboratory determinations of decreasing soil permeabilities with depth down valley-side soil profiles, closely related to texture differences.

Direct observations of throughflow and measurements by dye tracing, especially in the Dalwallinu area (Conacher, 1975) and near Moora (Sherwood, 1969), confirm the above interpretations. In one Dalwallinu valley, valley-side soils had salinity concentrations as low as 55 mg l^{-1}; throughflow water in percolines yielded values of around 1600 mg l^{-1}; streamflow water further downvalley had a salt content of 21 000 mg l^{-1}; and perched, subsurface waters in the flat valley floor were extremely saline at 96 000 mg l^{-1}. These data were interpreted as reflecting an accumulation of water and a concentration of salts downvalley, responding to time, distance, and high evaporation rates associated with capillary action and suction in summer (Conacher, 1975). Throughflow rates varied widely, ranging from approximately 10 cm in 30 minutes to 10 cm in 24 hours during or shortly after rainfall events.

Subsequent fieldwork in the Pingelly, Brookton, Darkan, Mt. Barker and Denmark areas and in the Darling Ranges (figure 1) has repeatedly shown ephemeral valley-side throughflow and seasonal, perched soil-water systems at the margins of scalded valley bottoms. Further, a farmer organisation known as WISALTS (Whittington Interceptor Salt Affected Land Treatment Society) has been responsible for the construction of interceptors on more than 300 farm properties throughout the wheatbelt since early 1978, on the valley sides either on or slightly off the contour. They are excavated by bulldozer, and cut through the surface soil materials into the higher clay-sized-content subsoils, which are pushed against the bank on the downslope side to prevent leakages. The purpose of the interceptors is to trap throughflow above the clay subsoil (and also overland flow), to prevent it from reaching the scalded valley floors, and so to dewater and hence rehabilitate the valley-floor salt scalds. Observations of numerous interceptors in many wheatbelt locations have repeatedly confirmed the ubiquitous presence of valley-side throughflow.

The locations of zones of concentration of throughflow in the Western Australian wheatbelt may be categorised as follows:

(a) topographic depressions down valley sides;
(b) concavities towards or at the base of valley-side slopes;
(c) stream heads, especially where coincident with a) and/or b);
(d) areas adjacent to stream channels, but especially marked where they coincide with types a) and/or b). (This fourfold typology corresponds fairly closely with that proposed by Kirkby and Chorley 1967);
(e) throughflow zones with no apparent topographic expression. Two are described in Conacher (1975), which were identified from auger holes during appropriate weather conditions. Others have been observed near Pingelly (Conacher, 1975);
(f) sand seams (a term in widespread use amongst WISALTS farmers), which are roughly linear, extending down valley sides, and also with little or no surface expression. They are distinguishable from the air during appropriate weather conditions (when the seams are wet

in relation to adjacent soils), and infra-red aerial photography is effective in highlighting their presence. At least two subtypes of sand seams occur:

Subtype A appears to be paleo drainage channels subsequently filled by coarse sands thought to be eolian (plate III). The channel shapes range from gully morphology to relatively wide (100 m +), shallow depressions. The minimal soil development suggests that infilling is recent.

Subtype B covers a range of other features. Tentatively, it is proposed that these are also eolian-infilled paleo channels, but of much greater age. Subsequent events have possibly included the infilling of relatively thin colluvial materials above and/or inter-mixed with the sands, and greater pedological alterations. Repeated field observations and reports from farmers indicate that throughflow discharges from sand seams are considerably greater than from the other five throughflow types.

Other researchers, working to the west and southwest of the wheatbelt, have more recently identified throughflow and/or perched water systems.

Williamson and Bettenay (1979) identified throughflow in sand-textured soil materials overlying higher clay-sized content subsoils in the 'subdued catchment' at 'Yalanbee', the CSIRO experimental farm near Bakers Hill. They also identified a perched water table in lateritic gravels in the cleared West catchment of their paired catchment study, and stated that the substantial increases of stream- and saltflow recorded after the West catchment was cleared in 1974 were 'probably due to an increase in perched groundwater

Plate 11.III Sand seam exposed by construction of a WISALTS interceptor near Mt. Barker (figure 1). The small, upslope-extending gullies in the sand seam are being caused by ephemeral throughflow.

movement into the stream through the banks of the deeply incised surface drainage lines'. Batini *et al.* (1977) identified perched groundwaters in the Helena catchment. In the South Dandalup catchment (figure 1), Shea and Hatch (1976) found that 'the surface and subsurface aquifers are often discrete from the groundwater'. Peck (1976) noted the presence of a perched water table on the less permeable kaolinitic layer in forested areas, and in 1979 Peck *et al.* suggested that such perched aquifers may develop more frequently and hold water for a longer period of time than before clearing.

Unfortunately most of these descriptions related to field observations of throughflow in soil profiles or of seepages in interceptors, or to monitored water table levels in observation bores. There are very few data on flow rates or discharges, and no direct data on changes following clearing.

The variability of throughflow velocities from dye tracing in the Dalwallinu and Moora areas has already been mentioned. A 12-month period of throughflow discharge measurements from a gauging station in an interceptor at Batalling Creek in the Wellington Dam catchment, showed that throughflow comprised 8.2% of rainfall for the year (Public Works Department, 1979). This figure is considerably greater than the per-rainfall-event distribution as overland flow reported by Puvaneswaran from the Wungong valley (table II); and throughflow at Batalling Creek was more significant than overland flow (2.1% of the year's rainfall) and deep infiltration (1.7%). The only other direct data on throughflow discharges are in Conacher (1975), where Whittington reported from the Brookton area that up to 56% of individual rainfall events was distributed as throughflow compared with up to 9% for overland flow.

Evidence of increased throughflow following clearing is indirect. From Williamson and Bettenay's (1979) data, the percentage increase of streamflow in the West catchment following clearing in 1974 ranged between 230% in 1976 and 789% in 1975 (the 1977 increase was infinite). Likewise saltflow in kg ha $^{-1}$ increased by between 195% and 814% - although streamflow salt concentrations were still very low, with a maximum flow-weighted reading of 63 mg l $^{-1}$. Their monitoring shows that deep groundwater could not have contributed to these increases, and that the amount of salts in the rainfall could not have accounted for the increased saltflow. The above data therefore probably indicate the percentage increases of throughflow and of salts leached from near-surface soil materials caused by clearing native vegetation and, in this instance, replacing it with pasture grasses. However these data come from only one small and, in terms of slope, atypical site.

Reported salt concentrations in throughflow water vary from up to 400 mg l $^{-1}$ at Batalling Creek and from the 'subdued catchment' at Yalanbee, to 1600 mg l $^{-1}$ (at the site near Pingelly and from the Dalwallinu area). With the high evaporation rates, the continued accretion of such water to near-stationary, perched water systems in valley bottoms eventually leads to much higher salt concentrations in surface soils. However where there are significant outputs of salts in streams, other sources of salts, such as deeper, saline groundwater, are important. If deep groundwater provides salts to streams, it clearly also influences the formation of salt scalds.

Increased infiltration of soil-water to deep groundwaters raising water tables

It has long been widely accepted that the salinisation of most soils and streams in the Western Australian wheatbelt has been caused by raising groundwater tables by the increased infiltration of water which used to be transpired by natural vegetation, and the release of saline groundwater into soils and streams through capillary action and seepage. Burvill (1947) described this model and wrote that 'the water level rose considerably in wells or is now known to be much closer to the surface than it was many years ago before clearing'. Smith (1962) stated that 'meagre historical evidence discloses that watertables...., in some cases, have risen from 24 m to within 1.8 m of the surface within 10-15 years of initial clearing' (data units have been converted from feet). This has been repeated in a number of places (Anon, 1962; Smith, 1966; Holmes, 1971; Peck, 1977; and Mulcahy, 1978).

The distinction between the level of water in a piezometer tube and the actual height of the water table is an important one. There is no doubt (Smith, 1962) that many wheatbelt valleys are characterised by the presence of deep confined or semi-confined aquifers, sometimes under considerable pressure (figure 3). The fact that potentiometric heads in valley-bottom locations react - within 24 hours - to rainfall events on the catchments shows that these aquifers are continuous and are open to rainfall inputs (Bettenay *et al.*, 1964). But, as pointed out by those authors, this hydraulic response reflects a pressure transmission rather than an actual movement of water. Likewise, a high potentiometric head indicates that a *potential* for upward flow of water in valley bottom exists; it does not necessarily follow that such upward movement actually occurs.

Substantial empirical evidence supporting the rising groundwater table model was not published until Williamson and Bettenay (1979) presented empirical data from the 'subdued catchment' at Yalanbee (figure 4). This showed that the increase in the potentiometric head must have corresponded with an actual rise of the groundwater table which was responsible for the increased stream salinity. To date, as far as is known this is the only unambiguous evidence of an actual rise of the groundwater table as distinct from a rise of the potentiometric head.

Monitoring being conducted by several agencies is likely to produce significant findings in the near future, but at present the results are mainly indicative. For example, from their paired catchment study, Williamson and Bettenay's (1979) data show a rise of potentiometric heads in the cleared catchment whereas potentiometric heads in the uncleared catchment generally remained unchanged (figure 5). Peck *et al.* (1979) also refer to this paired catchment, and present results from three experimental sub-catchments in

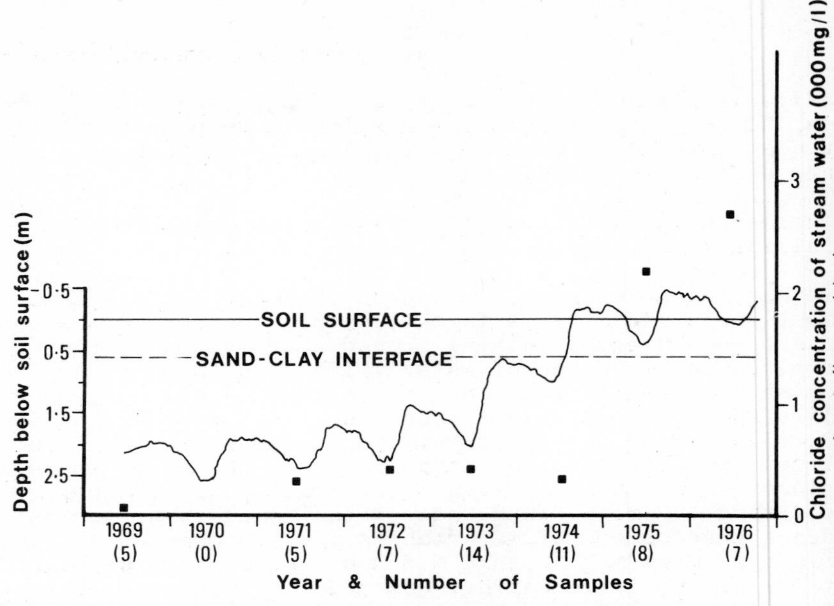

Figure 11.4 Increased potentiometric head of deep groundwater following the completion of clearing in 1963 of the 'subdued catchment' at Yalanbee, near Bakers Hill (figure 1). Streamflow salt concentrations increased after the potentiometric head crossed the interface between the perched water system (in sand-textured surface soil materials) and the clayey subsoils. From Williamson and Bettenay, 1979, Figs. 3 and 4

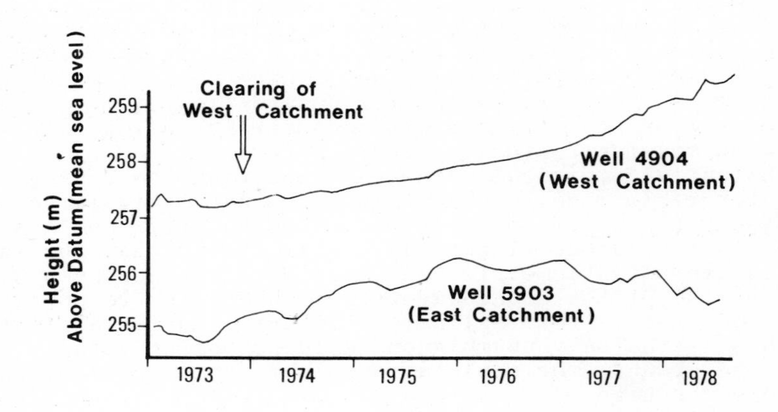

Figure 11.5 Comparison of changing potentiometric heads of deep groundwater between the uncleared (East) and cleared (West) catchments, from the paired catchment experiment at Yalanbee, near Bakers Hill (figure 1). From Williamson and Bettenay, 1979, Fig. 5

the Wellington Dam catchment. They conclude that, following clearing, potentiometric heads are rising in 80% of the bores which intersect permanent aquifers in the deeply weathered soils of the Darling Range, at rates of around 0.8 m per annum in a 750 mm per annum rainfall zone, and 2 m per annum in a 1100 mm per annum zone. Groundwater monitoring in bauxite-mined areas in the Darling Ranges have so far produced inconclusive results. 'There is only one site at which an appreciable rise in the groundwater level has been recorded' (Steering Committee, 1978a, p.28). Further south, monitoring of groundwater tables in clear-felled coupes in the woodchip license area centred on Pemberton (figure 1) had also produced inconclusive results up to 1978 (Steering Committee, 1978b, pp. 18-19): 'There are some indications that there is a slight increase in groundwater level following clearing ... . In Moralup coupe six bores which had previously been dry throughout the year were found to contain water during 1977.'

Contrary evidence, albeit of a different and perhaps less precise nature, has been reported in Conacher *et al.*, (1972), Conacher and Murray (1973) and Whittington (1975). In 1970 and 1971, none of the 50 farmers with soil salinity problems interviewed in the York-Mawson area could give evidence of a long-term rise of the groundwater table in farm bores and wells (Conacher and Murray, 1973). In 1972 20 more farmers were interviewed in the Dalwallinu area, where the extent of salt scalding had increased rapidly in the previous 15-20 years. Here water levels had risen in five bores and wells, fallen in three, had remained unchanged in 13, and in 30 cases the farmers did not know (Conacher *et al.* 1972). The authors concluded that there was no conclusive evidence of a widespread rise of the groundwater table, a conclusion reinforced by responses to questions concerning long-term changes in streamflow regimes and salt lake water levels. Near Brookton, Whittington's experience was that at the time (in the 1930s) that soil salinity and surface waterlogging was spreading across the valley flat at 'Springhill', a fresh-water well and several springs dried up and the streamflow regime became much more erratic (Whittington, 1975).

Irrespective of whether groundwater tables have actually risen after clearing, three further questions need to be asked.

a) What is the depth over which capillary action will be effective in bringing salts to the surface from a groundwater table?

b) What is the rate of upward flow of water from groundwater exhibiting a high (above ground surface) potentiometric head?

c) Do all salt scalds fulfil one or both of these criteria?

Capillary Action

There is uncertainty as to the effective height of capillary action in the wheatbelt. Peck (1978b) has discussed the 'critical depth' at which a saline water table presents a salinity hazard in dryland agricultural conditions, and concluded that this may be greater than the 1-2 m commonly assumed on the basis of established irrigation technology. Nevertheless, in the absence of data, it seems that a depth of ± 2 m may be used as a guide.

Rate of upward movement through a confining layer of groundwater under pressure

Holmes (1979) applied a modelling approach to this question for the Western Australian wheatbelt (figure 6), in which

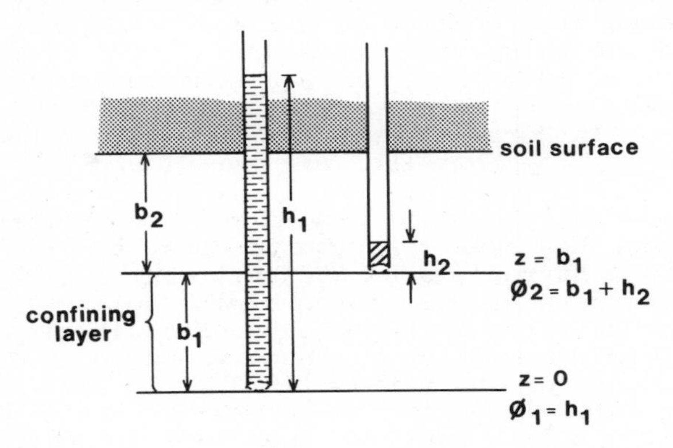

Figure 11.6 Schematic diagram of upward groundwater flow through a confining layer of thickness b_1 and hydraulic heads of $\emptyset_1$ at the lower boundary and $\emptyset_2$ at the upper boundary. Modified from Holmes, 1979, Fig. 2

the rate of upward flow through the confining layer of thickness b_1 is given by

$$v = K_1 \frac{\emptyset_1 - \emptyset_2}{b_1} \tag{1}$$

where $\emptyset$ at $z = 0$ is $\emptyset_1 = h_1$, and

$\emptyset$ at $z = b_1$ is $\emptyset_2 = b_1 + h_2$

and K_1 is hydraulic conductivity of the confining layer.

Holmes (1979) used 1 mm day $^{-1}$ for K_1 as 'suggested by recent measurements'. From Whittington's 'Springhill' scald near Brookton, the following additional values are obtained: $b_1 = 6$ m; $b_2 = \pm 1$ m; $h_1 = 7.61$ m; and $h_2 = 0.62$ m (Conacher, 1975). This value for b_1 - the thickness of the confining layer - falls within the upper part of the range found by Negus (pers. comm., 1972) in several wheatbelt locations (2.2-6.3 m, discussed further below) and by the Department of Agriculture from the Dalwallinu area (3.5 - 7.3 m - George, pers. comm., 1972). Note that in both instances a depth of 1 m has been assumed for b_2, the thickness of soil materials overlying the confining layer b_1.

Applying these values to equation (1),

$$v = 1 \times \frac{7.61 - 6.62}{6} \text{ mm day}^{-1} = 0.165 \text{ mm day}^{-1}$$

or 60 mm yr^{-1}, which is considerably less than the approximately 2000 mm yr^{-1} potential evaporation rate at Brookton.

Holmes (1979) also applied the above method to calculate the movement of salts upwards from the groundwater. Using the above data:

$$v = 60 \times 10^{-3} \text{ m yr}^{-1}$$

$$= 600 \times 10^{-3} \text{ m in 10 years, or}$$

$$0.6 \text{ m}^3 \text{ per m}^2 \text{ in 10 years.}$$

If all this water is evaporated and the salts which were transported by the upward flow are distributed in the top 100 mm of soil, and if the salt concentration of the groundwater is 10 000 mg l^{-1}, then 0.6 m^3 of water will have contained 6.0 kg of salt. If the bulk density of the soil is assumed to be 1500 kg of soil per m^3, the salt concentration in the postulated 100 mm of the top soil would become

$$\frac{0.6 \times 10}{1 \times 10^{-1} \times 1500} \text{ kg kg}^{-1}$$

after 10 years, or 0.04 kg salt per kg of dry soil. From the data presented earlier in this paper, using a range of 3000 - 40 000 mg l^{-1} (excluding surface salt crusts where present), actual salt concentrations in the top 100 mm of salt scalds range between 0.02 and 0.27 kg salt per kg of dry soil.

From these calculations, therefore, it appears that the upward movement of saline, semi-confined groundwaters through the confining layer can explain the presence of salts at or near the surface; but cannot explain the widespread presence of perched soil-water aquifers and surface waterlogging. The calculations may, however, be incorrect in two respects. First, actual evaporation rates are likely to be lower than the potential rates. Surface salt crystals have a high albedo and will reflect a significant proportion of incoming solar radiation in summer; and the actual evaporation rate of saline water, due to its greater density, is lower than the evaporation rate of fresh water. Nevertheless neither of these factors are likely to reconcile the massive difference between the potential evaporation rate and the calculated upward flow rates.

The second possible error is more important. The hydraulic conductivity figure of 1 mm day^{-1} may be incorrect, especially if the confining materials contain continuous cracks or fissures, or old root channels. An observation of such root systems has been recorded by Nulsen (1978).

Peck (1977) has noted that 'the range of solute velocities in non-uniform or highly structured soils presents particular problems which have received little attention'. Peck and Watson (1979) have subsequently modelled hydraulic conductivity and flows in non-uniform soils. They show that in a situation such as an old root channel penetrating a material of low hydraulic conductivity, where the length of the channel is considerable in relation to its diameter (by a factor of up to 10^3), then the flux within the channel is primarily controlled by the conductivity of the channel material rather than that of the matrix. They state that the ratio of the two conductivities may be as large as a factor of 10^4 or more. Further, 'a relatively small spatial density of channels containing high-conductivity material can make a substantial contribution to the total flow of water through soils of very low conductivity'.

These theoretical findings are important in the wheat-belt because many species of the natural vegetation have long taproots which enable the plants to draw on groundwater during summer. Since clearing of this vegetation has been relatively recent, it can be assumed that many roots or root channels are still intact; and in a number of locations water has been observed seeping at the surface from old tree stumps (plate IV). Nevertheless, further work is required to determine the conductivities of the materials in such channels, as well as the significance of what are essentially point-sources of seepage (plate IV) in the context of the overall development of perched water tables and surface waterlogging on salt scalds.

Plate 11.IV Seepage emerging from old tree stumps on a salt scald in the York-Mawson area (figure 1). The significance of such point-source seepages in overall scald development is not known. However the edge of this scald has extended at least 50 m from downslope of the stump-seepages to a zone well upslope

The extent to which salt scalds fit the criteria of groundwater tables within ± 2 m of the surface, or with hydraulic heads at or above the level of the perched water table

The 'subdued catchment' at Yalanbee (Williamson and Bettenay, 1979) fits both criteria, and Batalling Creek fits the latter criterion (Public Works Department, 1979). Another salt scald at Yalanbee has an hydraulic head at least 3 m above the ground surface, and therefore it also fits the latter criterion. However Negus' (pers. comm., 1972) data from holes drilled in 13 salt scalds in a number of wheatbelt locations yield varying results. Groundwater aquifers were intercepted at depths ranging between 3.2 and 7.3 m below the surface in 11 of the 13 holes and are therefore below the 'critical' capillary depth. In four of the 11 holes the groundwater was reported to be under pressure. Negus provides data for only three of these four holes, and the piezometric heads ranged from 0.61 m above ground surface (Whittington's 'Springhill' scald) to 1.1 and 1.5 m below the surface. Thus only one (*possibly* four) of the thirteen holes definitely meets the latter criterion. In the remaining two holes no deep groundwater was encountered.

Likewise, in a valley system near Dalwallinu, the Department of Agriculture sank 22 nests of open-well piezometers, with four holes drilled to different depths at each nest. The deepest hole at each nest was drilled to the main aquifer, which was encountered at depths ranging between 4.5 and 8.3 m below the surface. The main aquifer was under pressure but the piezometric head stood at about 2.5 m below the surface - possibly within the capillary fringe, but certainly incapable of providing an upward flow of water to the waterlogged surface (George, pers. comm., 1972).

From the data presented above, at most there are seven of 17 salt scalds for which data are available and where the piezometric head of the main aquifer is either within the capillary fringe or (in two cases) above the ground surface. In two of the seven scalds (the 'subdued catchment' and Batalling Creek), salt budget calculations show that the amounts of salts being exported by streamflow cannot be explained by throughflow alone (discussed further in the following section). On the other hand, eight of the 17 scalds have the piezometric head and/or the groundwater table too far below the surface for salts (and water) to reach the surface by upward flows or by capillarity; and in the final two instances no groundwater was present. One would, however, be very hesitant to ascribe these 7:8:2 ratios to all wheatbelt salt scalds, and more data from different, defined situations are required.

For those salt scalds not influenced by deep groundwater, the evidence already presented strongly suggests that increased throughflow plus, where relevant, increased streamflow which is in turn mainly supplied by throughflow, are the predominant sources of water and salts. In such situations the WISALTS interceptor approach would seem to be a logical means of tackling the problems. However where salt scalds are being influenced by groundwater as well as by throughflow (and streamflow) then some interesting questions arise.

Increased throughflow and increased infiltration to deeper groundwater, leading to a mixing of perched- and groundwater systems: practical implications

The two sites from which there is clear evidence of the operation of both inputs are the 'subdued catchment' at Yalanbee (Williamson and Bettenay, 1979) and Batalling Creek (Public Works Department, 1979).

At Yalanbee there is evidence of a rising water table after clearing (figure 4) and the mixing of saline groundwater (± 10 000 mg l^{-1} of chlorides*) with fresh (<250 mg l^{-1} chloride) perched waters, leading to increased salt concentrations in streamflow. The total chloride content of catchment soils to a depth of 0.75 m was calculated from 30 sites in 1968-69 and totalled 610 kg ha^{-1}. Between 1972 and 1976 inclusive, the chloride output from the catchment, excluding the input from rainfall, was 975 kg ha^{-1} (Williamson and Bettenay, 1979). It is thus clear that throughflow cannot account for the net outflow of salts from the catchment.

At Batalling Creek there are no measurements of changing hydraulic heads of groundwater following clearing, but there are measurements of throughflow discharges. From the monitored hillside, in 1978 the percentage of rainfall distributed as throughflow was 4.8 times more than that distributed as deep groundwater seepage, while the quantity of salts contributed to the stream by deep groundwater was 19.4 times greater than that contributed by throughflow (Public Works Department, 1979). As at Yalanbee, it is clear that groundwater is providing the bulk of the net salt outflow from the catchment. The Public Works Department (1979) measured a net outflow of chlorides from the Batalling catchment of 1 135 000 kg in 1978, and from their monitored site estimated that throughflow would have contributed between 41 000 and 78 000 kg depending on the scaling factors used.

The above evidence comes from streamflow although at both sites the valley floor is salt-scalded (since 1974 at the Yalanbee site). The 'Springhill' site near Brookton presents interesting findings - and questions - from a salt scald where conditions suggest that both throughflow and groundwater mechanisms would be operating. Data from this site, where the hydraulic head of the deep aquifer is 61cm above the ground surface, were used to calculate the upward flux of water movements from equation (1); although the groundwater salt concentrations here are less than usual, being only 6817 mg l^{-1}. It was also at this site that Whittington made the throughflow measurements referred to previously.

Whittington (1975) has constructed a comprehensive system of valley-side interceptors on the contour, extending from interfluve to valley bottom throughout the catchment, in order to intercept throughflow and to dry up the perched waters in the valley bottom (plate V) (described in Conacher (1974, 1975). Whittington claims that natural vegetation in a small water reserve on the valley flat is regenerating since interceptor construction started in the 1950s; that pasture has returned to previously-bare areas; and that crops

*Chlorides comprise roughly between 50% and 60% of total soluble salts.

Plate 11.V Contour interceptors holding water at 'Springhill' near Brookton (figure 1)

have been grown in the 1970s in paddocks uncropped since the 1940s: in sum, that 56.7 ha of 60.7 previously salt-scalded hectares have been rehabilitated.

Field observations since October 1970 have shown the spreading of barley grasses and some clovers, improved trafficability, and the drying up of the perched water systems in the worst-affected part of the property. Yet in September 1979, water from the deep aquifer was still trickling over the top of a windmill standpipe approximately 60cm above the ground surface, and located in the centre of this area. The dewatering of the perched water system can therefore only be a response to the interceptors, which are preventing the bulk of valley-side throughflow (and overland flow) from reaching the valley flat. If the perched water was solely or primarily a response to the high potentiometric head of the deep aquifer, then the valley-side interceptors would have had no effect. Indeed, since the interceptors are contour or absorption banks, not drains, and therefore divert some intercepted water to deeper levels (presumably to the deep aquifer), they would have made the situation worse by increasing the deep aquifer's hydraulic pressure.

Thus it appears that a deep aquifer under pressure does not necessarily lead to salt scalding or to mixing of perched- and ground-water systems. Alternatively, while a properly constructed interceptor programme is likely to result in dewatering of valley-bottom perched water systems, slow, upward movements of saline water from the deep aquifer may still occur. Even if this upward movement is considerably less than evaporation rates (as previously discussed), salts will still be brought up and deposited at the drying front. If, following dewatering of the perched water system, the drying front is located below the root zone of pastures and crops, then there may be no real problem. This may have occurred at 'Springhill'. Alternatively, the drying front may be located within the root zone or at the surface. If, through minimum tillage and mulching, soil structural and organic properties can be improved as attempted by Whittington, dewatering of the perched water system may have permitted leaching by winter rains to have removed salts below the root zone. The partial success of attempts by a number of farmers to improve salt scalds by adding mulch - in the form of hay, straw, stubble or coarse sand - suggests that there is merit in this argument. However most farmers have mulched without controlling waterlogging and their experiments have therefore failed after two or three seasons. The control of waterlogging, by dewatering the perched water system, seems to be the key.

Another related question concerns the inputs to the deep groundwater. Bettenay *et al.* (1964) identified these inputs as being located around the base of the fairly extensive granite domes which characterise the Belka valley area near Merredin (figure 1). As was pointed out by Conacher and Murray (1973), however, such a source cannot be generally applicable for the wheatbelt as a whole since the granite domes are not ubiquitous. More recently, Peck (pers. comm., 1978) has suggested that upland areas of broken lateritic duricrust may be the sources. Williamson and Bettenay (1979) and Peck *et al.* (1979) have identified throughflow or perched water aquifers in such locations. To these probable input areas should be added the various types of throughflow zones which were classified earlier in this paper. Water accumulates for longer periods in these zones than elsewhere in the landscape, and may therefore provide a source of deep infiltration to groundwater.

It follows that a catchment system of low-gradient, interceptor drains (as distinct from contour or absorption banks), with the uppermost interceptor located as close as possible to the top of the interfluve, would trap the greater part of overland flow and throughflow from all such input areas, transfer the water more directly to stream channels (or to farm dams), and dewater deep groundwater as well as perched water systems in valley bottoms.

The alternatives are to find high-transpiring, deep-rooted commercial crop or pasture species which will continue to grow through the hot and rainless summer and transpire groundwater from salt-scald catchments, or to revegetate substantial wheatbelt areas (and water catchments) with native vegetation. No suitable crop or pasture species are known at present, and the latter possibility would not find a great deal of favour with farmers. The method is, however, being implemented in the Wellington dam catchment by the Public Works Department, at considerable expense since farm resumptions are involved.

Further research is clearly required into situations where both deep groundwater and throughflow contribute water and salts to salt scalds and to streams. As pointed out by Williamson and Johnston (1979): 'It is unlikely that either the rising groundwater or throughflow hypotheses for salinity development... could be shown to be exclusive. In many catchments, a flow model which includes the two pathways in a dynamic form may be more realistic'. They go on to present evidence showing that seasonal changes in the ionic composition of catchment waters would provide a useful tool for analysing hydrologic responses.

CONCLUSION

Salt scalds may be distinguished from other badlands by their relatively undissected topography and the presence of high concentrations of soluble salts at the soil surface. It is this latter characteristic together with waterlogging and the development of anaerobic conditions, and not erosion, that has been responsible for destroying vegetation. Nevertheless their essentially bare surfaces render salt scalds highly susceptible to erosion by water and wind. The extent and continuing expansion of these areas in southwestern Australia, and the closely associated problem of water salinity, are causing serious concern.

The fundamental causes of salt scalding are related to changes to the hydrological regime following clearing of the natural forests and woodlands for agriculture, and not to climatic changes. In this respect salt scalds may differ from other badlands. However the question of the relative importance of overland flow and throughflow as causative mechanisms seems to be common to many badlands. The incomplete evidence from Western Australia strongly suggests that an increase of ephemeral throughflow after clearing, and the development of seasonal, perched aquifers, have been much more important in this respect than increased overland flow.

However a major factor which complicates the understanding of the mechanisms responsible for salt scalding, is the widespread presence of saline groundwaters in the kaolinised, 'pallid zone' of the deep-weathered soils. Indeed for some 50 years it has been widely believed that the raising of saline water tables, due to increased infiltration of water which was previously transpired by the natural vegetation, is the main cause of soil salinisation. The evidence now available indicates that groundwater tables - or the hydraulic heads of the deep aquifers where they are under pressure - are not always close enough to the surface. Where they are, the relatively much fresher, ephemeral throughflow (and streamflow in appropriate locations) seems to be the dominant input of water, but slow, upward flows of groundwater provide the dominant input of salts. Even where groundwaters are contributing salts in this way, the evidence further suggests that dewatering perched water systems in valley-bottom locations, allied with good soil management, may substantially rehabilitate salt scalds. Comprehensive systems of throughflow interception and diversion on a catchment basis may also reduce infiltration to deeper groundwaters, lowering pressure heads and hence reducing the contribution of salts to the surface from these saline aquifers.

There are a number of areas where considerable further work is required. These relate to improved quantification of the relative roles of overland flow, throughflow, streamflow and deep groundwater as mechanisms responsible for soil salinisation in different locations; the establishment of high-transpiring, deep-rooted crops on catchments; and the effectiveness of catchment interceptors and salt scald soil management on the dewatering of perched water systems and the rehabilitation of salt scalds.

12 Surface morphology and rates of change during a ten-year period in the Alberta badlands

I.A. Campbell

INTRODUCTION

Much discussion in geomorphology has centred on the manner in which landforms change through time and the rate at which these changes occur. This is particularly true in the case of slope studies which has long been a focus of considerable geomorphic investigation, speculation and controversy. This situation is compounded by the relative scarcity of field studies and measurements of the type necessary to resolve many of the points of argument.

Comparatively few studies in geomorphology have been concerned with long-term direct measurement, rather than inferential observations, of rates and the operation and effects of geomorphic processes on surface morphology and slope forms. This is in contrast to the many investigations on closely related topics such as fluvial geomorphology or hydrology, where often several decades of data and records are available for particular rivers. Some notable exceptions to the general scarcity of long-term direct field measurements on slopes may be found in the work of Rapp (1961) in an eight-year study of the rates of removal of slope material from a valley in northern Sweden; Schumm and Lusby (1964) monitored a seven-year series of slope changes in Colorado; Brunsden (1974) reported a five-year period of measurement on clay slopes in Dorset, and Luckman (1978) has observed creep and solifluction movement on alpine talus slopes over a ten-year period. Such studies provide the kind of measurements that are necessary for the accurate assessment of rates and variability of weathering and erosion processes and changes in slope morphology.

The reasons for the comparative lack of detailed long-term investigations do not stem from lack of concern but are understandable responses to the inherent difficulties associated with the design, establishment and maintenance of a reliable observation system that is required for the collection of an accurate sample of spatial and temporal data. Additionally, there is the awareness that in many, if not most, natural landscapes geomorphic processes operate at a relatively slow rate. Commonly, the landforms exhibit no perceptible change over protracted periods of time. This requires that, where field measurements are involved, the

degree of precision be very high, frequently in excess of normally available surveying or measurement techniques. It also means that the minute real changes may be disproportionately affected by errors in measurement the magnitude of which may be unknown.

Yet, as in other observational sciences, long-term studies are required in geomorphology for the information they provide about process rates and the problems involved in their extrapolation over geologic time. In addition, and of more importance, these long-term data are required for the verification of various geomorphic theories such as dynamic equilibrium (Hack, 1960); frequency and magnitude of geomorphic processes (Wolman and Miller, 1960); complex responses and thresholds (Schumm, 1973) and allometric change in landforms (Bull, 1975). Without the ability to test these and other theories against direct field measured data they remain essentially speculative however well-founded in deduction and logic and regardless of their support by inferential observations about the behaviour of landforms under the effects of various processes.

In recognition of the relative slowness of operation of most geomorphic processes, several investigators have examined landforms which undergo rapid evolution in terms of geologic time. These include steep, barren slopes (Schumm, 1967); badlands terrain (Schumm, 1962; Smith, 1958) and geotechnically created landscapes such as spoil heaps (Schumm, 1956; Bridges and Harding, 1971). The advantages of such areas for the detailed study of rates of change and processes are readily apparent. The lack of a dense vegetation cover, or even its complete absence, facilitates direct surface measurement. In the case of spoil heaps the unconsolidated nature of the deposits means that the effects of such processes as rain splash and surface flow are usually accelerated to rates well above the geologic norm. As a result, slopes evolve quickly and it becomes possible to collect a considerable amount of accurate data within a comparatively short span of time. But this attention to rapidly evolving landscapes, especially man-altered terrain, raises questions about the geomorphic applicability of the results to the general situation. It has been pointed out (Bloom, 1978) that under accelerated erosion conditions not all processes (such as soil formation) are accelerated proportionately and the results cannot be interpreted as representing true landform development over geologic time. By extension, such landforms cannot be used as surrogates for natural landscapes. The same argument pertains, though with greater force, to the use of scale models for the study of processes (King, 1972).

Naturally created badlands however have the advantages of widespread geographic occurrence under a wide range of environmental conditions and, while arguably representing extreme examples of normal landscapes, are the product of a complete set of interrelated natural processes. The materials of which they are composed are the product of geologically normal lithification and sedimentation processes and their structure, density and degree of consolidation are representative of the rock classes to which they belong. Badlands then, for various reasons, offer near ideal opportun-

ities for the study of the processes and rates of weathering and erosion and the associated patterns of change in surface morphology (Campbell, 1974).

THE PRESENT STUDY

This paper examines ten years of field measured data from an area of badlands in Dinosaur Provincial Park, Alberta (Fig. 1, Plate 1), and describes the changes in surface morphology that occurred within the period 1969-1979. The data were collected from nine one-metre square surfaces (Plate 2) on a representative sample of slope and lithological units (Campbell, 1974) using a portable frame device (Campbell, 1970a). Twenty-five measurements from within each 1 m^2 area were taken twice each year for ten years to give a total of 4500 measurements. Each measurement is exact to ± 0.5 mm; surface changes are therefore precisely monitored, and it is possible to assess the rate and form of surface changes by both degradation and aggradation in an open system.

The Study Area

The basin used for this study (Fig. 2, Plate 1) has a total relief of about 100 m and mean slopes of approximately thirty degrees. The lithological units consist mainly of a mixture of interbedded shales and sandstones of the Upper Cretaceous Oldman Formation. Their detailed characteristics are discussed by Campbell (1970b, 1974) and by Hodges and Bryan (this volume). The regional climate has a pronounced winter freeze-thaw period and a summer convectional-rainstorm pattern (Longley, 1972). This requires measurements to be taken at the close of each of these two distinct phases of varying geomorphic activities. Hence, the measurement times are usually in May (end of the winter period) and October (end

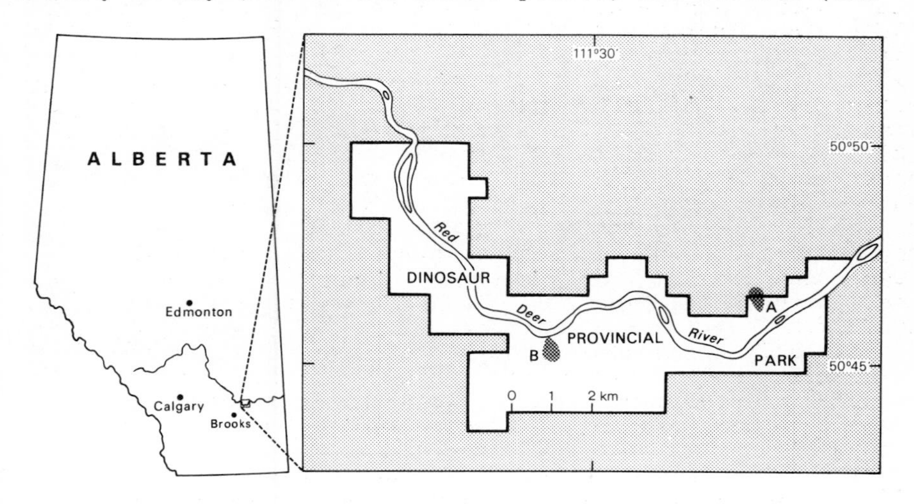

Figure 12.1 Location map showing position of basin 'A' (this study) and basin 'B' (study by Hodges and Bryan in the present volume)

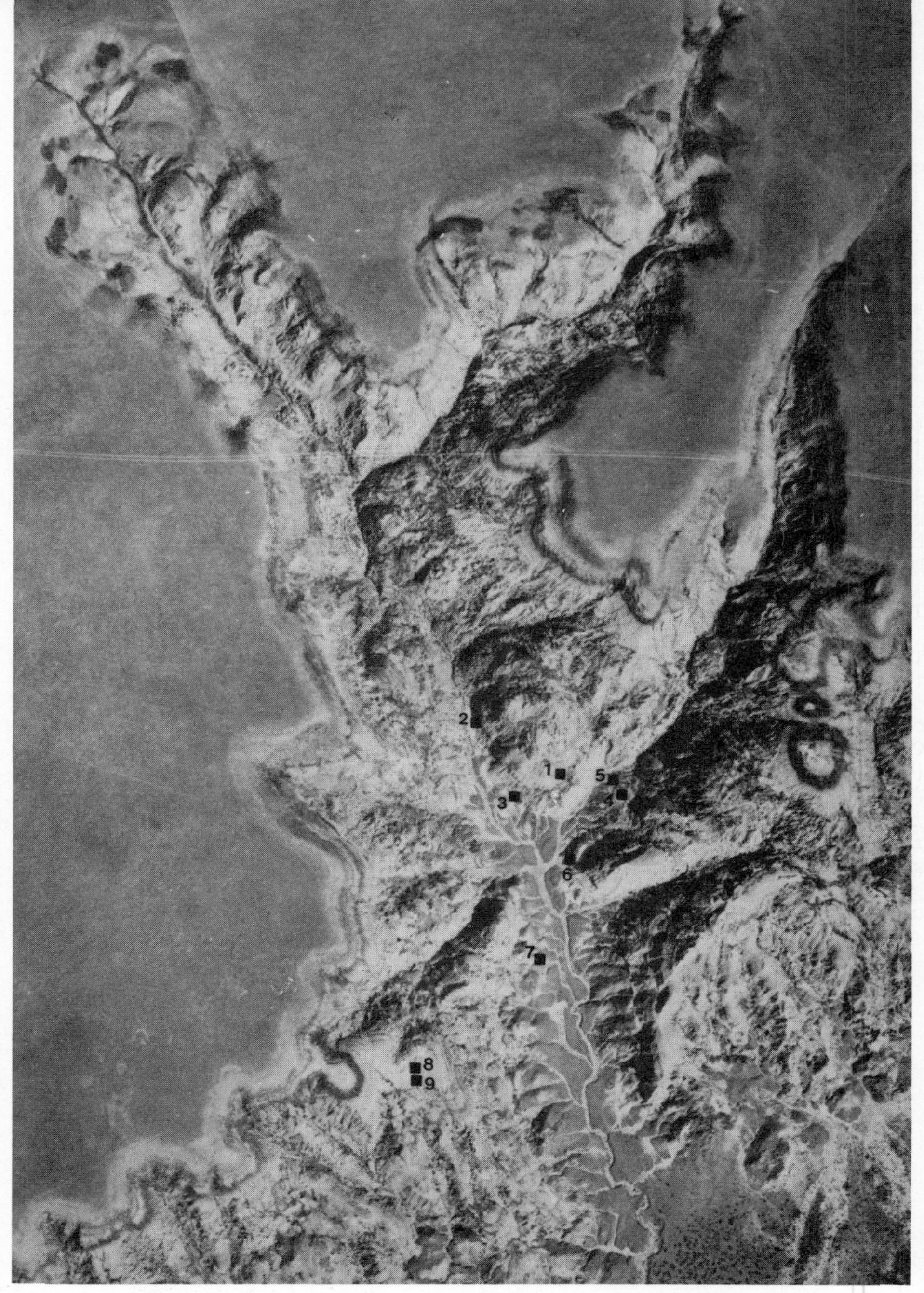

Plate 12.1 Aerial photograph of study basin showing plot locations

---

Plate 12.2 The nine plots as they were in August, 1979. Plot 7 shows the method of frame-mounting for field measurements. Plots numbered as shown

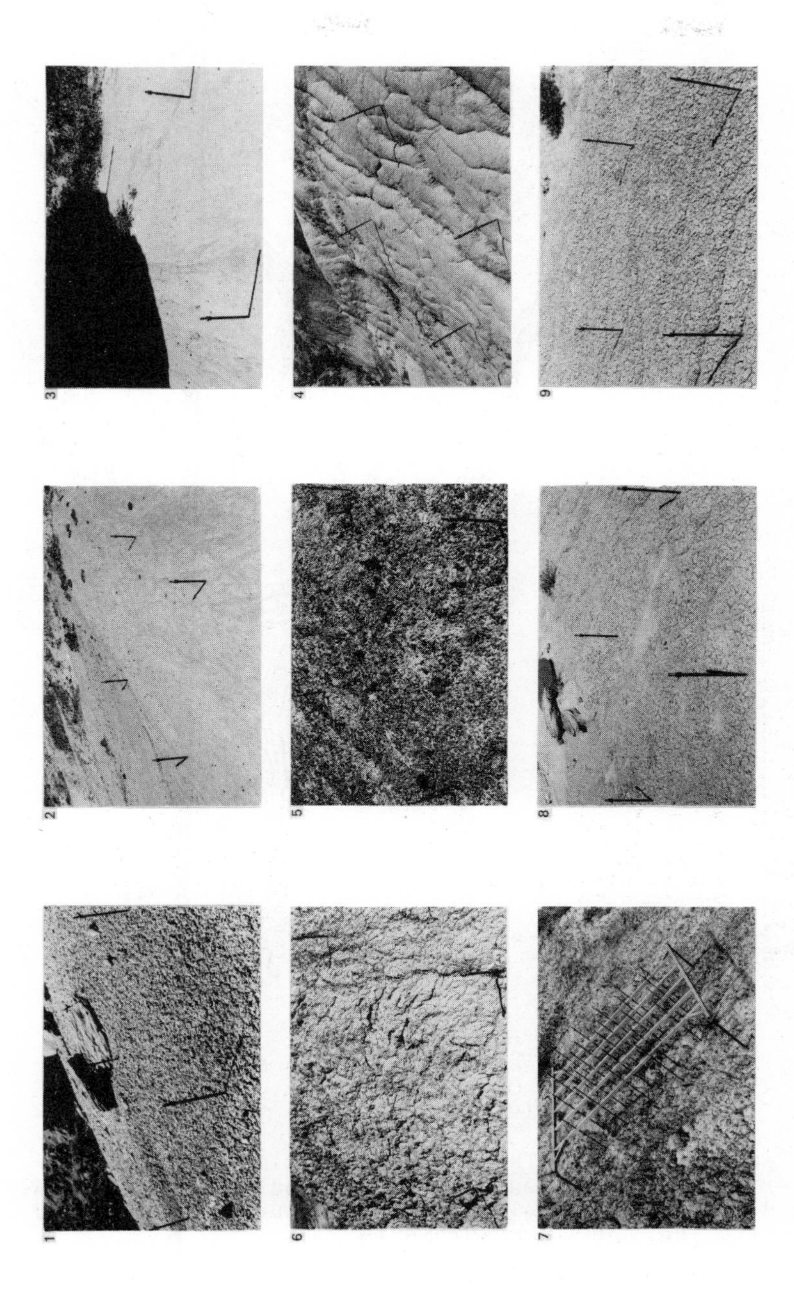

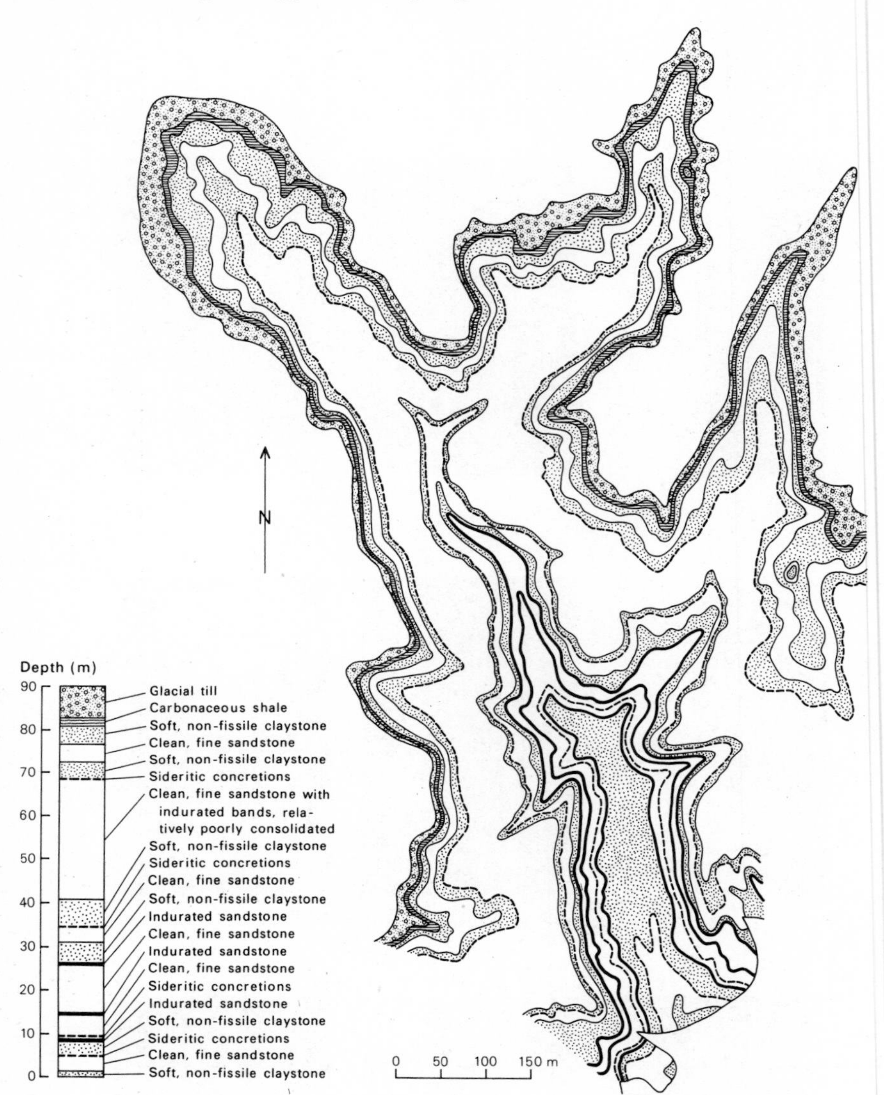

Figure 12.2 Geology of study area basin showing outcrop pattern of Oldman Formation. Source: Faulkner (1970)

of the summer rains). The measurement times are listed, together with the mean surface elevational changes that were recorded since 1969 (Table 1). Each value is the mean of twenty-five individual measurements.

Time	Month (cumulative)	Year and month	Plot number 1	2	3	4	5	6	7	8	9
1	7	7 69	0.0	0.0	0.0	0.0	0.0	0.0	0.0	0.0	0.0
2	11	11 69	0.69458	-0.25376	0.05364	-0.24536	-0.17807	0.66177	-0.09264	0.92255	0.14490
3	16	4 70	-0.14781	-0.23543	-0.13725	-0.18961	-0.19636	0.83337	-0.10466	1.05964	0.22636
4	22	10 70	-0.67253	-1.94126	-1.59521	-2.40905	-2.24544	-1.24289	-2.05347	-0.65284	-2.35913
5	29	5 71	-1.75967	-1.91133	-1.69466	-2.49687	-2.31271	-0.88596	-1.59943	-0.94058	-3.62944
6	34	10 71	-1.17323	-0.83397	-0.19095	-1.53285	-1.41636	-0.18166	-1.11740	0.02183	-2.79233
7	41	5 72	-1.14183	-0.94769	-0.39781	-1.42308	-1.40544	0.45572	-1.09233	0.45243	-2.99136
8	46	10 72	-0.99402	-0.90979	-0.45549	-1.73843	-1.93453	0.28044	-1.35178	0.11906	-2.64403
9	53	5 73	-0.68731	-0.44891	-0.19890	-1.74043	-1.84181	0.46369	-1.22354	0.24407	-2.62842
10	58	10 73	-2.83240	1.02948	0.43361	-2.40906	-2.27453	-0.44457	-1.49771	-1.89901	-4.70462
11	65	5 74	-2.41483	1.42850	0.59273	-2.47492	-2.15271	0.18483	-1.33704	-1.26601	-4.46265
12	70	10 74	-2.77697	1.40057	0.70412	-2.52881	-2.20907	0.18006	-1.31493	-2.22444	-4.26167
13	77	5 75	-1.75893	1.36466	0.60666	-2.47492	-2.15998	0.21830	-1.23975	-1.48826	-4.24216
14	82	10 75	-1.52244	2.33628	1.09596	-3.18346	-2.63453	-0.11792	-1.86478	-2.36533	-5.71540
15	89	5 76	-2.11552	3.74682	1.35653	-3.14355	-2.51998	0.29957	-1.44908	-2.62409	-6.07054
16	94	10 76	-2.17095	2.92283	0.50721	-3.77225	-3.11452	-0.54655	-2.07265	-2.71656	-6.15444
17	101	5 77	-2.35029	2.69501	0.50516	-3 96372	-3.29441	-0.38353	-2.10631	-2.88341	-6.11882
18	106	10 77	-1.77938	2.52543	0.43753	-4.53854	-3.74168	-0.96354	-2.76523	-2.32184	-7.46328
19	114	6 78	-1.01631	2.27604	0.80550	-4.67027	-3.79441	-0.95717	-2.34805	-1.26617	-7.02813
20	118	10 78	-2.08423	2.33790	1.11977	-5.01955	-4.04713	-0.85838	-2.57950	-1.82774	-7.60767
21	129	9 79	-2.64776	1.81518	1.03822	-5.42472	-4.21259	-1.19937	-2.74901	-2.08372	-7.80866
RESULTS OF REGRESSION ANALYSIS											
MEAN			-1.492944	0.876027	0.218399	-2.637116	-2.270774	-0.200180	-1.521871	-1.130496	-4.205309
STANDARD DEV.			0.965150	1.682176	0.806453	1.548504	1.204420	0.633884	0.803114	1.263889	2.500659
RATE PER MONTH			-0.017583	0.036836	0.015517	-0.038882	-0.029562	-0.008354	-0.017874	-0.026674	-0.063884

Table 1 Mean elevation changes (cm) and results of regression analysis (rates in cm/month)

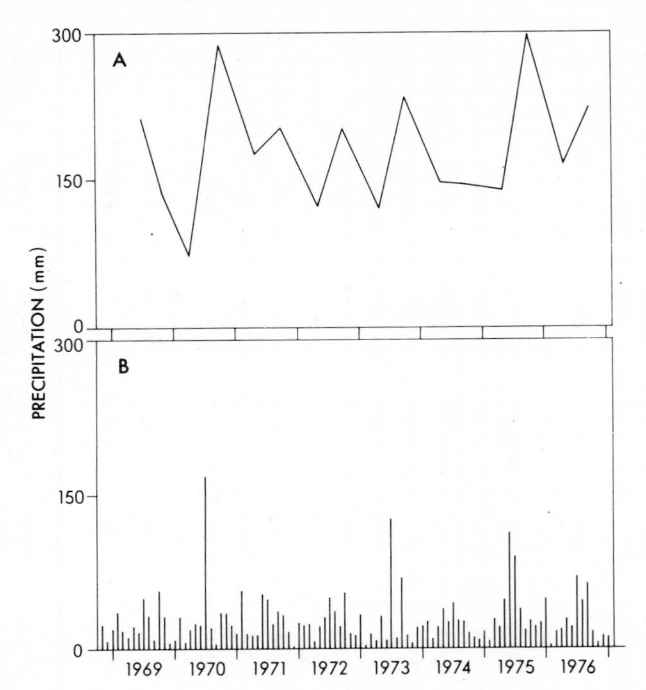

Figure 12.3 Precipitation patterns, mean of six badlands area stations, 1969-1976. 'A' shows cumulated amounts between each plot measurement times; 'B' shows monthly amounts

Rainfall and Erosion

The precipitation regime of this portion of Alberta is one with a marked summer maximum especially in the June-August period when it falls from locally generated convectional thunderstorms which are of great spatial and temporal variability (Longley, 1972).

It is these summer rains that probably represent the most important erosional mechanism in the badlands; in some years spring snow melt may contribute to erosion but this is extremely variable from year to year and frequently there is little or no spring runoff from the badland surfaces.

Under such circumstances some relationship is expected between rainfall amounts and erosion rates. Given the fact that the erosion data is cumulative over each season (between the two annual measurement periods), it can be compared with seasonally accumulated precipitation.

Monthly climatological studies are published for six stations within a distance of 35 km of the study area; four at Brooks and one each at Duchess and Empress (Atmospheric Environment Service, various dates). The precipitation data correlates well between these stations so a mean value (Fig. 3) probably gives a good representation of precipitation on the badlands. The data are, unfortunately, only currently published for the period 1969-1976, nevertheless they serve to illustrate the general purpose of this analysis.

The most concentrated rain in the period 1969-76 occurred in June 1970, with a mean (six station) total of 166 mm. The summer period of 1970 accumulated 272 mm. During the summer of 1970 all the plots, except number 1, underwent a strong erosional phase (Fig. 4). Plots 6 and 7, for example, also showed an increased surface roughness at that time, possibly reflecting rilling conditions (Fig. 5).

Two more episodes of heavy rain occurred in the summers of 1973 and 1975 (Fig. 3). In these periods, plot 1, for example, responded strongly in 1973 but weakly in 1975. The 1975 summer rains, though high in total (mean total 282 mm) were widely spread over the summer period. This probably inhibited a strong erosional response.

It is evident that although a general relationship can be observed between the erosional behaviour of certain plots and the precipitation pattern at specific periods there is no overall response characteristic that can be determined. In other words, it is a highly non-linear relationship. This tendency is also evident in the observation of runoff-generating rainstorms (Bryan and Campbell, 1980), which shows that different badland surfaces behave with a considerable degree of variability.

Elevation Changes

The pattern and rates of elevation changes (degradation and aggradation) are shown on Figure 4 and Table 1. Each plot is shown individually with each individual measurement time being recorded. A zero base line from the starting time and elevation is shown as a horizontal dotted line across each individual plot record. The twenty-five recording point positions for each plot's measured set of data have been converted to a vertically oriented coordinate system with the origin in the top row of points. The mean of these twenty-five relative positions is used for the elevation variable. Since the starting elevation is arbitrary for each plot, the mean elevation for the initial measurement (month 7) is subtracted from all values (Table 1). all of the time series begin with a zero value at that time and continue with mean elevations. Lowering of the surface through loss of material is indicated by a decrease in the elevation.

Plots 4, 5 and 7 show consistent surface lowering, though the rates vary both within each plot and between these three surfaces. Plots 2 and 3, which are located on small alluviated surfaces (Plate 2), both recorded consistent degradation until month 53 (May, 1973) since which time they have undergone a fluctuating sequence of aggradation that reached peak values in both cases in month 89 (May, 1976). The two surfaces clearly behave in almost identical fashions, though with variations in amplitude, even though they are widely separated (Plate 1). Apart from these two (plots 2 and 3) the general pattern has been one of surface degradation in which some similarity of behaviour can be observed. The time period between months 16 and 24 is represented on all plots, 2 and 3 included, as a phase of rapid degradation, which was followed by an equally dramatic phase of recovery.

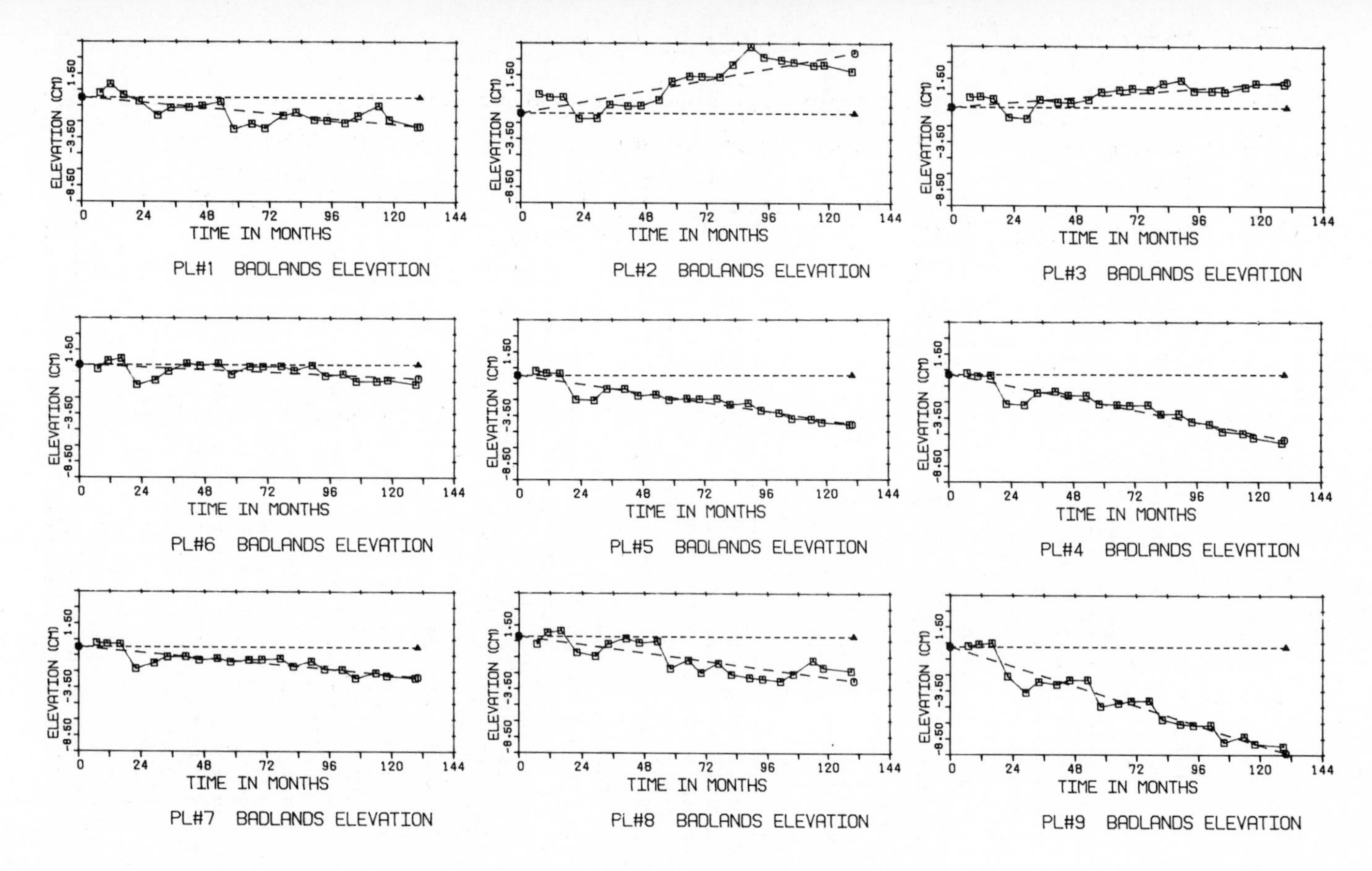

Figure 12.4 Mean elevation changes on badlands plots, 1969-1979 (cm)

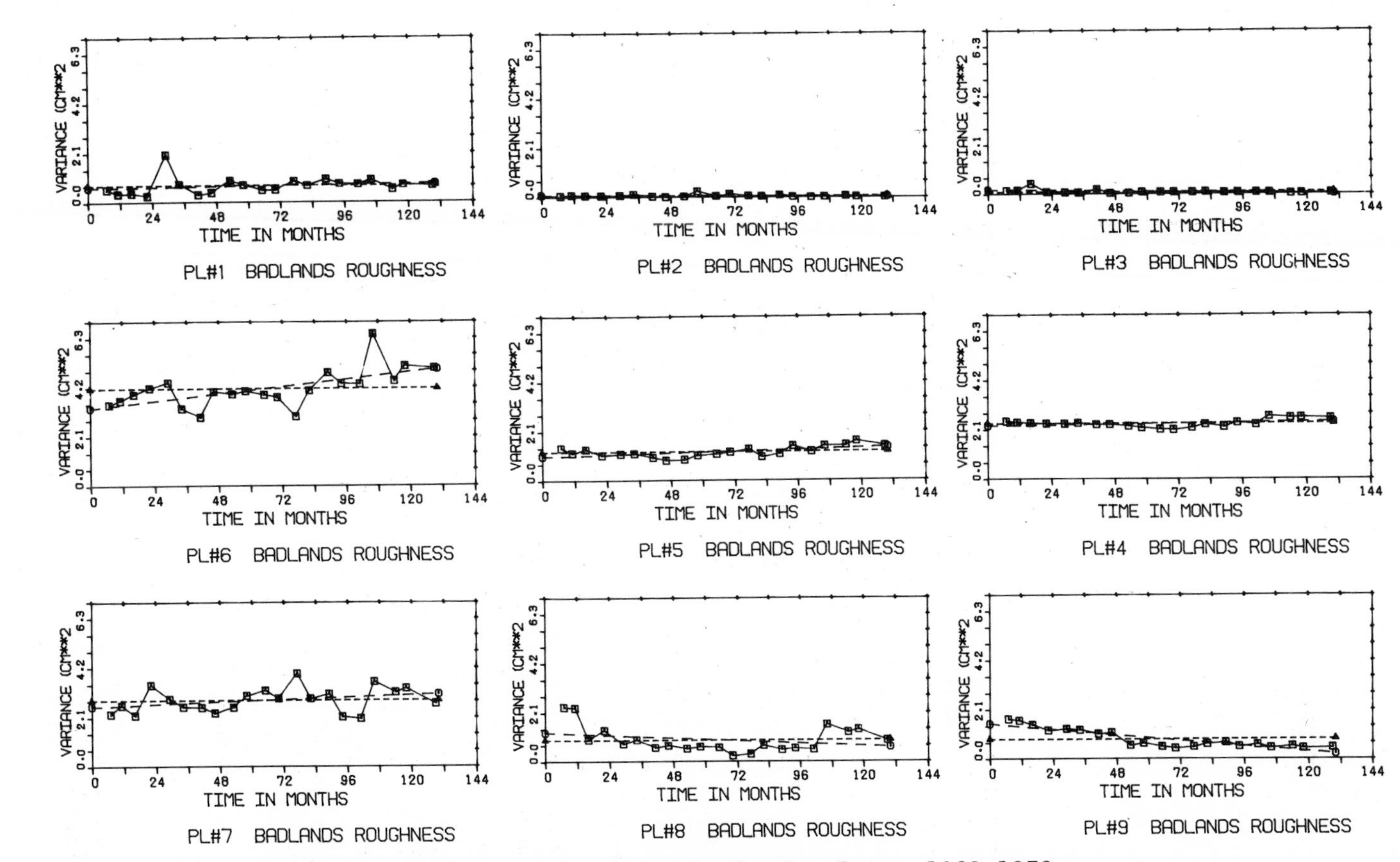

Figure 12.5 Trends in surface roughness on badlands plots, 1969-1979

Some threshold value of erosion must have been exceeded on a widespread basis throughout the system (Bryan and Campbell, 1980). Thereafter, an alternating pattern of aggradation and degradation, which is damped by low amplitude in most plots, but shown clearly in plots 1, 8 and 9, especially the latter, is apparent. Table 1 shows for plot 9, for example, a continual degradational trend but one in which distinct phases of accelerated lowering alternate with periods of slower degradation. In fact, five such accelerated phases, together with their 'recovery' periods may be discerned. It is important to note that plots 8 and 9 form a pair, 8 slightly upslope of 9, on the same lithological unit. They differ only in position and angle (Table 2).

Table 12.2 Plot angles, standard deviations and rates of surface change, 1969-1979

Plot	Mean Slope (°)	Standard Deviation (°)	Rate of surface change, cm/month (+aggradation; -lowering)
1	21.48	2.83	-0.017
2	4.85	0.53	+0.036
3	5.98	0.60	+0.015
4	32.99	2.66	-0.038
5	26.53	2.69	-0.029
6	40.34	3.97	-0.008
7	46.41	4.24	-0.017
8	7.63	2.76	-0.026
9	12.56	2.27	-0.063

Throughout the ten-year period no discernible change in slope angles was recorded. There is a tendency for the steeper plots to exhibit a greater standard deviation from their mean slope angle, a fact which probably reflects the relationship between the angle of repose of the surface materials and their relative ease of entrainment by slope processes (Table 2).

Rates of surface lowering are not related to slope angle. Plot 9, which is losing surface height most rapidly is at slightly more than 12°, whereas plot 1, on lithologically comparable material and which is almost twice as steep (21°), lowers at less than one-third the rate of plot 9 (Table 1).

The overall rate of surface lowering on a projected monthly rate (plots 2 and 3 excluded) varies from a minimum value of -0.008 cm/month (plot 6) to a maximum of -0.063 cm/month (plot 9), with a mean rate of -0.028 cm/month (Table 2). Over the decade of measurement the three most rapidly eroding surfaces (plots 9, 4 and 5) have lowered by amounts of 7.80, 5.42 and 4.21 cm respectively. The other degrading surfaces (1, 6, 7 and 8) lie within a lowering range of 1.19 - 2.74 cm.

Overall, the trends of rate and angle change confirm the findings made in 1968 (Campbell, 1970b) and 1974

(Campbell, 1974). There is no evident correlation between erosional rates, angular change and lithological variation.

Roughness

In an attempt to further analyse and describe the changes in form, a measure of surface roughness can be applied (Fig. 5). This shows trends in surface roughness of the plots. A simple planar trend surface through the twenty-five measurement points is used as a reference plane from which the variance of the distance of the points defines the roughness value, i.e. the variance of the trend residuals. It may be expected, for example, that major erosional phases, which may be accompanied by rill-cutting or other form changes would have the effect of increasing surface roughness and hence the form of the surface. Figure 5 shows, for example, that the two alluviated surfaces remain very smooth as may be expected since on these types of surface irregularities are being eliminated or smoothed by continued sedimentation and sheet flow processes (Bryan, Yair and Hodges, 1978).

It is notable that plots 1, 4, 5 and 9 also remain relatively consistent in their surface roughness. That is, no trend towards increased roughness is evident since the slope of the regression line and the zero base line are almost coincident. In fact, plots 8 and 9 show, overall, a trend towards increased smoothness (decrease in variance). Only plots 6 and 7 show a trend to increased surface roughness. These two were identified earlier (Campbell, 1974) as being notably aberrant surfaces in terms of their overall behaviour and there is some evidence that internal mass movement processes (slumping) may be at work which is perhaps a reflection of the steepness of their slope angle (Table 2). There is no evident relationship between roughness (surface form variation) and erosion rates or phases of aggradation and degradation.

Normal-slope relief ratios

As a final measure of form change in this study a technique was devised to measure the manner in which mass change within the plot was reflected in slope form. In an earlier study (Campbell, 1974) the hypsometric integrals (Strahler, 1952) had been calculated for each plot in an attempt to ascertain whether the plots, which form part of slopes, had attained a geometry accordant with prevailing geomorphic processes. The hypsometric integral values for the plots ranged from a 0.52 to 0.57. It has, however, been pointed out (Honsaker, pers. comm.) that it is more appropriate to determine such values from a plane surface normal to the slope rather than from an hypothetical horizontal plane. This is because the plots are portions of slopes, not true basin slopes, and it is distributions of material (aggradation and degradation) within the plot which determines the form of the plot surface. The hypsometric integral used previously (Campbell, 1974) is too insensitive to these small surface changes.

The normal-slope relief ratio (Fig. 6) is analogous to the elevation-relief ratio (hypsometric integral) except that the residual distances from the trend plane are used

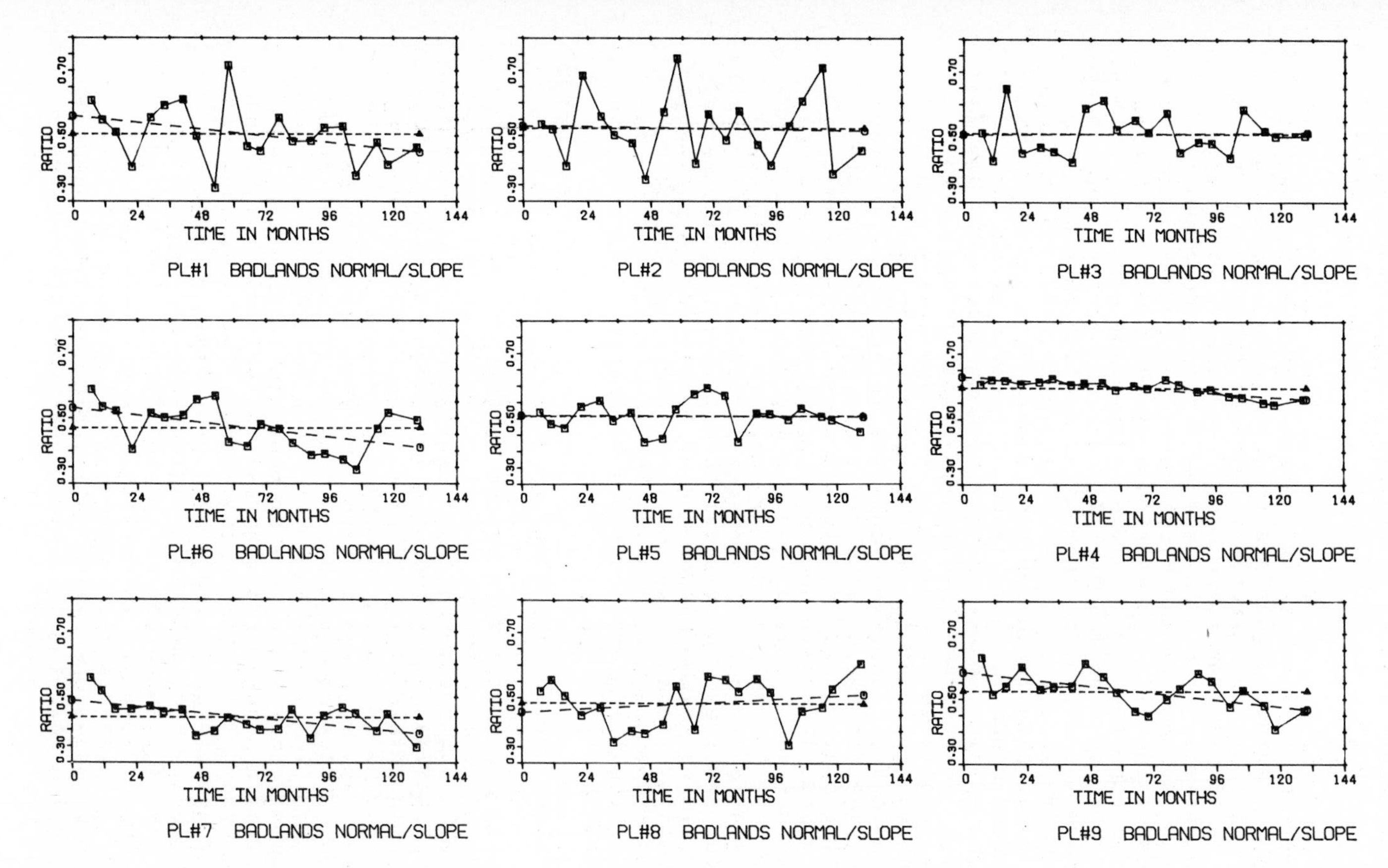

Figure 12.6 Normal-slope relief ratios for the badlands plots, 1969-1979

(c.f. roughness). The elevation ratio $\frac{\text{mean-minimum}}{\text{maximum-minimum}}$ (Pike and Wilson, 1971) calculated from these residuals is a measure of the amount of dissection of the surface and its area-volume distribution.

The non-dimensional ratio must lie between 0.0 and 1.0. Values near 1.0 indicate an essentially flat surface with narrow rills and/or cracks. Low values indicate flattish surfaces with residual narrow lumps or highs (small rock fragment). Values that group around 0.5 are intermediate; either smoothly sloping or with fairly regular undulations of any magnitude. In a larger scale interpretation, as in the case of Strahler's (1952) hypsometric integral, the latter are considered to be in the equilibrium or steady state condition (Schumm, 1956).

Both aggradational surfaces (plots 2 and 3) show long-term trend values close to the 0.5 value. It should be noted that the amplitude range is calculated individually for each plot and hence high values on one plot are relative only to the trend surface of that plot. The range therefore, for plots 2 and 3, while apparently large, actually represents very small relief amplitude values. The strong, rilled sandstone surface (plot 4) also has a consistent trend of values near 0.5. Plots, 1, 6, 7 and 9, show a relatively rapid evolution to lower ratios that is more-or-less consistent with their erosional behaviour. So, although plots 4 and 5, for example, are eroding rapidly, as is plot 9, the form of 4 and 5 remains consistent while plot 9, while smoothing (c.f. roughness, Fig. 5) overall, is leaving residual highs (Fig. 6). Plot 8, conversely, is apparently becoming more dissected, though this may only be a reflection of a tendency towards rougher surface conditions in the period between months 106 and 118 (1978).

Measurement Procedures

The measurement techniques used in this study were modelled on a device used to record frost-heave processes in the Canadian subarctic (Heywood, 1961). Similar devices have been used by Curtis and Cole (1972), Lam (1977) and Barendregt and Ongley (1979).

The frame used here has the advantages of being lightweight, robust and, because it and the measuring rods between which the measurements are taken both have machined surfaces, easily allows measurements to be taken to within 0.5 mm. Furthermore, the data derived from the frame facilitates three-dimensional reproduction (contouring) as well as up and downslope profiles.

As such, the frame possesses several advantages over more commonly used nails and washers, stakes, or similar techniques for single point measurements. These devices by their very nature may cause localized erosion or deposition. On the other hand, the frame, at least in its optimum size (one square metre) requires a minimum surface area for installation and, although the surface can be reasonably uneven, undulations of a large amplitude (>30 cm) would entail the use of inconveniently long mounting posts. The frame would likely also be impractical on near vertical, rapidly

eroding stream banks. Under these circumstances the simple erosion pin is likely to have greater utility.

Ideally, photogrammetrically-produced miniature contour maps would provide the best technique for recording surface and form changes. These would create no terrain disturbance and the maps could be drawn at any scale and contour interval to suit the study. There is, however, no indication in the literature that such a technique has been used.

SUMMARY

It should be stressed that the patterns and changes described above show only the effects of processes, not the processes themselves, and that they are applicable only to the plots as slope segments rather than to entire slopes. The data, and the figures do, however, provide a method by which to compare the behaviour of these contrasting surfaces. Different slope segments clearly behave quite differently in their erosional rates (or depositional rates) and in the manner in which these changes are reflected in their forms. In some cases over the ten-year period of measurement the behaviour of the plots in elevation, roughness and normal-slope relief ratio is highly consistent (plots 2 and 3, and plots 4 and 5). While all plots except 2 and 3, have shown surface lowering the rates at which this has occurred differ markedly and without regard to slope angle. There is some indication of a cyclical pattern in erosion (plots 1, 8 and 9, Fig. 4), and this is discernible to some extent in the aggradation pattern of plots 2 and 3 (Fig. 4).

It is evident that only long-term multiple measurements provide the data required if such a study is to have any significance. Any two or three year period selected at random would fail to reveal, for example, the true nature of aggradational-degradational behaviour of the two alluviated surfaces (plots 2 and 3). In like manner, the pronounced rhythmical pattern that is revealed over the long-term on plot 9 would not become evident within a two or three year period. Data from individual erosion pins are likely to be particularly misleading given the variations within plots apparent in this study.

Yet, some plots show a great deal of consistency of behaviour as if they were in complete adjustment to the long-term nature of the processes operating on them. Plot 4 exhibits minimal change in surface form (roughness and normal-slope values) while eroding rapidly, whereas plot 5, which erodes at a similar rate and also retains minimal roughness values, undergoes relatively large variations on its normal-slope values.

There is an indication that some of the plots (2, 3, 4 and 5) have developed a remarkably consistent pattern of external form adjustment to the geomorphic processes that operate on them. They are essentially smooth (Fig. 5) and have apparently achieved equilibrium conditions (Fig. 6) while either eroding, or degrading. Other plots, i.e. 6 and 7, while lowering at fairly steady rates, undergo major form changes. Plots 1 and 9, while eroding, show little variation in roughness (Fig. 5) but reveal major equilibrium adjust-

ments. Some elements in the badlands landscape are clearly in a rapid phase of evolution from one form to another and are very sensitive to external changes in geomorphic processes, others are not and exhibit little variation in their external characteristics over the decade of study.

Acknowledgements

The research for this paper was funded partially by the Alberta Research Council and the Natural Sciences and Engineering Research Council of Canada. The cooperation of Alberta Parks and Recreation and Mr. C. Beasley of Beasley Ranches Ltd., is gratefully acknowledged. The invaluable assistance provided by Dr. J. Honsaker and his thoughtful advice on the computer analysis was essential to the completion of the study. All drafting and photographic work was performed by staff of the Department of Geography, University of Alberta.

13 Quaternary evolution of badlands in the southeastern Colorado Plateau, U.S.A.

S.G. Wells and A.A. Gutierrez

The evolution of badland landscapes is governed primarily by denudational processes of bedrock hillslopes (Scheidegger *et al.*, 1968). Most geomorphic studies of badlands have incorporated simple and sophisticated types of field instrumentation to measure these denudational processes (Schumm, 1956b; Campbell, 1970, 1974; Bryan *et al.*, 1978). Such studies consider processes and rates of erosion over time spans of years and less, during which hillslope hydrology, hillslope morphology, and drainage network are independent variables (Schumm and Lichty, 1965). Over longer time periods, such as thousands of years, these variables become dependent upon total runoff and sediment yield, relief above base level, climatic conditions, and bedrock geology. The evolution of badland landscapes over long time periods, such as the late Quaternary, has received little attention.

The primary purposes of this paper are: (1) to delineate the evolution of badland landscapes during the Quaternary in the southeastern portion of the Colorado Plateau; (2) to evaluate the relative roles of base level, climate, and bedrock geology on regional badland evolution; and (3) to date the origin of regional badland development in this portion of the Colorado Plateau. These badlands have not been described previously, although they cover much of the 4000 km^2 area north of the Chaco River (figure 1). Long-term erosional and depositional processes associated with badland evolution in the Chaco drainage basin are elucidated by four major methods: (1) analyzing regional trends in badland morphology, (2) analyzing stratigraphic relationships of Quaternary deposits associated with badland evolution, (3) dating of these Quaternary deposits by radiocarbon techniques and by cultural artifacts, and (4) comparing these results with detailed studies within two instrumented badland watersheds. This paper utilizes both geomorphic data on badland landscapes and on landscapes surrounding badlands to determine long-term badland evolution.

The dominant landforms of the southeastern Colorado Plateau are pediments and badlands (Hunt, 1974). Within the Chaco drainage basin, these landforms create stepped topography in which gently sloping upland surfaces (pediments) are capped by Quaternary sediments and are dissected by badland areas (plate 1). Tributaries draining these badland

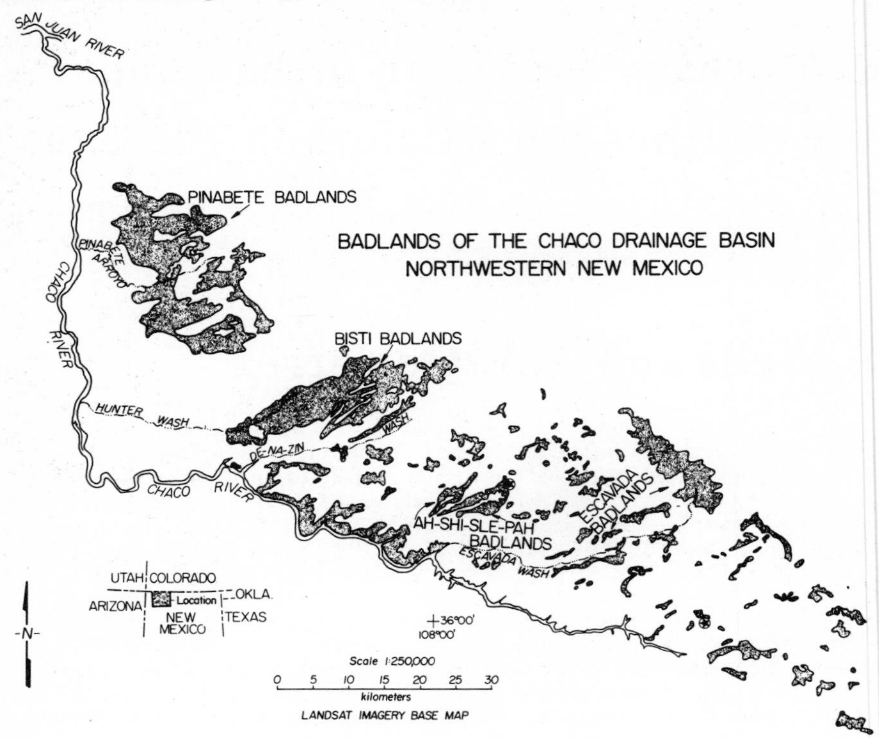

Figure 13.1 Distribution of badlands of the northern Chaco River drainage basin, southeastern Colorado Plateau. Badlands indicated by pattern, and two instrumented watersheds indicated by stars. Quantitative measurements of badland processes based on data from these instrumented watersheds

areas flow into the major drainage system, the ephemeral Chaco River, which drains northwestward into the San Juan River (figure 1). Remnants from periods of base-level stability are preserved along the Chaco and its tributaries as pediment and terrace remnants (Scott *et al.*, 1979; Love, 1980; Wells *et al.*, 1981).

Badlands are developed on the Cretaceous Kirtland and Fruitland Formations which dip northward and northeastward less than 5°. These formations are composed of friable sandstones, thin coal beds, and thick mudstones and are commonly covered with thick Quaternary fluvial and aeolian sediments. These less resistant bedrock units rest upon a clastic sequence, the Cretaceous Pictured Cliffs Sandstone, and are overlain by another clastic sequence, the Tertiary Ojo Alamo Sandstone. These resistant units form broad cuestas or hogback ridges and only have thin veneers of Quaternary surficial deposits.

The present climate of the Chaco River drainage basin is arid with a mean annual precipitation of approximately 220 mm. Past climates of this area have included mixed-

Plate 13.1 Quaternary deposits capping Cretaceous Fruitland Formation. Unconformity indicated by solid line. Badland development occurs in bedrock areas exposed on the dissected edges of these upland surfaces. Aeolian deposits rest unconformably upon the older fluvial deposits in upper right hand corner (just above subject)

conifer communities around 10 000 yrs. B.P. to pinyon-juniper woodlands around 5000 yrs. B.P. (Betancourt and Van Devender, 1980). Shifts in vegetational communities from late to mid-Holocene are attributed to drying of the climate (Betancourt and Van Devender, 1980; Love, 1980). Hall (1977) indicates that the present arid climate of this area was established by mid-Holocene (approximately 5800 yrs. B.P.) with a slight decrease in aridity at approximately 2400 yrs. B.P. Additional evidence for this climatic trend is the period of widespread aeolian deposition throughout this region between 5600 and 2800 yrs. B.P. (Wells *et al.*, 1981; Schultz and Wells, 1981).

REGIONAL BADLAND DEVELOPMENT

Variations in badland occurrence and characteristics

The regional distribution of badland topography in the Chaco River drainage basin is illustrated in figure 1. The areal extent of badland topography varies from the southeastern portion of the study area near the headwaters of Chaco River to the northwestern portion along the distal reaches of the Chaco River. The southeastern portion of the Chaco drainage basin has a high frequency of small badland areas, many of which are too small to show on figure 1. The northwestern portion of the Chaco drainage basin has a low frequency of badland areas (figures 1 and 2). A systematic downstream increase in average badland size for major tributary drainages of the Chaco River is illustrated in figure 2. The southeastern portions of the drainage basin have badland sizes averaging 5 km^2 and less. Measurements of frequency of badland areas per tributary watershed are given in figure 2. In general, tributary watersheds in the southeastern areas have 30 or more badland areas; whereas, the distal, northwestern tributary watersheds have 5 badland areas.

Both the frequency and size of badland areas are a measure of badland development. A well-developed badland area is one of large areal extent and is continuous over that area, such as the Pinabete or Bisti badlands (figure 1); whereas, a less-developed badland area has several small, isolated badland areas. Thus, the badlands north of the Chaco River can be divided into two groups: (1) well-developed badlands of the northwestern Chaco drainage basin, and (2) less-developed badlands of the southeastern Chaco drainage basin. The geomorphic parameters of these two badland areas are summarized in table 1. The increase in bad-

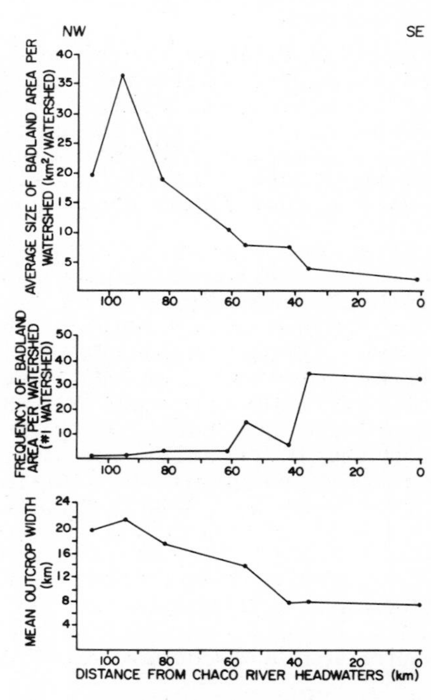

Figure 13.2 Changes in badland area size and frequency per tributary watersheds along the Chaco River, and variations in mean outcrop width of the Cretaceous Kirtland and Fruitland Formations with distance down the Chaco River

land development from the headwaters (southeastern) portion of the Chaco drainage basin toward the distal (northwestern) portion appears to be related to two regional geomorphic parameters: (1) increasing outcrop width of the Cretaceous Kirtland and Fruitland Formations with distance down the Chaco drainage basin, and (2) increasing relief above base level due to the greater amount of base-level lowering in the downstream portions of the Chaco River.

Influence of bedrock, climate, and base level on regional badland development

The mean outcrop width of the Cretaceous Kirtland and Fruitland Formations increases from the southeastern portion of the Chaco drainage basin toward the northeastern (figure 2). In addition, the course of the Chaco River is controlled by the strike of the Cretaceous Pictured Cliffs Sandstone and thus follows the outcrop pattern of these Cretaceous units (figure 1). Increasing bedrock width provides more surface area over which badlands can develop; thus, the increase in badland area per tributary watershed parallels the increase in outcrop width. A linear regression analysis of the change in badland area and mean outcrop width with distance down the Chaco River is summarized below:

(1) $Y = 1.48\, e^{0.03X}$, $r = 0.98$, significant at 95% level where Y = average badland size/tributary watershed, X = distance from Chaco headwaters;

(2) $Y = 5.55\, e^{0.02X}$, $r = 0.88$, significant at 95% level where Y = mean outcrop width of Kirtland and Fruitland Formations, X = distance from Chaco headwaters.

The similarity between these two regional trends is expressed quantitatively by the similarity of the two coefficients, 0.03 and 0.02, in equations 1 and 2.

In areas of wider outcrops of the Kirtland and Fruitland Formations more areally extensive Quaternary deposits cap the Cretaceous bedrock. In order for badlands to develop, these capping deposits must be eroded to expose the underlying bedrock. The widespread removal of these Quaternary deposits is a function of the amount of dissection which is controlled by base-level lowering along the Chaco River. Changes in the regional base level, Chaco River, are related to climatic fluctuations (Andrews *et al.*, 1975; Carrara and Andrews, 1973) which affected both the San Juan and Chaco Rivers (Love, 1980). Tributaries along the distal reaches of the Chaco responded earlier to climatically induced base-level lowering of the Chaco River than did the headwaters in the southeastern portion of the study area. Thus, fewer episodes of base-level change are preserved in the headwaters of the Chaco as compared to the distal regions (table 1).

The integration of small, isolated badland areas during the dissection of the Quaternary deposits and underlying bedrock increased with increasing relief above base level. A comparison of the frequency of badland areas per tributary

Table 13.1 Comparison of geomorphic parameters of badland areas within the southeastern and northwestern portions of the Chaco River drainage basin

GEOMORPHIC PARAMETER	SOUTHEASTERN CHACO DRAINAGE BASIN	NORTHWESTERN CHACO DRAINAGE BASIN
Average badland size per tributary watershed (km^2/watershed)	<10	>10
Frequency of badland areas per tributary watershed (#/watershed)	>10	<10
Frequency of base level remnants per tributary	2 - 3	4 - 5
Average relief above local base level (m)	30	100
Average length/width ratios of badland watersheds	1.5:1.0	3.5:1.0
Orientation of badland watershed axes	random	preferred orientation N63°E to N87°E

watershed of the Chaco River, the relief above base level, and the number of base-level remnants per tributary (table 1) illustrates this regional influence of climate and base level on badland development.

Badland watersheds within the northwestern and central portions of the study area show a general northeast-southwest elongation. Badlands in the southeastern portion of the study area are more circular and do not trend in any preferred orientation (table 1). The axes of elongation for the badland watersheds in the northwestern and central portions of the study area have a mean of N70°E with a range of N87°E to N63°E. A similar trend, approximately N70°E, is expressed in the major aeolian dunes which occur in this portion of the study area (Schultz, 1980); whereas, the aeolian deposits of the southeastern portion of the Chaco drainage basin are primarily sheet deposits and lack any preferred orientation.

The orientation of badland watersheds in the central and northwestern areas is influenced by the preferred orientation of the aeolian dunes on the upland surfaces. During the initial development of badland watersheds, the Quaternary deposits are fluvially eroded along the edges of the upland surfaces (plate 1). Runoff on the upland surfaces is channelized between the elongate dunes in the less-permeable interdunal areas and is directed down the regional slope of these surfaces into the badlands. Badlands extend headward into the upland surfaces in these interdunal channels. Thus, landscapes outside of badland watersheds influence the evolution of badland landscapes.

LANDFORMS AND PROCESSES WITHIN BADLAND WATERSHEDS

Badland watersheds within the Chaco River drainage basin display a characteristic succession of landforms and surficial deposits between the watershed periphery and basin axis or mouth. These landforms are distributed concentrically from the watershed boundaries to the center and mouth of the watershed (figure 3). The occurrence of these landforms and their relative positions within a badland watershed are independent of watershed size, morphology, and location within the study area (figure 3). Seven zones are defined which depict the major landforms and their associated surficial deposits and geomorphic processes. The major characteristics of these zones are given in table 2. Zone A represents a zone of erosional stability with some aeolian deposition. Zones B through F are characterized by fluvial dissection and hillslope erosion; whereas, zone G is dominated by fluvial deposition.

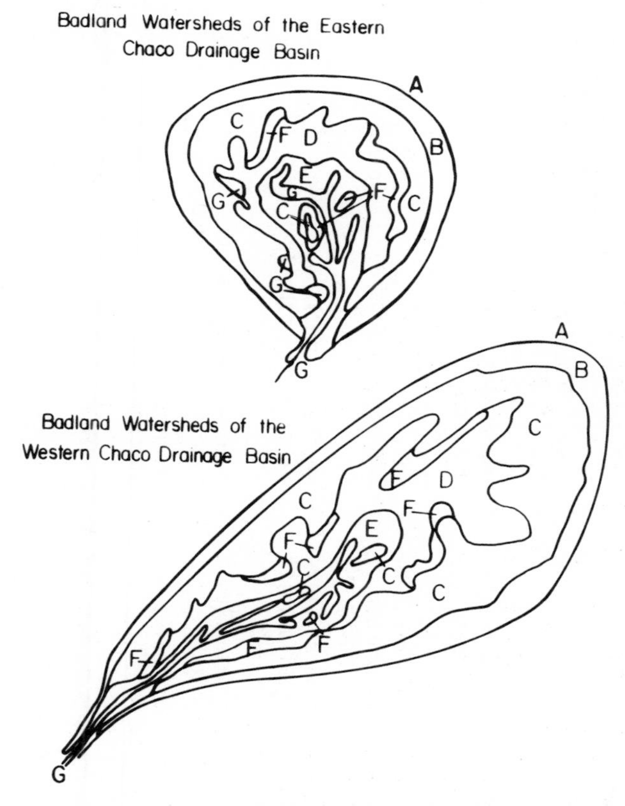

Figure 13.3 Major landform zones with badland watersheds for the eastern and western portions of the Chaco River drainage basin. See table 2 for description of zones A through G

Table 13.2 Major characteristics of landform zones within badland watersheds

ZONE	LANDFORM	SURFICIAL MATERIAL	DOMINANT GEOMORPHIC PROCESS	HILLSLOPE EROSION	
				Type	Rate
A	subplanar geomorphic surface	Quaternary fluvial, colluvial & aeolian sand, silt & clay, gravel lenses; 0.5 - 7 m thick	erosional stability & aeolian deposition		
B	convex drainages dissecting geomorphic surface deposits	reworked Quaternary deposits	hillslope erosion (zones B to E/F)	rilling and sheet wash	erosion rate 0.3-2.0 cm/yr
C	convex, rounded interfluves, with bedrock channels	weathered mantle, 0-60 cm thick		creep, sheet flow	(sandstone) 0.5-2.0 cm/yr (mudstone)
D	convex, rounded interfluves with alluvial channels	weathered mantle & alluvium in channels		(same as above)	(same as above)
E/F	low-relief interfluves; micropediments	thin weathered mantle with veneer of alluvium		sheetwash	0.5-0.6 cm/yr
G	alluvial channels, point bars & overbank areas	alluvium of variable thickness	fluvial deposition		

Two badland watersheds, which represent the two stages of badland development in the Chaco drainage basin, were selected to measure the processes within these geomorphic zones. In addition, these two badland watersheds were instrumented to determine the combined rate of hillslope and fluvial erosion, here referred to as the denudation rate. These watersheds are the Ah-shi-sle-pah and Windmill watersheds located in the western and eastern portion of the study area, respectively (figure 1). The upper Ah-shi-sle-pah watershed is a portion of well-developed badlands which is 111.2 km^2 in size and has 125 m of total relief. The Windmill watershed is an incipient badland watershed with an area of 0.025 km^2 and a total relief of 15.5 m.

Three methods are used to measure the denudation rates of these watersheds. The most extensively used method is that of erosion-pin technique described by Miller and Leopold (1963). The erosion pins were placed in the ground at 50-cm intervals across hillslopes and channels. Erosion rates are averaged for the entire length of the transect according to those procedures described by White and Wells (1979); this method provides an overall denudation rate that is representative of hillslope and fluvial erosion. Additionally, the erosion-pin transects were installed to obtain denudation rates characteristics of the major geomorphic zones, as well as the headwater, mid-basin, basin mouth, and total basin areas.

The second method of measuring denudation rates, a technique designed by Campbell (1970), utilizes a 1-m^2 grid with 25 measuring points at 20-cm intervals. With this devise changes in land-surface elevations are recorded over time for selected stations which are representative of the major geomorphic zones. In order to provide an independent check on these two methods, sediment traps were installed to measure total runoff and sediment yield from selected plots ranging from a few square meters to tens of meters in size. Additionally, the U.S. Geological Survey operates an automatic sediment sampling station in the Ah-shi-sle-pah watershed, and data from this station are utilized. The total sediment yield from these watersheds is calculated by multiplying the denudation rates by the areal extent of their representative geomorphic zones. Channel aggradation is measured by chain transect techniques similar to those given by Emmett (1974). Precipitation quantity and intensity are monitored by recording rain stations operated by the authors and by the U.S. Geological Survey.

Headwaters of badland watersheds (zones A and B)

The headwaters of badland watersheds are developed in most cases along the edges of dissected upland surfaces. A nickpoint occurs between the relatively flat surface and the steep, dissected edge (plate 1). Zone A is usually covered with relatively dense stands of vegetation (grasses and shrubs). Sandsheet deposits and dense vegetation on the surface result in high infiltration rates (20 cm/hr). In Zone B, erosion of the unconsolidated Quaternary deposits occurs by aeolian and fluvial processes. Ridge dunes (dated at 1500 yrs. B.P. and younger) form at the edge of the upland surfaces where sediment is blown up and out of the badland watersheds (Schultz, 1980). In most cases the ridge dunes occur along the contact between zones A and B, and therefore, are available for subsequent fluvial transport back into the basin. In many cases, sediment derived from these Quaternary deposits is temporarily stored downslope of zone B as small alluvial fans of thin sheet deposits. Bedrock hillslopes in these zones or directly downslope are covered with thin veneers of sediment (plate 1). Rills and discontinuous channels develop in zone B where the Quaternary deposits are slightly consolidated by pedogenic clay or calcium carbonate.

Bedrock hillslopes and channels (zones C through F)

Bedrock channels developed within the badland watersheds are subparallel to semirectangular in pattern. These channels are commonly meandering with very steep sideslopes. A typical longitudinal profile of a channel draining the badland watershed is given in figure 4. Within zones C through F, channelized flow is locally diverted underground through pipes. This process is common in areas where sandstones overlie mudstones (figure 4). Below the pipe entrance, debris flows and creep obliterate abandoned channel segments (Gutierrez, 1980).

Hillslopes within badland zones C and D are covered with thin veneers of weathered mantle ranging from 0 to 60 cm in thickness. Weathered-mantle characteristics such as thickness and surface relief vary with topographic location, bedrock type, and slope aspect (Gutierrez, 1980). South-facing slopes have thicker weathered mantles than north-facing slopes, and sideslopes and interfluve crests have thicker mantles than noseslopes. The bases of badland hillslopes typically are flanked by micropediments with poorly defined channels, which range in areal extent from 1 to 100 000 m^2. In many cases these occur as transitional zones between bedrock hillslopes and depositional areas in the valley floors.

Intrabasin alluvial channels and valley floors (zone G)

The central portions of badland watersheds are occupied by valleys filled with sediment derived from zone B and zones C through F. Clay mineralogy of sediments in zone G indicates that a major source area for these sediments is zone B. Sediments from zone B are characterized by kaolinite, feldspar, and quartz. Sediments in channels within zone G have greater quantities of kaolinite than montmorillonite-mixed types which are derived from the weathered mantle on the bedrock surfaces. Sediments derived from the bedrock hillslopes within the badlands are stored in the upper portions of the valleys until major runoff events can remove and transport the fine-grained debris (Gutierrez, 1980). Sediment transported out of the badland watershed is sand-size debris derived from zone B or friable sandstones within zones C through F.

Sediment that leaves the badland watersheds is deposited as alluvial fans or sheet deposits on the valley floors (plate 2). Depositional processes on these fans involve

Plate 13.2 Active sedimentation (light areas in upper third of figure) in valley floors at bases of badland hillslopes. Most of sediments are derived from topographically higher Quaternary deposits shown in lower third of plate

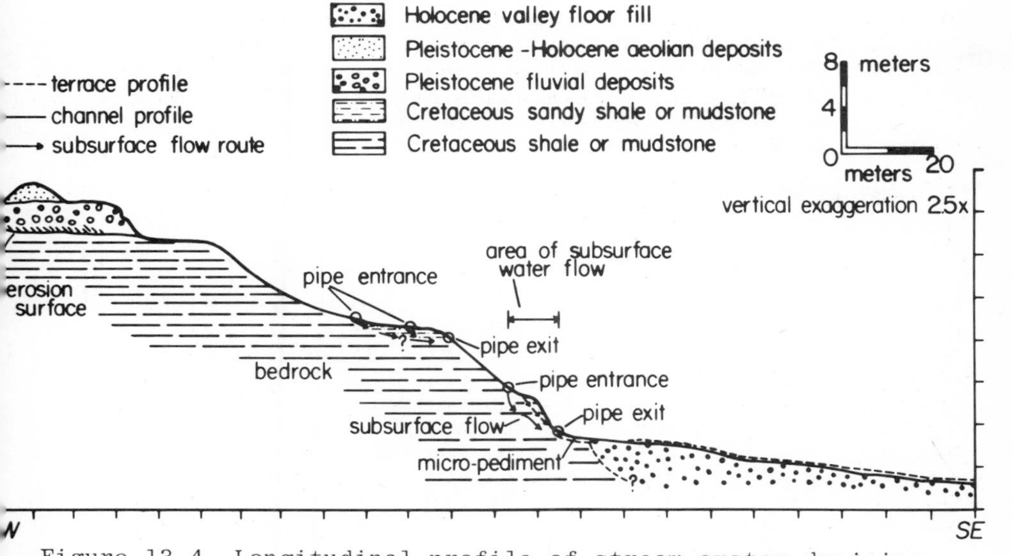

Figure 13.4 Longitudinal profile of stream system draining a badland watershed

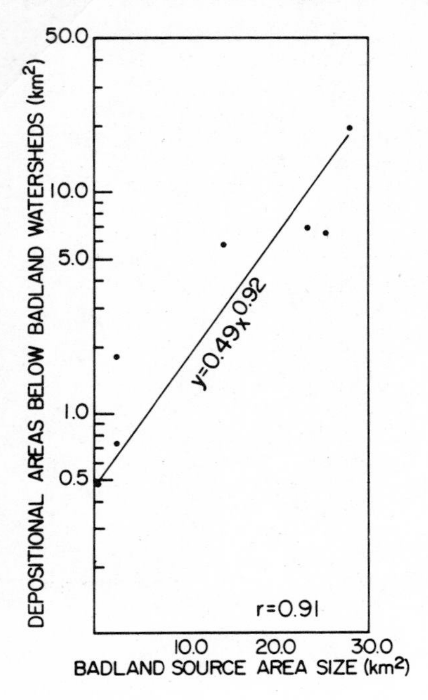

Figure 13.5 Relationship between size of badland catchment-area and size of depositional area below the catchment area

both mudflow and streamflow. Large concentrations of clays are mixed with sands and gravels, which are derived from the Quaternary sediments of the upland surfaces, to form mudflows. These sediments are confined, in many cases, to incised channels near the badlands but overtop the stream banks away from the badlands to form sheet deposits (plate 2). The size of these depositional areas is proportional to the size of their respective source areas, the badland watersheds (figure 5). The relationship between badland source-area size and depositional-area size is expressed by the following equation:

(3) $Y = 0.49\ x^{0.92}$, $r = 0.91$, significant at the 95% level where Y = size of depositional area,
X = size of contributing badland watershed.

This relationship is similar to that found by Bull (1964), Denny (1965), Hooke (1968), and Wells (1977) where alluvial-fan size was compared to size of contributing drainage area for desert basins in the southwestern United States. The exponent for the badland/catchment area relationship is 0.92 which is similar to the mean exponent of other alluvial fan/drainage basin area relationships (0.90) (Mabbutt, 1977). Larger badland watersheds provide more areas for temporary storage of the sediment, thus the exponent for equation 3 is less than unity. Systematic relationships between size of depositional lobes and size of contributing catchment areas suggest an equilibrium condition (Denny, 1967; Wells, 1978). Not all badland drainage basins and depositional areas show such relationships because of processes which disturb equilibrium between source areas and depositional areas. Stream piracy and shifting drainage basin boundaries are common in badlands due to rapid rates of erosion; thus, piracy can increase and decrease catchment size without concomitant adjustments in depositional-area size. Other factors which influence catchment/depositional-area relationships are the frequency of rainfall-runoff events which control the storage and removal of sediment from badland watersheds (Gutierrez, 1980), and the availability of sediment within a basin. Badland watersheds which have thicker deposits on the geomorphic surfaces for a given drainage basin size will have larger depositional areas.

Rates of processes in badland watersheds

The types and rates of hillslope erosional processes within each badland zone are summarized in table 2. Areas with the highest erosion rates include zone B (0.3 - 2.0 cm/yr) and friable sandstones in zones C and D (0.5 - 2.0 cm/yr). Cumulative water and sediment discharge are greater in zone B than in subwatersheds of similar size in zones C and D (figure 6). More sediment is yielded from zone B than zones C and D due to the relative availability of sediment, which can be transported out of these areas. Quaternary deposits of zone B require little or no weathering to prepare them for transportation; whereas, bedrock hillslopes must undergo weathering before material can be eroded from zones C and D.

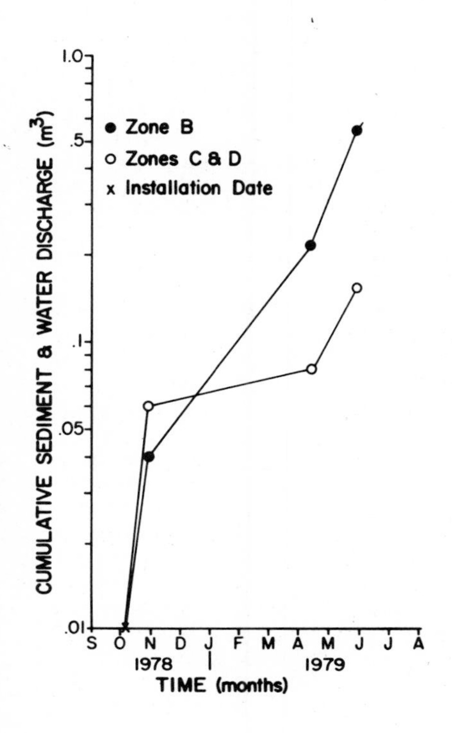

Figure 13.6 Total sediment and water discharged from instrumented plots within the Ah-shi-sle-pah badlands

Although the weathered mantle on bedrock hillslopes has greater areal extent, the mantle is usually a few centimetres thick. Quaternary deposits commonly are several metres thick and are constantly available for erosion as dissection of zone A proceeds.

The mean denudation rate for these two watersheds incorporates the three types of denudation data and the respective zones of sediment yield. The mean denudation rate for the Windmill watershed is 360 m^3 of sediment per year, and the mean rate for the Ah-shi-sle-pah watershed is 381,384.7 m^3 of sediment per year. The greater sediment yield for the Ah-shi-sle-pah watershed is due to the larger drainage basin size and greater relief. It should be noted that studies of western U.S. watersheds indicate that sediment yield decreases with increasing basin size due to the increased capability of long-term sediment storage (Schumm, 1977). However, in our study area, little sediment is stored within badland watersheds for long periods of time; rather, most of the sediment is transported out of the watershed into the valley floors during large precipitation events (Gutierrez, 1980). Such precipitation events are typical of the summer months (May-September) when rainfall intensities average 10.16 mm/hr. Rainfall intensities for the winter months (October-April) average 1.8 mm/hr and do not produce the sediment yields that summer events achieve. Suspended-sediment concentrations are greater for summer rainfall events and are usually in excess of 80 000 mg/l; whereas, winter rainfall events produce suspended-sediment concentrations in the range of 15 000 to 60 000 mg/l.

Correlation between catchment-area sediment yield and relief ratio has been established for western drainage basins (Schumm and Hadley, 1961); however, sediment yields in badland watersheds of the Chaco River drainage basin have not previously been analyzed. Figure 7 illustrates the relationship between sediment yield and relief-ratio for selected badland watersheds in the Chaco study area and in other western watersheds (Schumm and Hadley, 1961). As the relief ratio increases, the amount of sediment yielded from a badland watershed increases. The badland watersheds of this study yield more sediment for low values of relief ratios than the western watersheds studied by Schumm and Hadley (1961). This difference could be related to the vegetation cover in other western watersheds as compared to the paucity of vegetation in Colorado Plateau badland watersheds. Vegetation serves as an important agent to increase sediment storage in most basins (White and Wells, 1979), but in badlands vegetation is insignificant.

Rates of aggradation on the valley floors have been determined both by field instrumentation and by datable material within the valley fill. Rates of valley-floor aggradation range from short-term rate determinations of 0.5 cm/yr to long-term rate determinations of 0.1 cm/yr. This short-term rate of valley-floor aggradation is comparable to that mean rate of channel aggradation determined by Emmett (1974) over a ten year period which is 0.76 cm/yr. Most valley floors flanking badland watersheds have aggraded during the past 1000 yrs (Wells *et al.*, 1981).

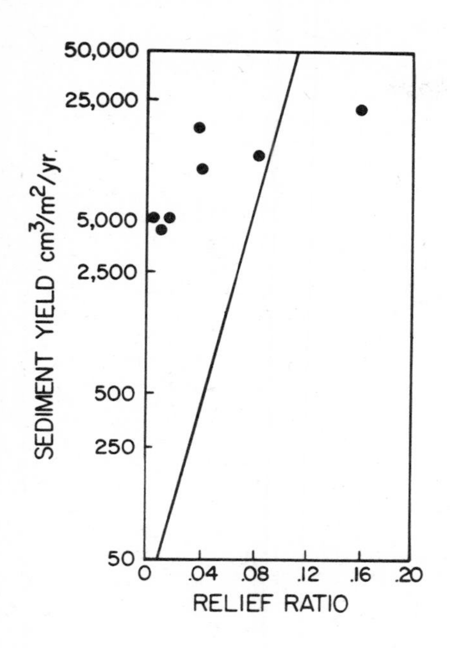

Figure 13.7 Relationship between sediment yield and relief ratio for badland watersheds in the Chaco study area (dots) compared to trend (solid line) established by Schumm and Hadley (1961)

ORIGIN AND EVOLUTION OF BADLANDS

Badland landscapes are only one of a suite of landscapes of the Chaco drainage basin. Other major landscape types include: (1) upland surfaces composed of Quaternary deposits capping pediment and terrace remnants, (2) late Holocene valley floors, and (3) late Quaternary aeolian deposits which occur both on the upland surfaces and the valley floors (figure 4). These upland surfaces and the valley floors along with their associated deposits are referred to as geomorphic surfaces (plate 1 and figure 8). Each geomorphic surface represents a suite of landforms and deposits which formed under relatively constant base-level and climatic conditions. Base-level lowering and concomitant dissection of these geomorphic surfaces expose the erodible bedrock. Badlands develop along the dissected edges of the older geomorphic surfaces and above the valley floors (youngest geomorphic surface) (figure 8). Drainage networks in the badlands head on the flanks of the older geomorphic surfaces and transport sediment from the geomorphic zones A and B (figure 3 and plate 1) to zone G (valley floors) (plate 2). Dissection of drainage network in badlands results in the formation of hillslopes developed on the bedrock (zones C through F).

The initiation of badland development at any point within the study area requires the removal of the Quaternary deposits of the older geomorphic surfaces. The removal of these deposits on a regional scale occurs by fluvial dissection in response to base-level lowering; local erosion of

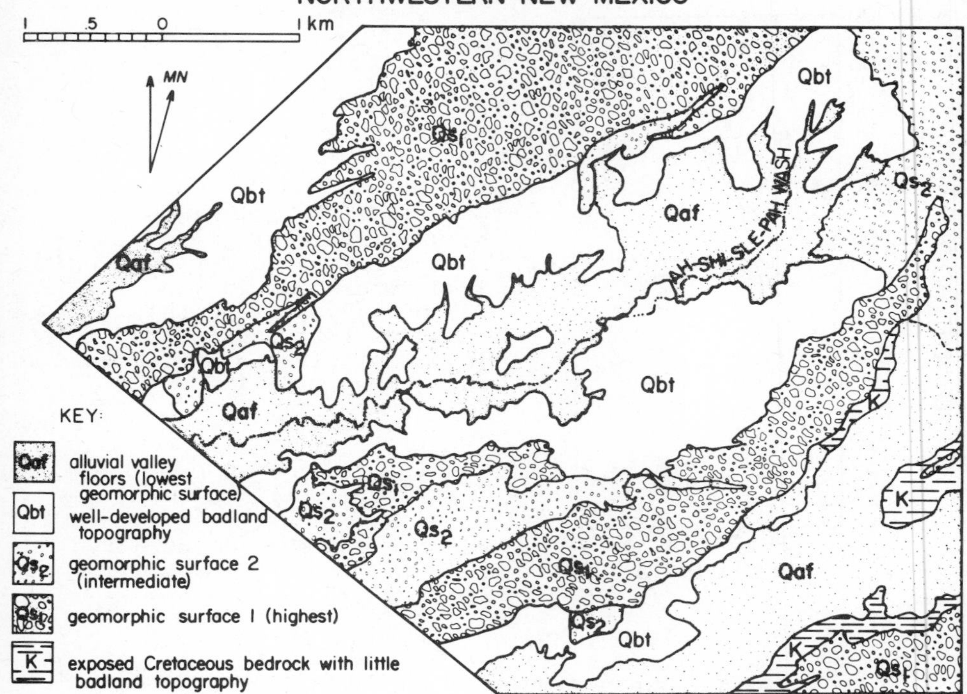

Figure 13.8 Generalized geomorphic map of the Ah-shi-sle-pah drainage basin. Geomorphic units Qs_1 and Qs_2 are stable with respect to erosion; whereas, Qbt units are areas of rapid erosion. Qaf units are areas of long-term aggradation

these deposits may occur by wind deflation. The initiation of badland development may be determined by dating those depositional events immediately proceeding dissection of the geomorphic surface. Additionally, the sediment derived from these badland areas and deposited in the surrounding valley floors can be dated. Both types of dates are useful in determining the origin of badlands and the time required for their evolution.

Age and long-term rates of badland development

The long-term evolution of badland watersheds in the Chaco River drainage basin is evaluated by extrapolating modern rates of basin denudation, reconstructing paleoenvironmental conditions and dating surficial deposits included in badland evolution. Reconstruction of geomorphic surfaces prior to dissection and badland development provides an approximate value of the amount of material removed during badland evolution. This volume divided by the mean erosion rates determined from instrumented watersheds is used to determine the amount of time required to form the present badland watersheds.

Both the Ah-shi-sle-pah and Windmill watersheds were used to date the period of badland development. Maps of

reconstructed paleotopography of the geomorphic surfaces were used to determine the ancient landscape level prior to badland development in both watersheds. The volume of material removed during badland development in the Windmill watershed is 108 250 m^3. The mean denudation rate for this instrumented watershed is approximately 360 m^3/yr. The calculated time required to remove 108 250 m^3 of Quaternary deposits and bedrock at this denudation rate is 300 yrs. The same procedure applied to the upper Ah-shi-sle-pah watershed yields a time of approximately 2800 yrs to remove 1.07 x 10^9 m^3 at a rate of 381,384.7 m^3/yr. These calculations give an approximate age of late Holocene for both types of badland watersheds.

The validity of extrapolating these denudation rates over such time scales appears reasonable in that the present arid climate was established by mid-Holocene; thus, similar processes (e.g., dominant summer precipitation-runoff events) probably have occurred over the past 3000 to 5000 yrs. Additionally, the denudation rates, which have been extrapolated over entire watersheds, are compatible with basin denudation rates determined from empirical equations in other studies (Wells and Rose, 1981).

To provide another measure of long-term badland development, areas were selected in which the material resting upon the geomorphic surfaces had been dated, such as the Bisti and Ah-shi-sle-pah badlands (figure 1) (Schultz, 1980; Wells *et al.*, 1981). Paleoenvironmental reconstruction of selected badland areas indicates that major aeolian depositional periods occurred prior to the development of the badlands (Schultz, 1980; Schultz and Wells, 1981). The last major aeolian depositional period has been dated by C^{14} radiocarbon procedures, estimates of ages of cultural artifacts, and estimates of the amount of time required for soil development. Stream valleys developed on the geomorphic surfaces and graded to the valley floors below badlands truncate the aeolian deposits dated at 5600 and 2800 yrs. B.P. (figure 9). Additionally, fluvial deposits inset into these valleys are dated at 1020 yrs. B.P. and younger (figure 9). Thus, a large portion of the Bisti badlands formed since this aeolian event which is dated at 5600 to 2800 yrs. B.P.

The average relief in the Bisti badlands is 15.7 m. Using this figure as an estimate of the average amount of surface lowering during badland development, the long-term rates of badland development are calculated at 0.3 cm/yr for the 5600 yr interval and 0.5 cm/yr for the 2800 yr interval. These values represent long-term denudation rates (both fluvial and hillslope erosion). These calculated rates are similar to rates measured by field instrumentation over the past three years in the Chaco drainage basin (table 2). In addition, estimates of 0.3 to 2.0 cm/yr are common for short-term field determinations of hillslope erosion rates and are compatible with hillslope erosion rates determined in other studies (Campbell, 1970; Schumm, 1956b):

Steveville badlands Alberta, Canada	Badlands of South Dakota
0.46 to 2.98 cm/yr	1.01 to 1.90 cm/yr

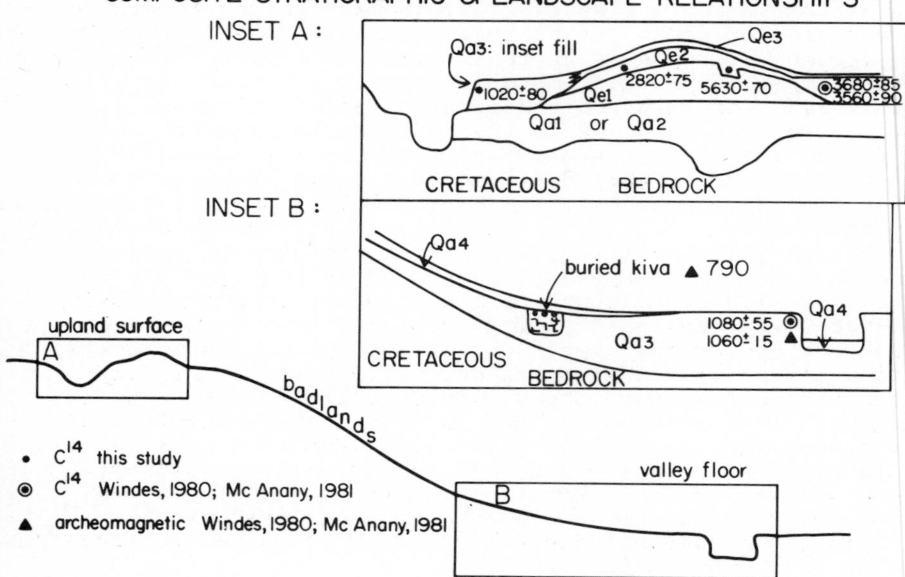

Figure 13.9 Stratigraphic-landscape relationships in the Chaco drainage basin. Upland surfaces are composed of alluvial deposits (Qa_1, Qa_2, Qa_3 with Qa_1 being the oldest) and aeolian deposits (Qe_1, Qe_2, Qe_3 with Qe_1 being the oldest). Valley floors are composed of alluvial deposits (Qa_3, Qa_4) and aeolian deposits (Qe_3, not shown in diagram). Selected radiocarbon and archeomagnetic dates are illustrated in their respective deposit. Data taken from Windes, T.C., 1980, Excavations at Pierre's Site 27-an arroyo hearth near the main site complex, unpublished report, Division of Cultural Research, National Park Service, 17 p.; and from McAnany, P., 1981, Geology and geomorphology of the Alamito coal lease, unpublished report submitted to Navajo Nation, 23 p.

Similar stratigraphic-landscape relationships illustrated in figure 9 are found throughout the study area indicating that regional badland development in the Chaco River drainage basin occurred in the late Holocene, with significant portions of the badlands forming in the last 3000 yrs. Climatic conditions during the Holocene changed from moister conditions in the early Holocene (Betancourt and Van Devender, 1980) to drier conditions after 5000 yrs. B.P. (Hall, 1977). Badland development during the late Holocene may have been aided by the increased aridity of the climate which would reduce stable vegetation cover on the dissected margins of geomorphic surfaces and permit rapid erosion of the

Quaternary deposits. Badland topography would develop subsequent to the exposure of erodible bedrock or regolith. Such extensive erosion during a relatively short time period would result in large quantities of sediment being introduced into the valley systems. With continued drying of the climate, less sediment would be transported out of the tributary drainage lines, and valley aggradation would occur. Dates on valley-fill deposits topographically below or inset into badland areas suggest a major period of aggradation beginning around 1000 yrs. B.P. (figure 3), which may represent the storage of sediments derived from badland watersheds.

Figure 10 summarizes the timing of badland development and other major geomorphic events in the Chaco River drainage basin. Although badland development has occurred locally during the late Quaternary, it was most extensive during the late Holocene (post 3000 yrs. B.P.). This period coincides with increased aridity in this portion of the Colorado Plateau and lush vegetation destabilization on geomorphic surfaces during this time as indicated by aeolian activity (figure 10). Major dune deposition during this time was in the form of parabolics which owe their origin to vegetational destabilization. Reduction of aeolian deposition between 3000 and 1500 yrs. B.P. (figure 10) may have been influenced by badland development. Any aeolian deposits on these geomorphic surfaces may have been removed with increased badland extension. High, local relief provided by the badlands

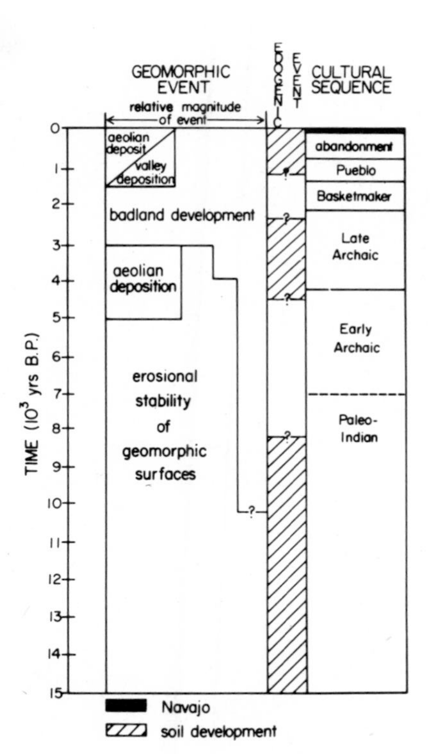

Figure 13.10 Chart Illustrating the general timing and relative magnitude of geomorphic events in Chaco River drainage basin. Cultural sequence adopted from Hall (1977)

may have prohibited major aeolian deposits from moving away from their source area on the cuesta of the Pictured Cliffs Sandstone (Schultz and Wells, 1981) to the geomorphic surfaces developed on the Kirtland and Fruitland Formations. Only small, localized dunes could form along the geomorphic surface margin, such as the ridge dunes. Finally, a decrease in badland development relative to alluvial-valley deposition over the past 1500 yrs. occurred as the valleys backfilled with sediment, covering the bedrock and reducing the areal extent of badlands.

Acknowledgements

The authors wish to express their thanks to Dr. R. Ingersoll, Mr. T. Bullard, Mr. L. Smith, Dr. R. Bryan, and Dr. A. Yair for reviewing the manuscript and to Dr. D. Love, Mr. J. Schultz, Mr. L. Smith, and Mr. T. Bullard for their contributions in field and laboratory. Cooperation of the Navajo Nation and the Bureau of Land Management is acknowledged. This research was supported by Grant No. 68R-3111 of the New Mexico Energy and Minerals Department and by a scholarship provided by the New Mexico Geological Society.

14 How old are the badlands? A case study from south-east Spain

S.M. Wise, J.B. Thornes and A. Gilman

We are currently studying aspects of some 40 archaeological sites in south-east Spain. In this work we have observed that on many sites the archaeological sediments and structures have survived 4000 years of denudantional history. In some cases this is because the sites lie on extensive flat plains or on limestone hilltops. In other cases, however, sites lie on the flanks of ridges in severely eroded badland terrain. Here the survival of materials *in situ* on steep hillslopes coupled with the configuration of the gully network system around the sites implies that both sheetwash and headward gully extension have been less active than the general terrain would suggest. This indicates that gully erosion with high sediment yields is not always associated with badlands.

Vita-Finzi (1969), in his pioneering study of circum-Mediterranean alluvial chronologies, pointed to the existence in several areas of two depositional units, an Older Fill and a Younger Fill. The Older Fill was thought to predate a period of channel erosion lasting from c. 8000 B.C. until post-Roman times, after which channel deposition again became dominant. Recent evidence (Davidson, 1980; Butzer, 1980) shows that the pattern is more complex. Our data suggest that badlands in the Guadix area of southeast Spain had been formed by 2000 B.C. and that later clearances probably did not significantly alter geomorphological patterns of activity. Possible explanations are discussed towards the end of the paper. In this case neither climate nor human activity need be the crucial variables and landforms are more likely to be a response to long term denudational instability.

A major problem in south-east Spain is the difficulty of establishing contemporary rates of erosion: events are not only of high magnitude and infrequent occurrence, but are also spatially discontinuous and greatly influenced by human activities. An event which at first seems to be effecting an enormous amount of work may recur only once every few centuries, so that the long-term overall process rate would be slow. The evidence presented here is therefore necessarily semi-quantitative and to some extent circumstantial.

Three sites are examined in detail: Cerro del Gallo, Cuesta del Negro, and El Culantrillo, all situated in badlands near Guadix in the province of Granada, south-east Spain. The sites are first placed in their regional geomorphological and archaeological context and then individually considered in more detail.

GEOMORPHOLOGICAL CONTEXT

Throughout south-east Spain there is a sharp juxtaposition of strongly folded metamorphic and crystalline rocks, which form the mountains of the Betic system (*sensu lato*), and of the material laid down in the cuvettes between mountain ridges -- poorly consolidated marls, silts, sands and conglomerates. The Tertiary basin to the north of Guadix lies between the Sierra Nevada to the south (figure 1) and the Sierra Harana and Sierra de Gor to the north-west and north-east respectively. The basin fill comprises marls and sands with conglomerate beds, which through progressive subsidence has reached a thickness of several hundred metres, with a maximum of 3000m (figure 1 and Vera, 1970). The lowest series within the Neogene are strongly folded, but the Middle Miocene deposits are only gently folded. The latter consist of:

1) Sands and conglomerates with calcareous or bioclastic cement which have an abundant marine fauna.

2) Marine marls and/or limestones, which together constitute the greatest thickness of the series.

The Guadix formation of Pliocene age is a conglomerate of dominantly fluvial origin capped by extensive caliche deposits presumed to be of Quaternary age.

Particle size analysis of bed-rock samples from the sites (figure 2) shows that the modal class clearly falls in the silt range and, although the samples show rather poor sorting, the 'tails' are relatively small. Samples 6 and 7 are taken from a terrace and river bed deposit and reflect the fact that most of the grades are easily transported.

The long term denudational history of the basin has been discussed by Birot and Sole Sábaris (1959), who conclude that it was closed until the breaching of the northern divide by the Guadiana Menor in the upper Tertiary. This is thought to have initiated the wave of erosion which dissected the Tertiary basin. Thus, three types of terrain can be distinguished (figure 1):

1) The hills, ridges and mountains of the Betic fold system.

2) The gently sloping, undissected upper surface of the Tertiary fill, capped by a calichified conglomerate. This surface passes marginally onto pediplains cut across folded limestones and metamorphics.

3) The badlands formed by the dissection of the Tertiary fill. Drainage densities here are very high, but the form of the slopes varies from the scarp edge, where available relief is high and many slopes are near-vertical free faces, towards the major channels

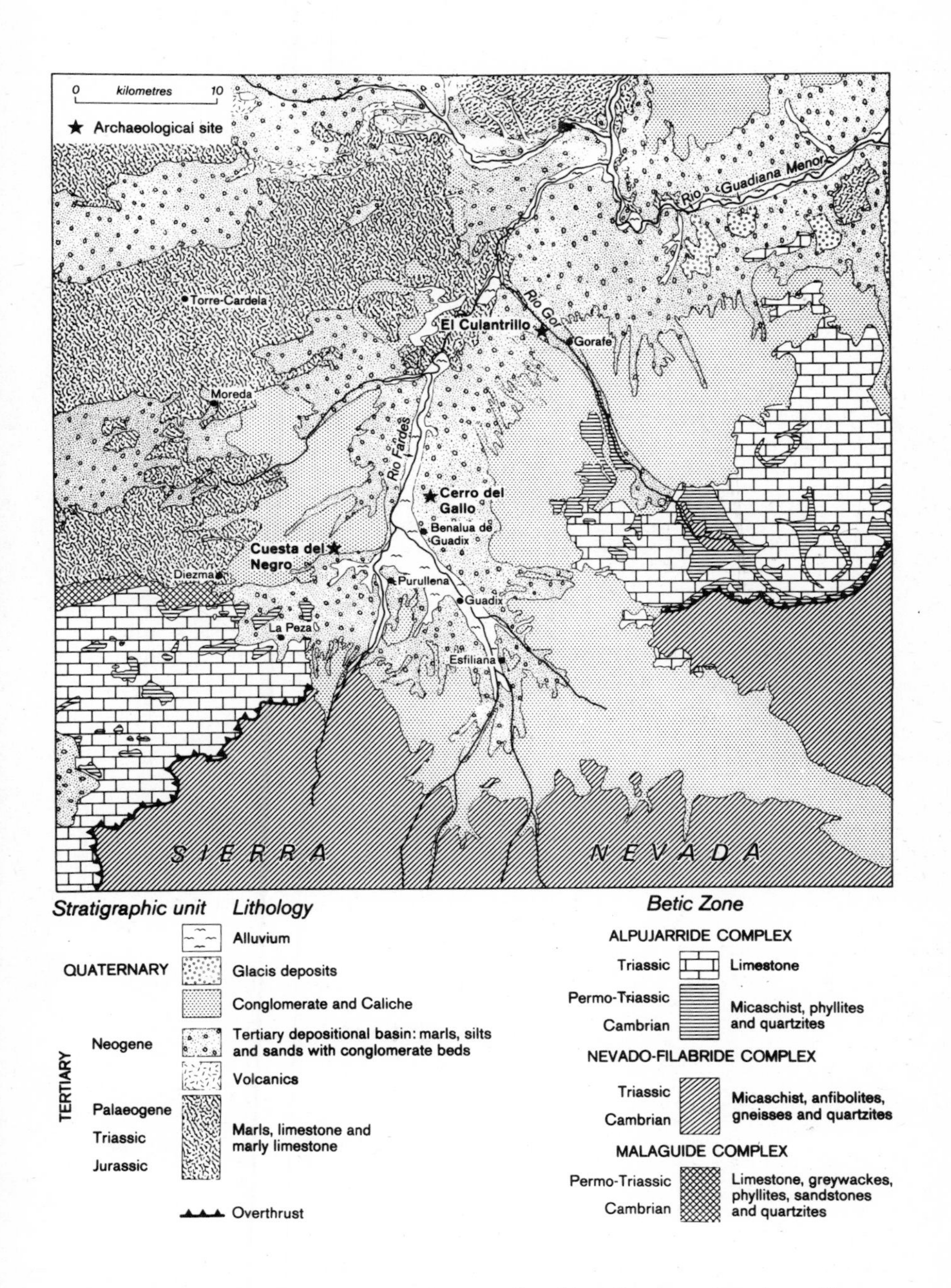

Figure 14.1 Geology Map of the Guadix Basin showing the location of the archaeological sites. Source I.G.M.E. (1970)

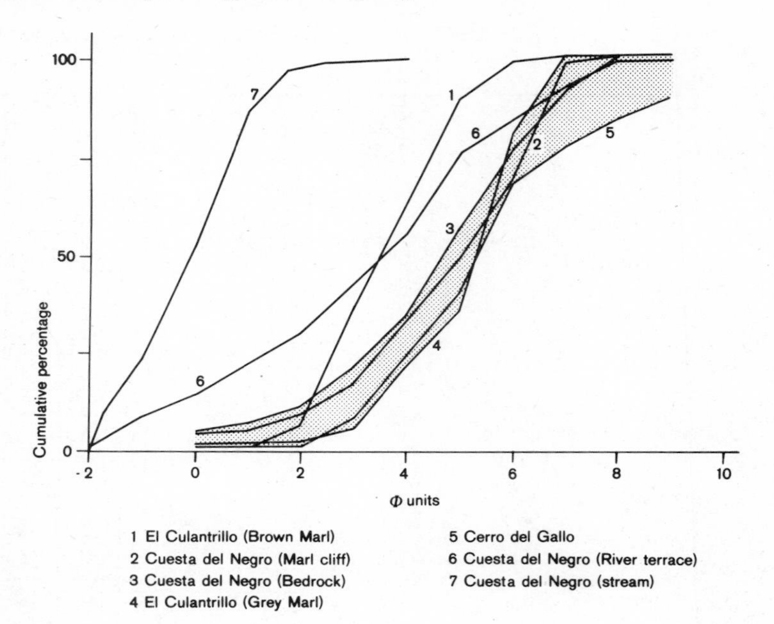

Figure 14.2 Particle size curves of sample of bedrock and stream sediment in the vicinity of the archaeological sites (Analysis carried out by Mr. P. Burrin)

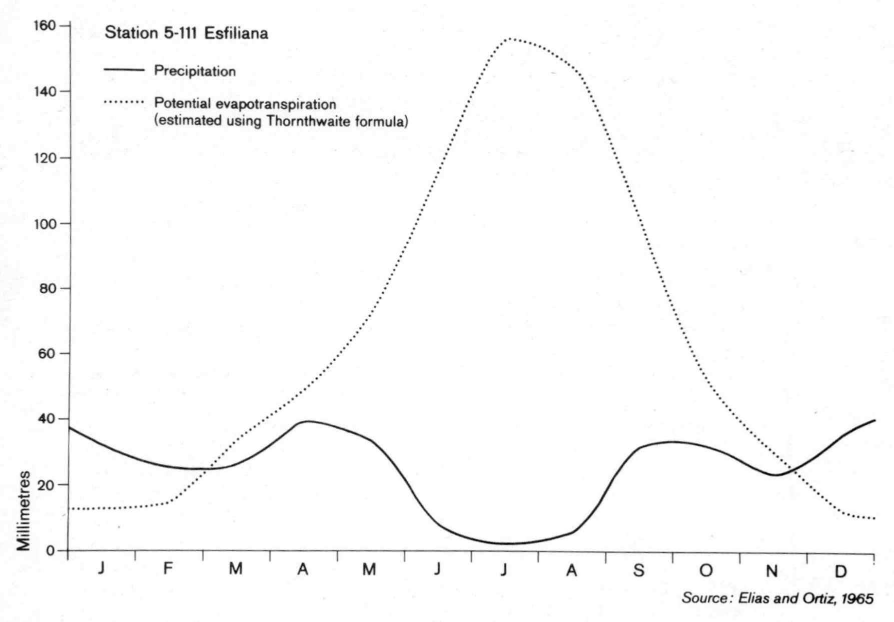

Figure 14.3 Rainfall and Potential Evapotranspiration at Esfiliana

along the valley axes, where dissection has reduced the relief and created more-isolated conical hills. One of the investigation sites is near the escarpment edge (Cuesta del Negro), one is in the middle zone (El Culantrillo) and one near the valley axis (Cerro del Gallo).

The climate of the basin is characterised by low rainfall due to the rain-shadow effect of the Sierra Nevada, with a very sharp increase in totals along the margin. Purullena and Benalua, in the centre of the basin, have mean annual rainfalls of 304 and 290mm respectively, while Torre Cardela and Moreda on the higher western edge have 525 and 643mm. Temperatures show large diurnal and seasonal ranges of about 25°C with severe winter frosts and snow. Summer temperatures in the centre of the basin reach 30-36°C during the daytime and vegetation cover is sparse, so that evaporation rates approach the maximum possible. Figure 3 shows the rainfall and estimated potential evapotranspiration using Thornthwaite's formula, for Esfiliana (Elias and Ortiz, 1963). These figures suggest that about 17.8% of the rainfall is 'effective', which agrees well with the runoff coefficients for the nearest gauging stations, on the Guadiana Menor, at Negratin (14.6%) and Posito (17.1%). 'Effective' rainfall for the basin edges is about 30% of the total. Rainfall intensities can be crudely estimated for stations near each of the archaeological sites: for a 30 min period with a 2.33 year return they are about 18-20 mm hr^{-1}, and for a 100 year return, about 90-150 mm hr^{-1} (I.C.O.N.A., 1979).

These results may be compared with infiltration characteristics for the marls of this area. Saturated conductivities based on 28 ring infiltrometer tests at the sites are about 10 cm hr^{-1} which is relatively high. A more meaningful figure in terms of saturated overland flow is to be derived from the total volumetric storage in the 'soil', which is estimated to be of the order of 28mm. Observations of the wetting front depth are consistent with the saturated conductivity values and these rates will be surpassed relatively infrequently by rainfall intensities, mainly in summer when the soils are very dry and the conductivities are 2-3 orders of magnitude lower. Recent work suggests, however, that saturation overland flow may be more common than Hortonian, even in this environment, so that these transmission figures may underestimate the occurrence of erosive events (Scoging and Thornes, 1980).

Stream sediment load figures are highly tentative, since they are based on 48 monthly 'grab' samples taken from the Guadiana Menor at the Posito gauging station. The Gor and Fardes Rivers, which drain the Guadix Basin, form jointly about half the catchment above Posito. Flow is perennial in both rivers, but is very low in August and very high in December and February. The 53 year record of daily flows for Posito shows strong interannual variations. The mean of annual mean daily flows is 19.09 cumecs, but the standard deviation is 7.47 cumecs and successive years can differ by an order of magnitude. The estimated sediment load is about 16 tonnes km^{-2}yr^{-1} (with upper and lower quartiles of 72 and

7 tonnes km^{-2}yr^{-1}). Although this is quite low, various empirical equations yield comparable results: estimates using the models of Fournier (1960), Carson and Kirkby (1972:219), Langbein and Schumm (1958), and Jansen and Painter (1974) group around a mean of 25 tonnes km^{-2}yr^{-1}. Corrected electrical conductivity readings permit one to estimate the dissolved solids load at 137.5 tonnes km^{-2}yr^{-1}, which indicates that, at present, most of the material is being removed in solution.

ARCHAEOLOGICAL CONTEXT

The major features of the cultural succession in south-east Spain from the Neolithic to the Bronze Age are summarised in Table 1 (see Gilman, 1976, for a more detailed discussion of the cultural evidence). The main outlines were established by Henri and Louis Siret (1887) on the basis of a technological/evolutionary seriation of materials from habitation and burial sites. It is only in the past 25 years, however, that there has been adequate stratigraphic confirmation and that its true dimensions have been established by C-14 determinations. There are still gaps in our knowledge of the sequence and significant regional variations in the cultural assemblages. In spite of its rudimentary character, however, the Neolithic-Bronze Age sequence is a coherent one and the sites around Guadix to which this paper refers can be clearly assigned to its Copper and Bronze Age phases.

The most distinctive features of the Copper Age Millaran Culture are its megalithic burial monuments and their grave goods. At the type site of Los Millares itself the large cemetery of dry-stone-walled corbelled-vaulted passage graves has yielded various utilitarian and ritual items of stone, bone, copper and pottery. The mortuary assemblages vary considerably between tombs, the differences probably reflecting both changes in ritual over time and social differences between the tombs' builders/users. Other megalithic passage grave cemeteries, such as those found in the Guadix area (the Gor/Gorafe, Fonelas, and Los Eriales tomb groups), are distinguished by simpler megalithic construction and a poorer selection of grave goods, but share idiosyncratic ritual items, such as shell bracelets and flat stone figurines, and the basic collective burial ritual. At Los Millares itself, charcoal from tomb 19 gives a C-14 date of 4380 ± 120 b.p., while at El Barranquete (Almagro Gordbea, 1973), tomb 7 yields dates of 4280 ± 130 b.p. (CSIC-81) and 4300 ± 130 b.p. (CSIC-82). C-14 dates are taken from the compilation -- with references -- by Almagro Gorbea and Fernandez-Miranda (1979). All dates are given here in uncalibrated radiocarbon years (b.p.), not solar years (B.P./B.C.).

Settlements are found in the vicinity of megalithic cemeteries and yield typologically similar utilitarian materials: the Los Millares site (Almagro and Arribas, 1963) typifies the juxtaposition, similar associations being found elsewhere. Several Copper Age habitation sites in south-east Spain have yielded reliably associated C-14 determinations: Los Millares - 4295 ± 85 b.p. (H-204/207); El Tarajal - a dozen determinations (CSIC 218-225, 227-230)

Table 14.1 Major Characteristics of the Later Prehistoric Cultural Sequence in Southeast Spain

Period	Early Neolithic	Later Neolithic	Copper Age	Bronze Age
Millenium B.P.	VII	VI	V	IV
Cultural Complex	Impressed Ware	Almerian	Millaran	Argaric
Settlement Type	caves & rock shelters	caves, rock shelters & permanent, open air villages	permanent, open air villages, some fortified	permanent, open air villages, often in extreme defensive positions
Burial Rite	not known	collective, in simple megaliths or in caves	collective in (sometimes elaborate) megaliths	individual in cists & large jars
Grave Goods		ultilitarian & ritual items	utilitarian & ritual items; differential wealth	personal goods (e.g. weapons, jewellery); differential wealth
Subsistence	sheep/goat herded; wheat, barley cultivated	mixed farming: cattle, pig, sheep/goat, domesticated; wheat, barley, legumes cultivated		

cluster around a mean of 4050 b.p. (Almagro Gorbea, 1976); Terrera Ventura (Gusi Jener, 1975) - 4240 ± 60 b.p. (CSIC 264), 4200 ± 60 b.p. (CSIC 265) and 4110 ± 60 b.p. (CSIC 267); Cerro de la Virgen (Schule and Pellicer, 1966) - five determinations (GrN 5593, 5595-5597, 5764) have a mean of 3873 b.p.; and Los Castillejos (Arribas and Molina, 1979) - 3840 ± 35 b.p. (GrN 7257). While a reliable periodization of the Millaran remains to be established, Copper Age settlements and burial monuments fall, broadly, into the third millennium B.C. here as elsewhere in Europe.

The Bronze Age in south-east Spain is represented by the Argaric Culture (Coles and Harding, 1979:216-226), a complex contrasting sharply with the Millaran. Settlements were generally placed in new locations (among well documented sites only Cerro de la Virgen and Los Castellones (Mendoza *et al.*, 1975) have both Copper and Bronze Age occupations), often in extreme defensive positions. Although burial in passage graves was not completely abandoned, the dead were usually buried individually in cists or large jars underneath the floors of settlements. Grave goods now predominantly consist of metal weapons and items of personal finery, such as jewellery and drinking chalices. The amount of grave goods varies greatly between tombs, reflecting class stratification. There are also pervasive changes in artifact styles throughout the material cultural repertoire.

THE ARCHAEOLOGICAL SITES

Topographical and geomorphological maps have been produced to illustrate the configuration of the three sites and the relationship of the archaeological deposits to the nature of the slopes. The topographic maps were drawn up from accurately measured spot heights, contours being interpolated with the aid of air photographs. Base maps for the geomorphological mapping were drawn up from air photographs at approximately 1:15 000, further expanded to give a working scale of approximately 1:2500.

Cuesta del Negro

This site lies on a long spur descending from the flat plain east of Diezma down to the Fardes River. The plain itself, with thick caliche horizons at its surface, represents the almost unbroken top of the Tertiary marls, conglomerates and sands which form the lithologies of the central part of the basin. It is impossible to date the surface, which is diachronous. In parts it has calichified conglomerates representing transport across the surface; elsewhere it transgresses onto limestones (e.g. around Diezma) and has bright red weathering products. There are no clearly defined channels on the surface; shallow depressions with smooth rounded cross-sections occur nearer the escarpment, where they are controlled in their descent to the Fardes by the many sandstone and conglomerate lenses found in the series. On weathering the conglomerates produce coarse, well-rounded debris in the granular and gravel grades; these contrast sharply with the silty-sands and calcareous marls which interdigitate with the hard lenses. The drainage pattern at the edge is dominated by those channels which have captured some of the drainage on the upper surface and hence obtained a competitive advantage. Vegetation is very sparse due largely to goat-herding rather than to slope instability.

The site itself has three distinct parts: a large mound of settlement debris at the upper end of the spur ('tumulus' on figure 4), an upper excavated area (Sites E and F) and a lower excavated area (Sites A-D). The upper area is on a rounded knoll with steep slopes to the south which also have *in situ* structures. The lower area is a shallow embayment crossed by a depression with house structures on each side. Steep gullies and evident signs of erosion occur on this south side. The site was excavated by the Department of Prehistory, University of Granada, in 1971 and 1972. Molina Gonzalez and Pareja Lopez (1975) have published part of the results from 1971 (Site A, figure 4) which indicate two main phases of occupation of the site. The earlier consists of

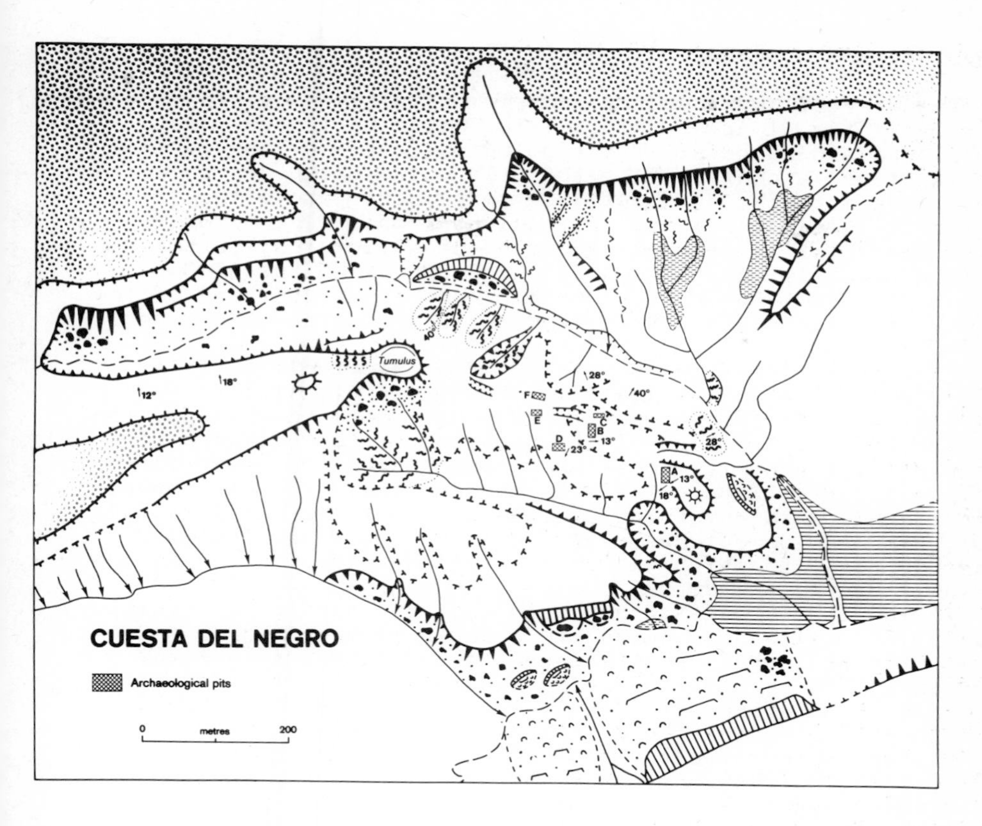

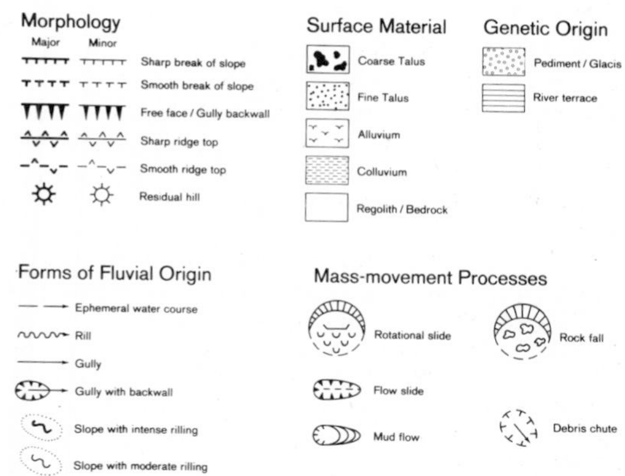

Figure 14.4 Geomorphological map of Cuesta del Negro (see key for symbols, which also apply to Figs. 5 and 7). The free faces shown are formed in the conglomerate beds which are contained within the marl, and which largely define the slope form. The full extent of archaeological deposit is not known.
Field mapping for all three geomorphological maps carried out by S. Wise in January 1980

hearths, settlement debris and pit burials, and yields artifacts of the El Argar culture, including carinated vases, chalices and daggers. The typological date for the earlier occupation is confirmed by the C-14 dates cited above, which date the material to about 4000 years B.P. Above the Argaric levels is an occupational horizon of late Bronze Age character, with rectangular stone structures, ceramics dated by excision and C-14 dates of 3095 ± 35 and 3160 ± 35 b.p. (GrN 7284-7285). The large mound at the head of the spur has also yielded Argaric materials (Carrasco Ruz, pers. comm.).

On the slopes the remaining archaeological debris is in places over 2m thick and is structured parallel to the slope, and although some materials have moved downslope, the *in situ* structures show that the shape of the hill and its superficial deposits and slopes have changed very little since Argaric times. The main changes have evidently been in the headward extension of some south-facing gullies. Much of this development has occurred along the line of a road across the site which was used until the mid-30's and is now significantly dissected.

The two streams adjacent to the site have well developed alluvial fills, now incised to a depth of about three metres, whose upper surface continues on to the Fardes as a river terrace. These comprise buff coloured silts with a slightly coarser mean grain size than the marls forming the bedrock (figure 2). No terrace dating has been carried out but their position with respect to the site suggests that they post-date it. They may represent sustained, slow aggradation or a single season of heavy storm deposition which was subsequently incised.

El Culantrillo

This site lies north-west of Gorafe and occupies an intermediate position between the escarpment edge of the dissected infill and the River Gor. The bedrock is marked by strong sandstone lenses which separate thicker bands of marls, which appear to be much more highly cemented, and the sandstone bands are further apart than at Cuesta del Negro. As a result they can sustain deep canyons with vertical walls 15-20m high. The site itself (figure 5) is a broad interfluve on either side of which the streams have cut deep v-shaped notches which rise to near vertical sandstone cliffs. The site is separated from the upper spur by a low col.

The site was excavated by Dr. M. García Sanchez in 1955 and published in 1963. The main dating points are provided by 12 burial pits yielding typical Argaric materials including chalices and daggers (García Sanchez, 1963) and typological dating indicates therefore that the site was occupied 4000-3500 years ago. The burials, containing *in situ* materials, extend down the entire spur. Tomb 1, which yielded three Argaric chalices, is particularly important. It is located on the steep southeast slope of the spur, below the point where the top ends in a sharp descent to the Rambla del Agua (below the 15m contour in figure 6). Since the tomb can only have been dug into the existing slope, it provides a *terminus ante quem* for the general configuration of the spurs.

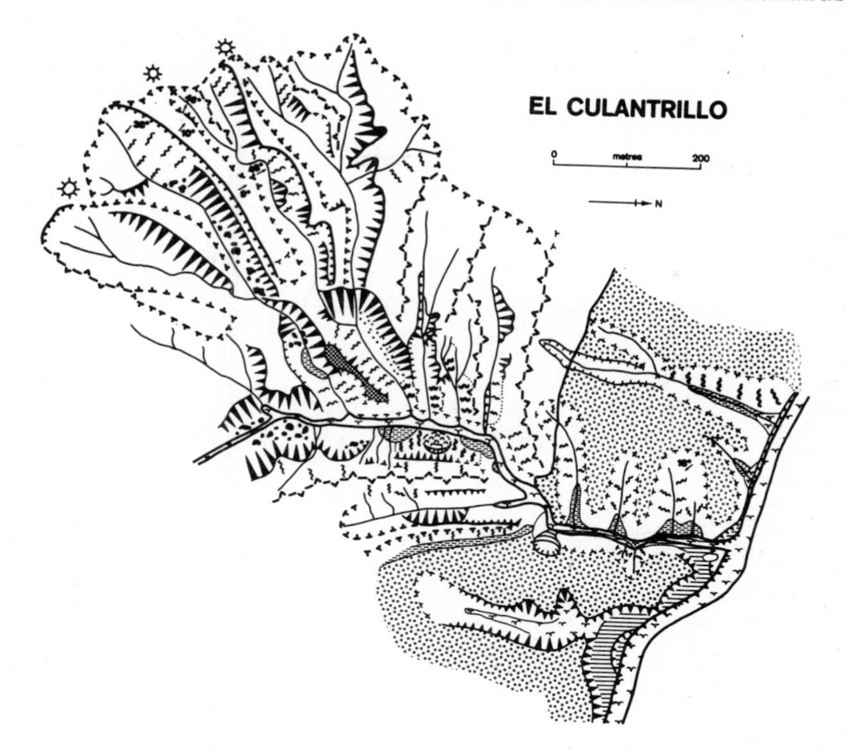

Figure 14.5 Geomorphological map of El Culantrillo. The approximate extent of the site is shown by cross-hatching. The small ridge on which it lies is shown in more detail in Figure 6. The pediment shown here is not the top of the Tertiary fill as in Fig. 4, but a pediment developed at the base of the badland slopes and cut in marl

The major stream at the foot of the slope has a small terrace and evidence of slight alluvial filling followed by entrenchment. It passes through a small gap and then joins the river Gor which has a strongly developed terrace 6m above the river which can easily be traced both upstream and downstream. At the junction of the stream draining the site and the river Gor, however, the two channels both cut into folded and steeply dipping bedrock. This suggests that there was slight incision at or before the site occupation, but little change since. As at Cuesta del Negro the bases of the nearly vertical marl cliffs are buried in talus suggesting that one of the main processes and principal sources of material is mass movement.

On the escarpment south of El Culantrillo (plate 1) is a fine series of megalithic tombs stretching along both sides of the Gor river between Gor and Gorafe and north to El Culantrillo (García Sanchez and Spahni, 1959). Typologically datable grave goods are scarce, but the tombs' contents

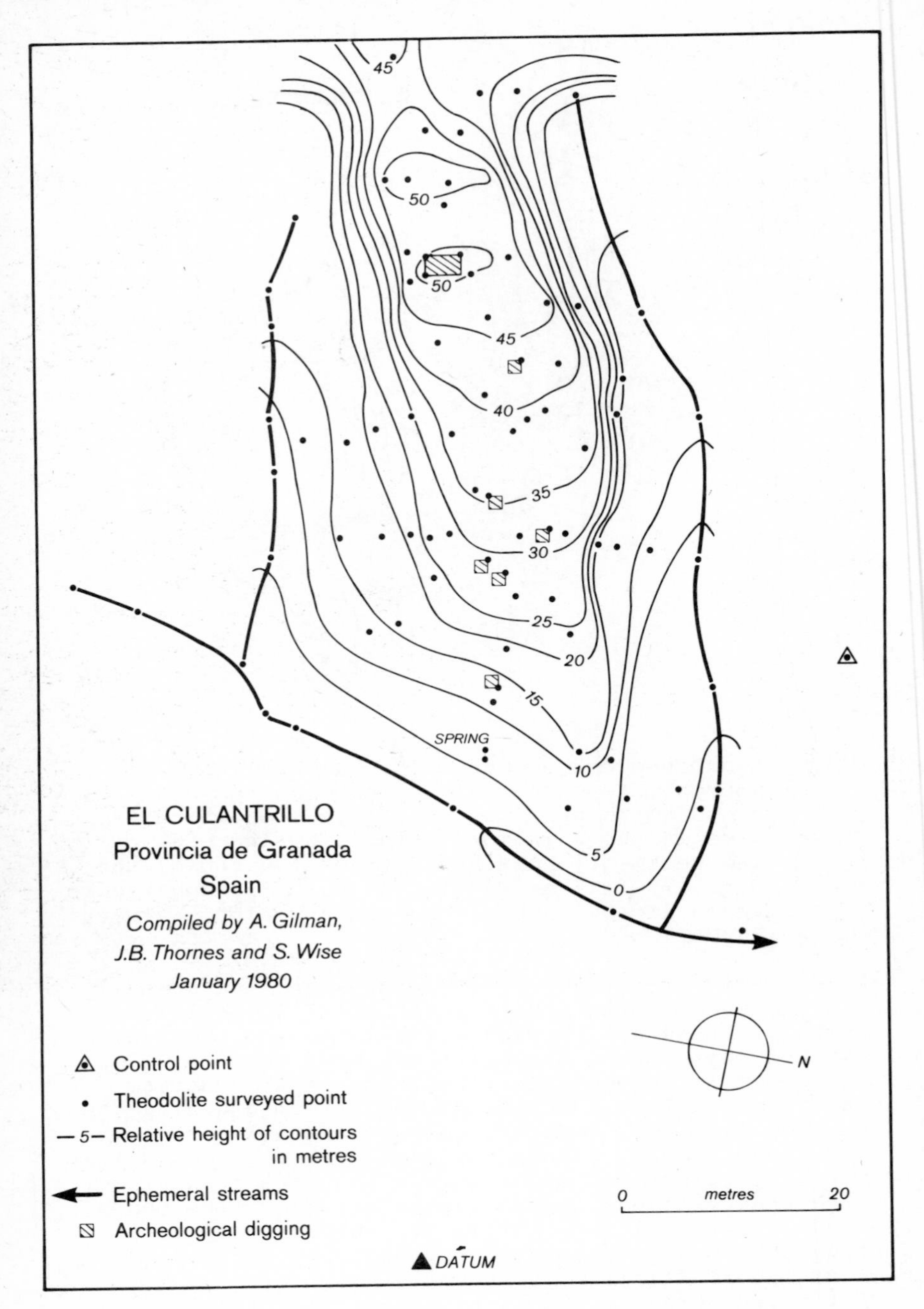

Figure 14.6 Topographical map of El Culantrillo

Plate 14.1 Air photograph showing the area around El Culantrillo (north is at the top of the photograph). The junction between the Gor and Fardes rivers is just to the north of this picture. The flat land in the south-east corner is the upper surface of the Tertiary material, which has been dissected to form the badlands as shown here. (Reproduced by permission of the Servicio Geografico del Ejercito.)

(e.g. shell bracelets) certainly indicate use in Copper Age times, i.e. as much as 5000 years ago. Many of the 198 tombs are located on the edge of the escarpment, an apparently intentional alignment which suggests that the upper parts of these precipitous slopes have remained stationary for five millenia. This is confirmed by megaliths found a few metres beyond and below the edge. Tombs 14 and 15 near the

Figure 14.7 Geomorphological map of Cerro del Gallo

settlement of La Sabina are located on slopes of 27-32°. The morphology here is complex, the streams running nearly parallel to the edge, but the stability of the escarpment since Argaric times at this location cannot be doubted.

Cerro del Gallo

This, the largest site, occupies a position near the centre of the basin about 5 km north of the village of Benalua de Guadix at the junction of the San Torcuato and Chamorro barrancos, which cut through the marl badlands from the Hoya de Guadix to the Fardes River (figures 7 and 8). It occupies several hectares on top of a broad interfluve running in a roughly east-west direction. The lithologies here are largely marls and silty clays (figure 2). Higher surrounding hills, such as that immediately to the west beyond the barranco Chamorro, are capped with calichified conglomerates. Although some thin caliche layers occur on the site there is virtually no coarse material except at the western edge.

There is a marked contrast between the north-facing and south-facing slopes on many of the ridges in the area, and at the site itself. On the south-facing slopes there is very little vegetation cover and slopes are intensively

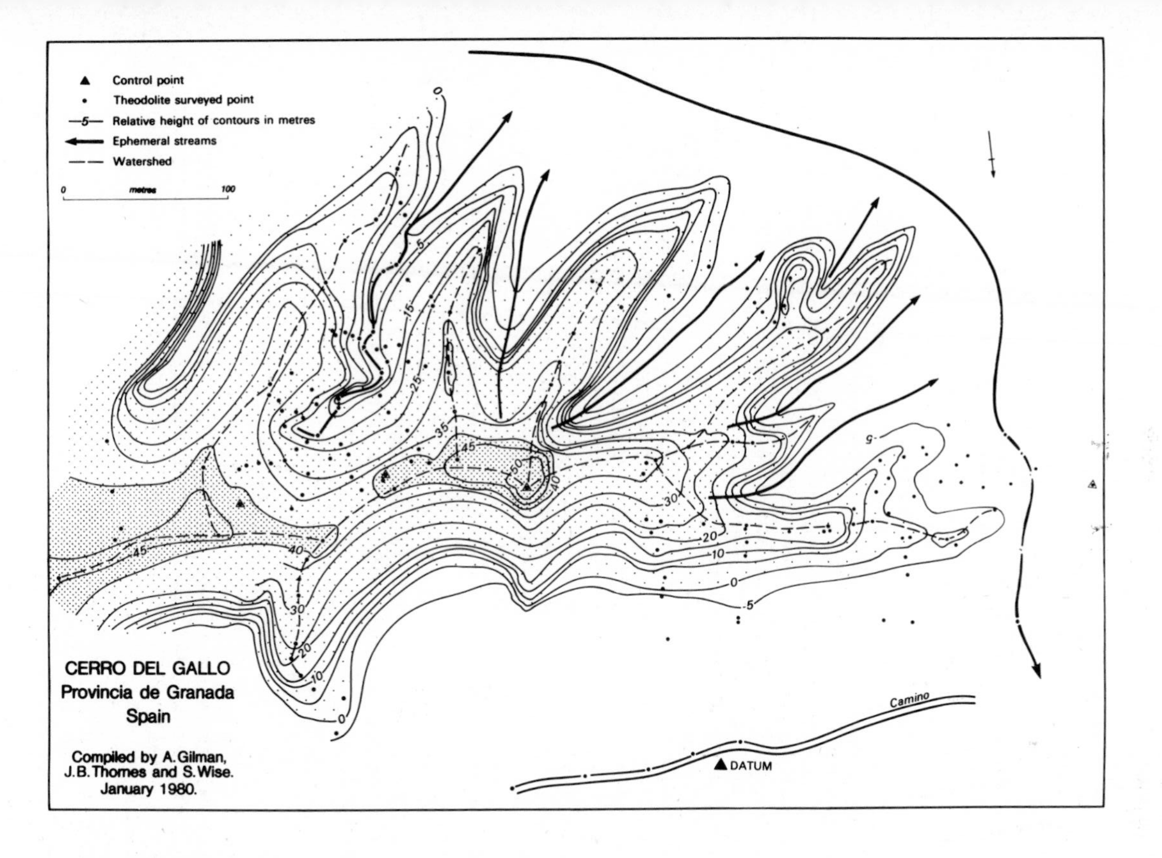

Figure 14.8 Topographical map of Cerro del Gallo

rilled and cut by deep gullies (figure 6). The north-facing slopes have an almost complete grass cover and an organic soil, and show almost no evidence of surface erosion. These slopes are also steep, due to arcuate failures along the base of the slope (figure 7). This contrast has been noted in other semi-arid areas (Strahler, 1950; Melton, 1960) and can probably be attributed to differences in the microclimate.

Cerro del Gallo has been the object of unauthorised excavations by amateur archaeologists. Torre Peña and Aguayo de Hoyos (1976) have published a general description of the locality and some of the artifacts found. The artifacts, which include chalices, carinated vessels and archer's bracelets, show that the site belongs to the El Argar culture.

Archaeological material is distributed over the entire interfluve (figure 7). The large embayment draining south between the two hilltops on the ridge contains thick deposits *in situ* (possibly 2-3m) indicating the existence of this embayment in Argaric times. This is being eroded by a steep-walled gully, which is currently removing *in situ* archaeological material. Hence there is clear evidence at this site that erosion has taken place since the site was occupied. This is localised along the drainage lines, however, and *in situ* deposits remain on many steep slopes. For example, on the western side of this gully (asterisk, figure 8) we observed the intact mouth of a hand-made (and therefore presumably prehistoric) pottery vessel exposed by a rill at a depth of 40cm. If this had been transported from above by erosion, it is unlikely that it would have retained a complete, unbroken orifice and so it is presumably *in situ*.

The footslopes of the steep spurs surrounding the site also show quite well developed gently sloping (c. 9°) glacis. These extend laterally into the gullies which penetrate the south and south-west facing slopes, and eventually merge with the fans from these gullies, passing without a sharp break onto the alluvial fill. Some of this deposition is certainly a recent phenomenon, but no dating has yet been attempted.

The following conclusions may be drawn from the evidence presented:

a) the existence and location of *in situ* archaeological deposits indicate that the drainage pattern and general slope configuration have remained unaltered since the sites were occupied, or over the last 4000 years.

b) all sites show clear evidence of erosion during this time, but this is localised along drainage lines and results from mass movement or headward gully extension, rather than overland flow.

c) erosion does not seem to be proceeding at a catastrophic rate, as survival of any archaeological deposit would then be unlikely.

DISCUSSION

At this stage it may be useful to summarise the evidence. First, there is the visual appearance of the Guadix Basin spectacular badland topography with high drainage densities and steep valley side slopes which are only sparsely vegetated. This, coupled with the semi-arid climate, leads one to expect very rapid rates of erosion. The estimates of erosion rates and the limited available sediment load data, however, indicate rates of erosion which are of the order of only 25 t $km^{-2}yr^{-1}$, which is moderate in global terms (Fournier, 1960). The *in situ* archaeological deposits also imply greater slope stability and less rapid erosion than the visual evidence suggests.

The visual impression is partially misleading because of the sparse vegetation cover resulting from a harsh climate and intensive grazing. Superficially, the apparent contrast between the 'hard' Palaeozoic and Mesozoic rocks and the 'soft' Tertiaries reinforces this impression, although closer inspection reveals that the marls are well cemented and capable of maintaining near-vertical slopes. On the other hand extremely high drainage densities (e.g. km/km^2) and steep slopes, often exceeding 30°, are difficult to reconcile with the low erosion rates.

The simplest explanation is that the erosion rates obtained are incorrect and should be higher. The values recorded are few in number, unreliable in quality and of short duration. Moreover, human activities, such as the construction of earth banks across ramblas to trap floodwater and sediments, may have suppressed the figures below those naturally occurring. (At the same time, the recent widespread neglect of terracing attendant upon the reduction of the rural population over the past 25 years may have promoted unusually *high* erosion rates in the period recorded at the Guadiana Menor gauging station.) The estimates based on published models agree very well, the total range being 15.8-40 t $km^{-2}yr^{-1}$, but this may simply reflect use of the same basic parameters in all models. The models are also inadequate as the parameters used are long-term climatic means, whereas in this area the large variations which occur in the short term are probably more important.

A second explanation is that rates were much higher in the past due to climatic changes. The theoretical models of Langbein and Schumm (1958) and Carson and Kirkby (1972) are very sensitive to small changes in values of the climatic parameters under semi-arid conditions. However, the limited evidence that exists implies relatively stable climatic conditions over the last few thousand years. The argument that widespread erosion was a result of forest clearance and similar early agricultural activities is incompatible with the archaeological evidence (see discussion in Davidson, 1980). This shows that more or less the present pattern of slopes, gullies and channels was in existence and that erosion rates during the last 4000 years in much of the landscape have not been very high.

Some support for this view may be obtained from the geomorphological maps, which show that in these areas mass-movements critically influence landform development through high relief and high drainage density. There is, at the end of the winter season, abundant evidence of high seepage rates and high pore water pressures near the valley floors, presumably due to the high infiltration capacities and the lack of piping. Near the escarpment (at Cuesta del Negro) and, to a lesser degree, in the middle zone at El Culantrillo, large slope failures on the channel boundary are the dominant forms. They may leave irregular debris piles at the foot of the slope or not, according to the efficacy of the stream undercutting and transporting activity. Towards the centre of the basin, at Cerro del Gallo, for example, relief is subdued and characteristic convex hillslopes with impeded basal removal occur and microclimatically induced assymetry is the dominant control of form. Even here the signs of erosional activity in the channels are much greater than on the interfluve slopes, implying a relatively weak coupling of the hillslope and channel morphologies. The dominance of mass-movement over wash processes contrasts strongly with the observations of Leopold *et al.*, (1966). Both processes are probably of small importance compared with solutional processes.

It is tempting to resort to lithological controls to account for the insensitivity of the interfluve areas. The strong cementing of the silts by calcium carbonate leaves a rock-like parent material which helps to control relief. Lithology also produces a strong capping effect of resistant sandstone lenses within the marls and of the strongly developed caliche horizons on the upper surface.

The most likely cause of these stability contrasts, however, probably lies in the longer term denudational history. Breaching of the endorreic basin in Tertiary times coupled with progressive uplift may have led to outward propagation of the network from the principal channels by headward growth. Parker (1977) has shown experimentally that this tends to occur by addition of first-order channels when there is continued basal incision. The contemporary restriction of erosion to minor extensions in headwater areas, the development of fills with only slight incision along the main valley axes and the relative atrophy of forms near the centre of the basin appear consistent with this hypothesis. Even if one assumes a pre-existing network which was subsequently entrenched in the headwaters, with complex response downstream, the incision of the high density network could lead to convex slopes with basal incision and removal. These conditions are considered conducive to slopes which are stable to run-off events and unlikely to be dissected (Smith and Bretherton, 1972; Kirkby, 1980). This explanation is consistent with the uplift which is believed to have occurred since Pliocene times (Vera, 1970). It is also consistent with the observation elsewhere in south-east Spain that the impact of the huge 1973 flood was predominantly on the channel margins, through landsliding, or in the channels themselves in upland areas (Thornes, 1976), suggesting poor hillslope-channel coupling. The main significance of high-magnitude, low-frequency events lies in removing the debris from collapse. The evolution of these badlands is only to be understood by considering the longer tectonic and denuda-

..ional history. Perhaps some other badland areas in south-east Spain should also be examined in the light of this proposition, for although in some areas reservoir sedimentation indicates high levels of erosion over the last century or so (e.g. Lopez-Bermudez, 1973), the historical empirical record alone will never be long enough to resolve the controls.

CONCLUSIONS

In the Guadix Basin there is a clear inconsistency between the high erosion rates which one would expect from the general visual appearance of the landscape and the persistence of appreciable interfluve areas untouched by erosion for 4000 years. This is accounted for by sharp variations in the slope form stability and the sensitivity of response in different areas of the basin, which are best attributed to tectonic activity and the long term history of denudation which together have caused continuing incision along the drainage lines since the Pliocene. As a result, interfluve areas have developed much more slowly and overall denudation rates are lower than one might expect on the basis of the prevailing climate. If erosion is intermittent, it is restricted to a few gully heads and parts of the major channels. Further field evidence must determine whether this conclusion may be applied to other instances of badlands in areas of tectonic uplifting. The results obtained in this work lead to somewhat similar conclusions to those of Yair *et al.*, (1980) concerning the rate of operation of processes in some badland areas.

Acknowledgements

We wish to thank Dr. Javier Carrasco Rus and Dr. Manuel García Sánchez for information useful in the preparation of this paper. The 'Prehistoric Land Use in Southeast Spain' project, of which this paper is a contribution, has been funded by the Tinker Foundation, the National Endowment for the Humanities, the Fundación Juan March, the Sociedad de Estudios y Publicaciones, the Fundación Universitaria Española, Northridge Archaeological Research Center, the Institute for Social and Behavioural Sciences (California State University-Northridge), and the California State University Foundation, Northridge.

15 Long term denudation rates in the Zin-Havarim badlands, northern Negev, Israel

A. Yair, P. Goldberg and B. Brimer

INTRODUCTION

Badland areas, widespread in semi arid and arid areas, are characterized by high to very high drainage densities (Schumm 1956a; Smith, 1958) with steep slopes devoid of vegetation. These features led to the assumption that badlands topography is indicative of low surface permeability entailing high runoff and erosion rates. The above assumption seems to be confirmed by short term experimental field research conducted in various areas where denudation rates of 1cm/year or more are reported (Schumm, 1956b, Hadley & Schumm, 1961, Campbell 1970b, 1974, Gerson, 1977 and Lam, 1977).

Furthermore some badland areas are relatively young on a geological time scale. The Steveville badlands in Alberta, Canada (Campbell, 1974) developed some 12-14 000 years B.P., after the last glacial period. The Mt. Sdom badlands, in the Dead Sea area, are carved in a diapir ridge uplifted some 20 000 years B.P. (Gerson, 1972). Schuldenrein and Goldberg (in press) report the development of badlands in the Jordan Valley at post late Neolithic time - ca 6000 years B.P.

However the generalisation of the high vulnerability of badland areas to runoff erosion seems questionable. Schumm (1956b) in his pioneer study conducted in South Dakota obtained a high permeability for the Chadron shales and consequently discarded runoff erosion as an important geomorphic process on this bedrock. The relative high permeability of some badland areas is also supported by the extensive occurrence of subsurface flow and piping (Mears, 1963, Parker, 1964, Heede, 1971, Barendregt and Ongley, 1977, Bryan *et al.*, 1978). More recently sprinkling experiments conducted in the Zin valley badlands in Israel (Yair *et al.*, 1980a), indicate that runoff generation in the study area must be infrequent under current climatic conditions. When runoff occurs its rate and yield are very low resulting in very low denudation rates varying from 0.03mm to 0.48mm during rainstorms highly extreme in their intensity and duration (Kutiel, 1978).

Badland Geomorphology and Piping

The discrepancy between the results obtained with the characteristic high drainage density of the Zin badlands, which represent a basically fluviatile landscape, raises the problem of the origin of these badlands, their age and the conditions under which they were formed and their average long term rate of denudation.

The present work attempts to elucidate the problem of the long term rate of denudation through a detailed palaeogeographic study of the area. Such a study is made possible by the occurrence, within the study area and in its vicinity, of prehistoric sites and fluviatile terraces whose stratigraphic, geomorphic and temporal setting are relatively well known.

THE STUDY AREA

The study area is located in the northern Negev highlands (Fig. 1) at an elevation of ca 500m. Average annual rainfall is about 90mm with extreme values of 34 and 167mm. The research site extends over the Havarim valley and the canyon box of the Zin valley into which the former valley drains. The Zin valley, with its area of 1540km^2, is one of the major arteries draining the northern Negev eastwards to the Dead Sea.

The Havarim valley, draining an area of 8.4km^2 is limited to the N.W. by the topographic and anticlinal Halukim ridge where Cenomarian and Turonian rock units crop out. (Fig. 2) These units, assymetrically folded, dip steeply to the S.E. where they are capped by the synclinal Avedat plateau where subhorizontal limestone and chalks are underlain by marls, soft shales and flint of the Paleocene and Campanian (Bentor, 1966). The Avedat plateau, deeply incised by the Zin valley ends to the north by a steep escarpment forming the southern limit of the Havarim drainage basin. North of the Zin valley and to the east of the Halukim ridge extends the flat Sde Boqer plain where a conglomerate of presumed Neogene age (Garfunkel, 1978) lies unconformably on top of the underlying shales. Remnants of the same conglomerate are found on the Avedat plateau itself. The Zin and Havarim valleys being incised below the conglomerate can therefore be considered as mainly post Neogene in age, their downcutting being related to the existence of a low base level in the Dead Sea area.

A strong assymetry characterizes the Havarim drainage basin (Fig. 1). The part extending north from the main channel covers some 70% of the whole basin. It has developed in limestone and other relatively resistant rocks. Its drainage density is relatively low. South of the main channel, where soft chalk and shales outcrop, a typical badland, rugged topography, has developed.

Fluviatile terraces and Prehistoric sites

Several levels of fluviatile terraces were recognized by Goldberg (1976) in the Zin and some of the adjacent valleys to the east. However the correlation of these terraces westward, with terraces remnants observed in the Havarim valley

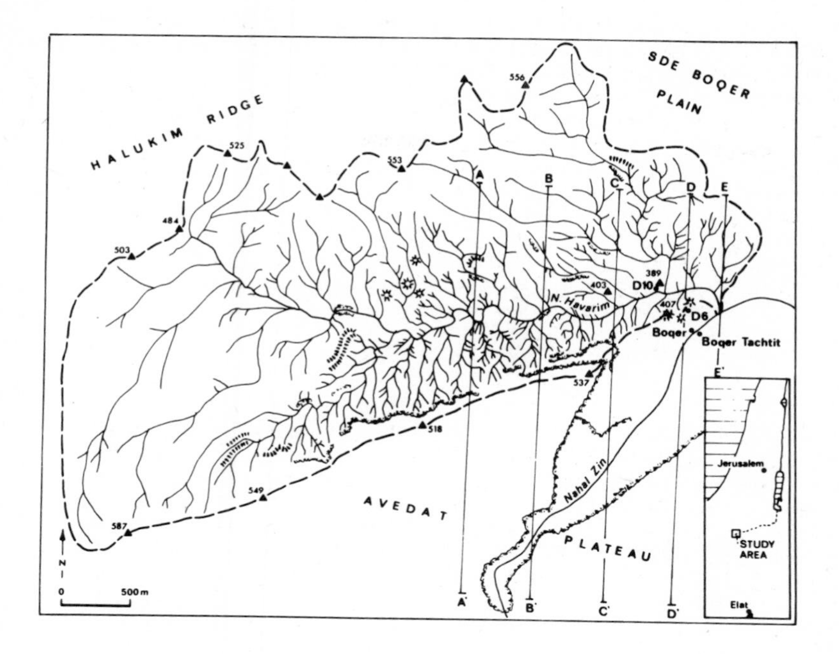

Figure 15.1 The study area

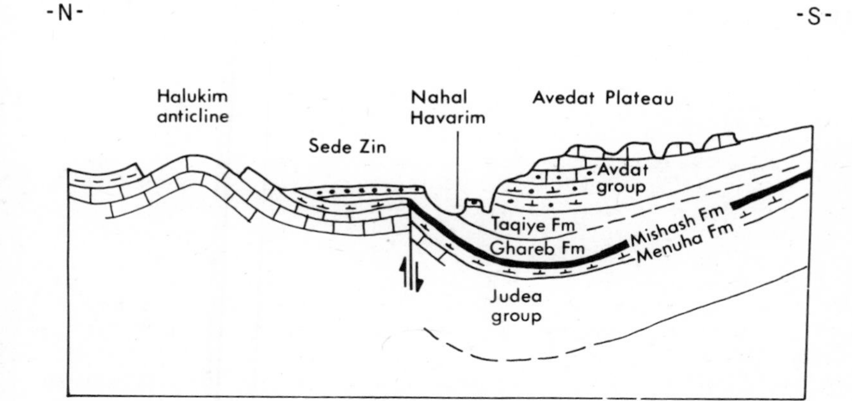

Figure 15.2 Geological setting of the study area (After Garfunkel, 1978)

(Yair *et. al.*, 1980b), was never conducted. The present study is primarily based on a systematic mapping of terrace remnants within the Havarim valley and their correlation with the dated terraces of the Zin system. The mapping was conducted in the field using air photographs at a scale of 1:1000. With the aid of a parallax bar and stereopairs topographic transects were constructed across the Zin and Havarim Valleys (Fig. 3). The accuracy of the elevation measurements was checked in the field and is estimated to be ± 75 cm.

On the basis of the altimetric measurements and geomorphological and sedimentological evidence three distinct levels of fluviatile terraces were recognized in the Zin-Havarim valleys.

The upper terrace

Remnants of an upper terrace are found on both sides of the Zin channel within its canyon like section as well as downstream. These lie at an elevation of 400-410m i.e. some 30-40m above the present channel. Some of the remnants preserve huge blocks of a well cemented conglomerate composed of well rounded limestone and dolomitic pebbles (Plate 1). This terrace can be followed into the Havarim valley along a stretch of 2km (Figs. 4,5). The alluvial fill, 1-2m thick, lies here directly on top of the underlying shales; and is composed of a mixture of fine and coarse gravels and sands

Plate 15.1 View of the fluviatile terraces in the Zin Valley

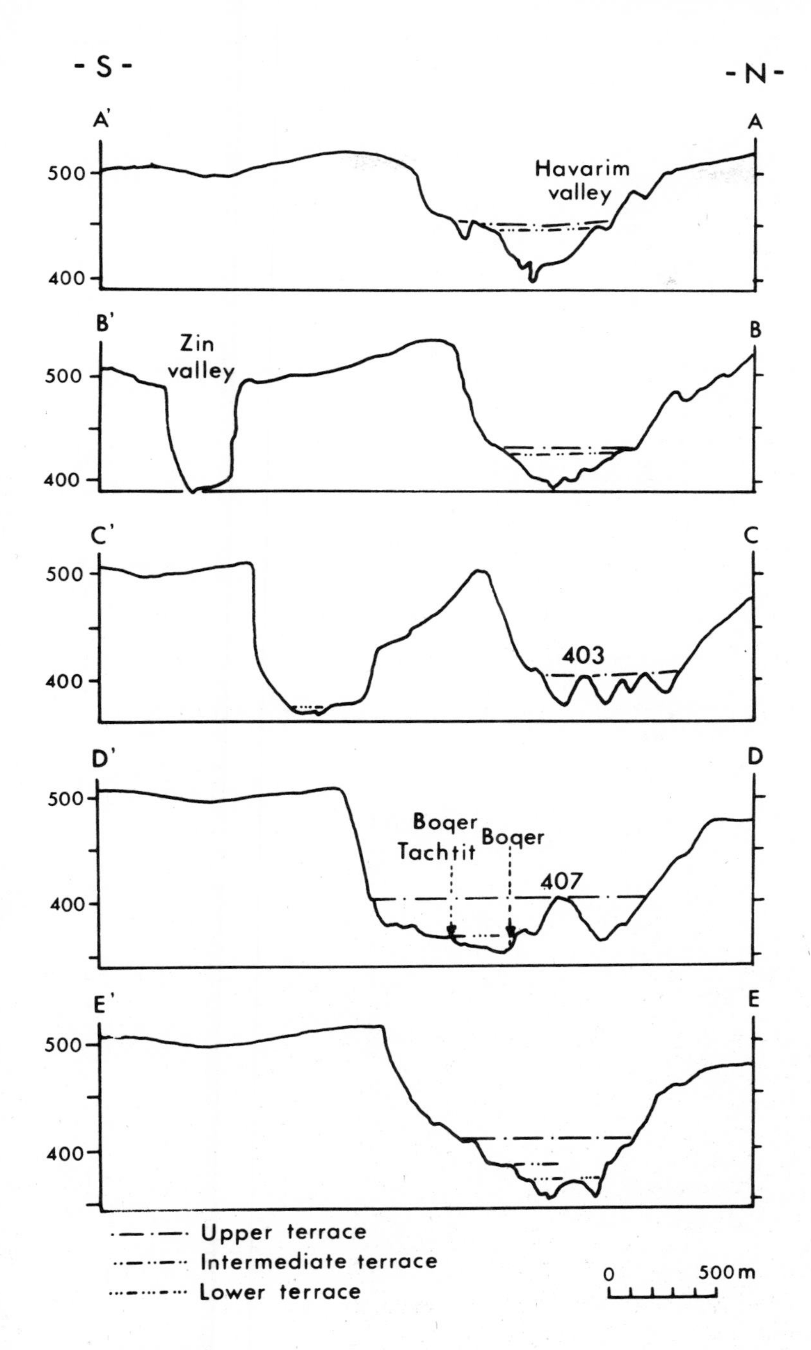

Figure 15.3 Topographic transects across the Havarim and Zin valleys

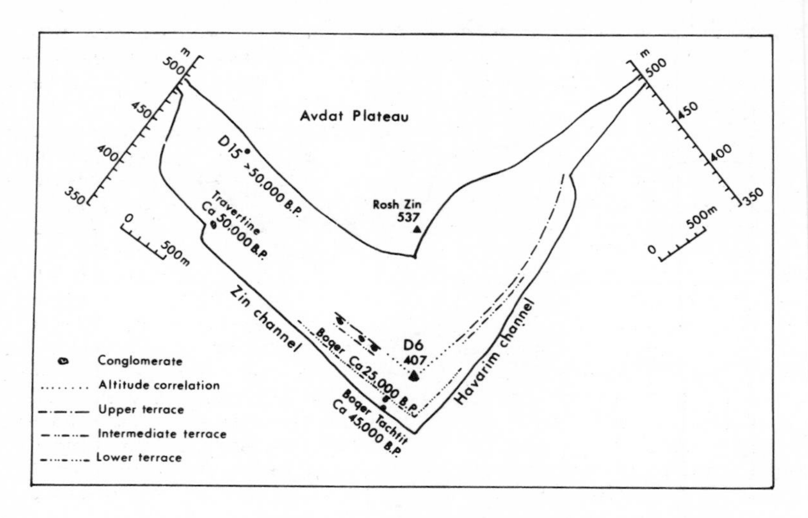

Figure 15.4 Altimetric correlation of the Zin-Havarim terraces

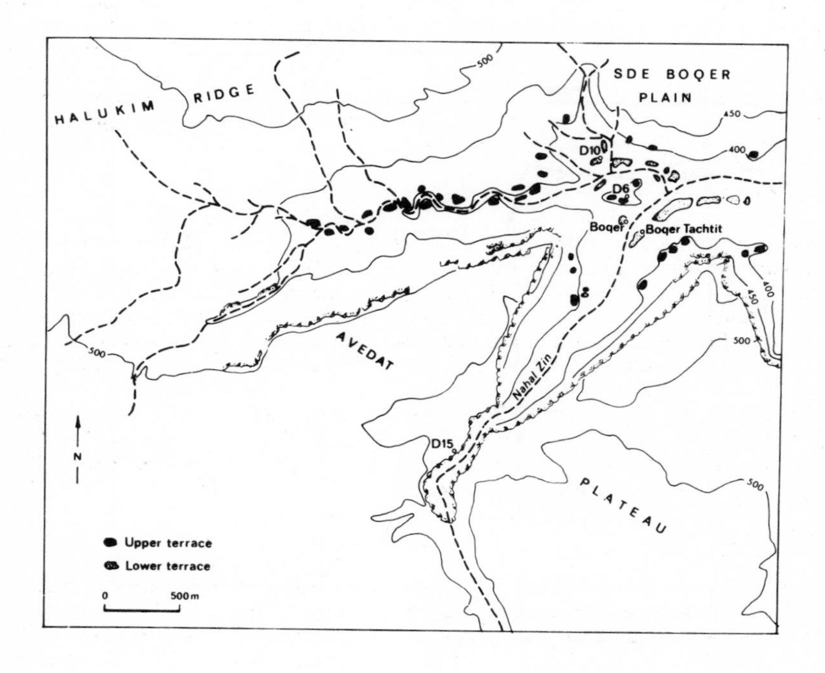

Figure 15.5 Spatial distribution of the upper and lower terraces in the study area

weathered in their upper part. No absolute dating exists for the upper terrace but its age is indicated by the prehistoric site of D6 located in the vicinity of the Junction of the Havarim and Zin valleys (Fig. 5). The site occupies a small detached hillock at an elevation of 407m, 40m above the present Zin channel. The top of the hill is capped by large and massive blocks of the well indurated conglomerate and is strewn by a large concentration of flints of early Mousterian industry (Marks *et al.*, 1971) whose age is estimated at 90 000 - 70 000 years B.P. (Goldberg, in press) A second Mousterian site - D10 - was identified within the Havarim valley at an elevation of 390m, some 400m north of site D6 (Marks *et. al.*, 1971). Field evidence (Fig. 5) suggests that the conglomeratic unit cropping out at the top of site D6 was formerly more widespread and filled the canyon box up to an elevation of 35m above the present channel. The Early Mousterian conglomerate was traced also for about 3km downstream to the east, where at the exit of the Aqev valley it links up with a similar conglomerate well dated by the Early Mousterian site of D.35 (Goldberg, in press). One can therefore imagine that by the end of the Early Mousterian period a widespread conglomerate covered the bottom of the Havarim and Zin valleys preventing a quick incision of the drainage system and therefore the formation of a badlands topography in the shales underlain by the conglomerate. A similar phenomenon probably characterized the steep slopes at the northern rim of the Avedat plateau where extensive scree mantles - partly well preserved up to the present composed of detached blocks of Eocene limestone-may have formed an effective caprock unit that prevented quick erosion of the underlying shales.

The intermediate terrace

A second terrace level parallel to that of the Mousterian conglomerate was identified within the Zin canyon and the Havarim valley. Remnants of this level are missing in the Junction area of the two valleys. (Fig. 5). The intermediate terrace is 5m below the Mousterian level and is characterized by a very thin veneer of unconsolidated gravels overlying the marly and shaly formations. This level indicates a standstill phase in the process of incision of the drainage network that took place after the early Mousterian. The intermediate terrace is devoid of prehistoric sites and its age, for the time being, unknown.

The lower terrace

Remnants of the lower terrace are limited to the lower parts of the Havarim and Zin valleys (Fig. 5). Very well defined terrace remnants flank the Zin valley some 150-200m east of site D6. The tread of the terrace is 12m above the present Zin channel which is incised to a depth of ca 2m into the bedrock (Fig. 6). The alluvial fill, 10m thick, is composed of interbedded calcareous silts sands and gravels. This terrace can be clearly followed for some distance into the Havarim valley where the depth of the accumulated material is 3-5m.

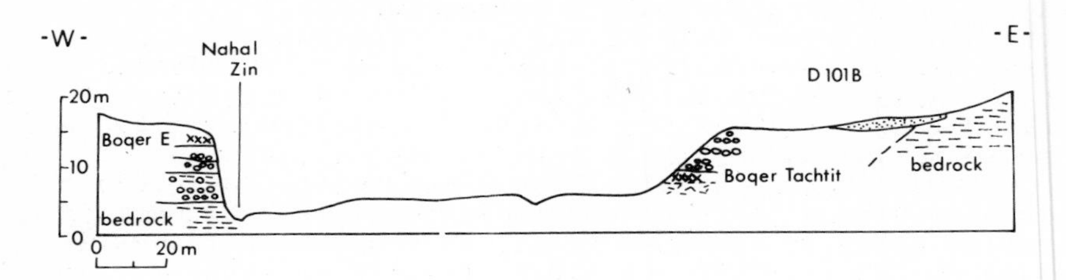

Figure 15.6 Transect across the Zin valley between prehistoric sites D100-D101

The sequence described above indicates three distinct evolution phases.

- The continuation of the post early Mousterian incision process down to approximately the depth of the present Zin channel.
- A depositional phase during which the thick gravelly alluvial fill accumulated in the Zin and Havarim valleys causing an increase in elevation of the local base level that controls - at least partly - the rate of the erosion processes in the study area.
- A second downcutting phase during which the Havarim and Zin valleys cut through the alluvial fill of the previous stage.

The tentative dating of the three phases described above is made possible by the occurrence of remnants of prehistoric occupation within the alluvial fill. Distinct occupation horizons, covering the transition period between the Middle and Upper Paleolithic were recognized by Marks (1977). The lowermost horizon yielded a C14 date of ca 45 000 y.B.P. This horizon is overlain by thick imbricated gravels and boulders of unknown age. But the tread of the terrace is capped by a thin unconformable veneer of colluvial silt containing an *in situ* Epipaleolithic site (D101 B. Fig. 6), dated by C14 to 13 500 y.B.P. (Marks, 1977).

An additional indication of the age of the uppermost layers of the lower terrace is given by the prehistoric site of Boqer (Marks, 1977) located on top of the terrace, west of the present Zin channel (Fig. 6). The upper layers of the terrace consist here of interbedded colluvial sands and silts. Scattered in space and time throughout these layers are a number of Upper Paleolithic sites which span an interval of at least 40 000 years B.P. to about 25 000 years B.P. The youngest occupation is covered by about 3m of colluvial silts and clays. Although these were not dated an estimate of about 22 000 y.B.P. for this final phase is assumed.
It is worthwhile noting that the sequence of events described above is not only of local significance as a very similar one was observed in eastern Sinai and northwestern Negev (Goldberg, in press).

Average long term rate of channel downcutting

The basic assumption in the present study is that the badlands topography has developed parallel to the incision of the main Havarim channel. This incision was practically completed by the end of the first incision phase some 45 000 y.B.P. The second incision phase, which spans the interval of 22 000 y.B.P. until the present may be considered as irrelevant to the purpose of our work as most of the incision occurred in the thick alluvial fill deposited during the period 45 000 to 22 000 y.B.P. The incision into the bedrock in the Havarim valley seems to have been limited to less than 1m in the second phase.

The development of the present badlands topography seems to have begun with the widespread Early Mousterian conglomerate that covered the relatively wide valley floor and capped the underlying soft shales. At that time the steep slopes connecting the valley floor to the Avedat plateau to the south and Halukim ridge - Sde Boqer plain to the north were probably covered by extensive scree mantles that inhibited, like the conglomerate, the incision of gullies into the underlying soft shales. Well preserved scree mantles can still be observed along the northern and western rims of the Avedat plateau.

The breaching of the Early Mousterian caprock surface started some 70 000 years B.P. or later. An incision of the drainage network took place. The calculation of the downcutting depth is based on the altimetric profiles constructed across the Havarim valleys (Figs. 1 & 3). Following the identification of the upper terrace level, on each of the profiles, the difference in elevation between this level and that of the present day channel was measured in the profiles. The figures obtained for the depth of downcutting varied from 11 to 41m, showing no systematic increase in the downstream direction. Taking into consideration a time lapse of 25 000 years (from 70 000 to 45 000 y.B.P.) for the incision process the calculated average annual rate of downcutting amounts to 1.3mm/year. When a longer period - 70 000 years - is considered the figure drops to 0.45mm/year.

The extrapolation of channel downcutting rates to denudation rates, that relate to a whole watershed, is highly problematic. Due to the high concentration of flow energy in the channel, the rate of channel incision probably represents the highest rate of denudation per unit area within a whole watershed. This may be especially true in badlands areas where the characteristic V shaped valleys offer the unique combination of very narrow channels, relatively steep longitudinal profiles and high depths of flow during storm events. Under such conditions energy concentration in the channels is very efficient and given the friable material denudation-incision rates very high. Theoretically if one assumes that the whole drainage system of the Havarim valley adjusted itself quickly to the downcutting process of the main channel and that within each of the small tributaries draining to the main valley the steep slopes were fully adjusted to the adjacent channel, then the rate of incision along the main channel can be considered as representing the denudation rate within the whole watershed. It is tacitly

Plate 15.2 View of a 'hanging valley' close to the main Havarim channel

assumed that during an incision phase the whole drainage network is hydrologically fully integrated and that any sediment amount delivered by the slopes to the channels is washed out of the watershed within a short lapse of time, storage time in the channels being almost negligible. Such a situation might have existed during the first post Early Mousterian incision phase i.e. from 70 000 to 45 000 years considered to be the decisive period for the sculpting of the present day badlands topography.

However, field evidence casts doubt on the idea that the small tributaries carved in the shales adjust themselves quickly to the main Havarim channel. Many 'hanging valleys' can be observed along the western rim of the Avedat plateau. The upper parts of these valleys exhibit well developed drainage basins with steep slopes and clear cut channels. The lower parts of these valleys narrow quickly in the downstream direction where the channels vanish gradually upon reaching very steep shale slopes adjacent to the Havarim channel (Plate 2). Discontinuities in flow lines are also indicated by the extensive occurrence of discontinuous gullies and pipes, especially on north/facing slopes (Yair *et al.*, 1980a). Such discontinuities can be regarded as indicative of the high infiltration capacity of the Taqiya shales. They support experimental data obtained recently by Yair *et al.*, (1980a) and suggest that the relative contribution of storm runoff, by overland and pipe flows, to the Havarim channel by the badlands area is rather limited even under extreme rainfall conditions. It may well be that the incision process of the Havarim channel should be primarily attributed to runoff originated within the northern part of the drainage basin. In this part, which covers some 70% of the watershed, extensive outcrops of limestone yield per unit area relatively high magnitude flows, probably the highest to be expected in the study area (Yair *et al.*, 1978; Yair *et al.*, 1980b).

A second indication of the high infiltration capacity and low erodibility of the shales is provided by the development of a thick and dense crust on a material originally very friable (Yair *et al.*, 1980a). The very formation of a crust, 30 to 40cm deep, in an area where average annual rainfall is estimated at 90mm requires, probably, stable conditions over a relatively long period and a high infiltration capacity of water to allow the cementation of the surficial layer. An indirect indication of the relative stability of the crust is provided by the occurrence of an extensive cover of soil lichens and mosses.

Conditions of stability are also indicated by the preservation of an Early Mousterian conglomerate on the interfluves, which are sometimes completely flat, for a period of 70 000 years. If one assumes a denudation rate of 1cm/year, often quoted in the literature, the area should have been lowered by some 700m!! and the preservation of the Early Mousterian conglomerate at an elevation of 40m above the present day channel completely inconceivable.

To summarize, field evidence of various types indicate that the denudation rate within the badlands area, during the period 70 000 to 45 000 y.B.P. must have been lower than the downcutting rate of the channel. A difference of one order of magnitude or more should not be excluded. Unfortunately the lack of precise methods prevents better quantitative evaluation of the long term average denudation rate. But the definition of the maximum possible rate of denudation - equal to that of channel downcutting - provides some useful information in a field where our knowledge is still largely deficient.

The second phase of evolution lasted from 45 000 to ca 22 000 y.B.P. channel downcutting in the main Havarim and Zin channels stopped leading to aggradation processes. During such a phase channel processes in the Havarim valley can no longer be considered as indicative of denudation processes; which probably continued within the small drainage basins. During this phase an unknown part of the sediment delivered by the slopes to the channels left the watersheds during extreme flow events. Another part was stored for long periods within the system as indicated by numerous small terraces and alluvial fans observed in second and third order streams in the badlands. The thickness of these terraces never exceeds 150cm. The majority cannot be correlated with the terrace identified along the main Havarim channel. The very occurrence of fluviatile terraces within small watersheds points to the low degree of integration of the hydrological system during this period. We have no way to evaluate the denudation rate during the depositional phase, but it must have been lower than in the previous period.

During the last phase in the evolution of the area, spanning the interval of 22 000 y.B.P. until present the downcutting process resumed. The maximum incision in the Havarim valley amounted to 5m (Fig. 4). Most of it occurred in the alluvial fill, incision into the bedrock being very limited and restricted to the lower part of the channel. The calculated average rate of incision is 0.25mm/year. This figure is quite low considering that during an incision phase, in a hydrologically integrated fluviatile system, the rate of channel downcutting represents the highest possible rate of denudation. However, for the reasons referred to in the analysis of the processes during the first post Early Mousterian phase, the average rate of denudation within the badlands during the last 22 000 years must have been lower than 0.25mm/year. The preservation of terrace remnants in small drainage basins and the fact that the present channel did not cut everywhere through the whole depth of the alluvial fill, together with the field evidence referred to in the first phase of incision indicate that the average rate of denudation within the badlands must have been significantly lower than 0.25mm/year.

CONCLUSIONS

An attempt was made to evaluate denudation rates in a badlands area over a period of 70 000 years. The attempt is based on prehistoric evidence, radiometric data; sedimentological and geomorphological evidence related to channel downcutting and aggradation processes. Data analysis lead to the following main points.

1) The major and decisive phase of channel incision in the Havarim drainage basin occurred in post Early Mousterian time from ca 70 000 to ca 45 000 years B.P. It amounted to ca 40m. Assuming a short standstill represented by the intermediate terrace, the average annual incision rate is 1.3mm Considering that the rate of channel incision represents the highest rate of denudation per unit area within a whole watershed and assuming a rapid and complete adjustment of the tributaries to the main channel, the figure of 1.3mm/year may be regarded as representing the highest possible average annual denudation rate in the study area. Such a figure is well below the rates obtained in spot measurements and even below denudation rates obtained for whole watersheds in Mt. Sdom (Gerson, 1977) and in Alberta (Bryan & Campbell, 1980).

2) However, field evidence that includes the occurrence of still existing 'hanging valleys' and of discontinuous systems of gullies and pipes suggests that some parts of the Havarim badlands are not yet fully integrated with the main channel. Under such conditions the extrapolation from incision rates to denudation rates is highly problematic. The only possible extrapolation is that the latter must have been lower than the former.

3) The formation and conservation of a thick regolith crust on north facing slopes (Yair *et al.*, 1980a) together with the preservation of thin conglomerates and terrace remnants for a period of 70 000 years are also indicative of relatively low erosion rates and stable conditions. The stability can be explained by the high threshold rainfall amount needed to activate runoff erosion in the study area (Yair *et al.*, 1980a). This high threshold suggests that erosion activity at the scale of a watershed is rather intermittent; occurring mainly under extreme high magnitude flow events whose recurrence interval is probably low to very low.

4) Data presented should modify the generally accepted assumption that high erosion rates are characteristic of all badland areas. They also cast doubt on the possibility to extrapolate from short term spot measurements of erosion to denudation rates. The latter rates usually relate to large areas - at least a whole watershed - over a long period of time.

16 Piping in the Big Muddy badlands, southern Saskatchewan, Canada

D.P.Drew

The Big Muddy Valley is one of an integrated network of glacial meltwater channels cut during the final retreat of the Wisconsin ice-sheet. The course of the valley is generally northwest to southeast across the southeastern portion of Saskatchewan, and the river is tributary to the Missouri River in northern Montana (figure 1).

The study area comprises a small segment of the southern flank of the valley some 19km south of the town of Bengough and 25km north of the United States border. At this point the valley is a wide, flat-floored feature at an altitude of 695m above sea level (a.s.l.). The Big Muddy River is a tiny underfit stream, meandering over the valley floor and dry for much of the year. Under present climatic conditions its erosional and transportational abilities are minimal. Valley walls are steep, ranging from 75m to 100m in height. Beyond is the flat prairie surface at c 800m a.s.l. covered, south of the valley, with a thin mantle of glacial drift and outwash deposits rarely more than 1m in depth.

The valley sides have been deeply dissected by a series of 'subsequent' gullies orthogonal to the main valley and formed by headward erosion and spring sapping. Some of these gullies have cut back for several miles into the prairie surface. Tributary valleys are separated by narrow, sinuous interfluves which often retain the original prairie sod on their summits. Generally interfluves decline in altitude towards the main valley floor and in places have been breached by secondary gullying to form isolated 'buttes' extending into the valley. One such interfluve (Main Ridge) with an isolated knoll (Castle Butte) beyond, was examined in detail. Castle Butte is completely detached from the ridge and is subject to erosion on all sides (plate 1). The capping of prairie sod has been removed from the butte, but remains in place on Main Ridge. The tributary valleys are completely dry except during the spring snowmelt and immediately following heavy rainfall.

The Big Muddy Valley is cut into bedrock, and numerous exposures occur both in the main valley and in the tributary gullies. The oldest formation represented in the area is the Whitemud, a clay-sand deposit of late Cretaceous age

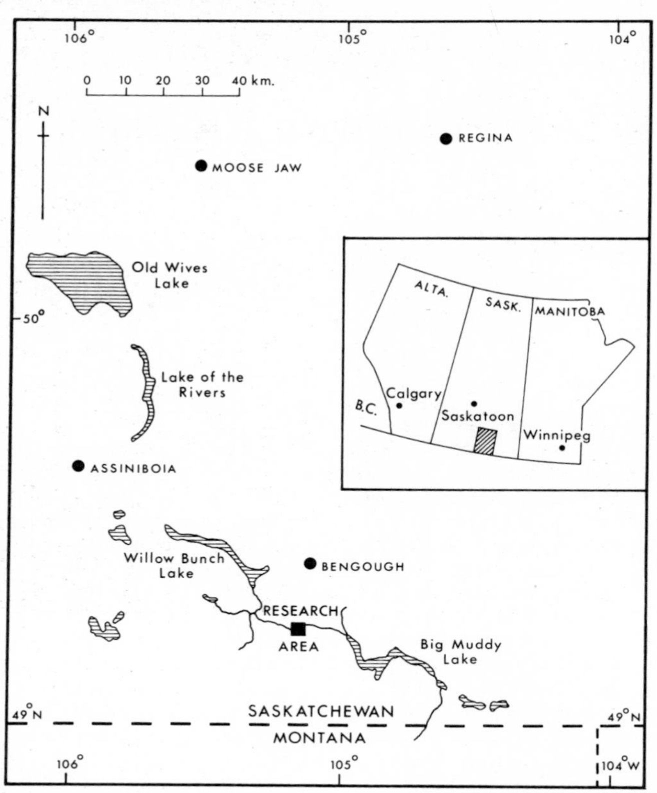

Figure 16.1 Location of the study area

(Byers, 1969). In many cases the sharp break of slope at the butte-valley floor interface (gravity slope-wash slope) corresponds to the upper limit of the Whitemud. The greater part of the valley wall is developed in the overlying unconformable Upper Ravenscrag formation (Houldsworth, 1941; Parizek, 1964) which consists of alternating layers of feldspathic sands, sandy clays, clays, shales, lignite and irregular lenses of ironstone (table 1). Variations in slope form, resulting from the differential resistance of various beds to erosion, are apparent. Dip appears to vary locally, but from observations made within the pipes, it rarely exceeds 10°.

Climatically the area is semi-arid with a mean annual precipitation of c 380mm and a potential evapotranspiration of about 760mm. Much of the precipitation falls as convectional rainfall during the short hot summer.

The vegetation of the valley and gully slopes is sparse and is xeric in areas where piping is intensive. Prickly Pear (*Opuntia cactus*), Moss Phlox (*Plox hoodii*), Sagebrush (*Artemisia cana*), Creeping Juniper (*Juniperus horizontalis*) and Ball Cactus (*Mamillaria*) predominate on exposed slopes whilst clumps of River Birch (*Betula occidentalis*) occupy the more sheltered and shaded gully floors.

Plate 16.1 Castle Butte showing piping. The main Big Muddy Valley is in the background

The soil in this area is a thin sandy regosol. The vegetal cover is denser on the light brown earths of the prairie top and largely consists of Spear Grass, Wheat Grass and Blue Grama. The prairie sod, with its high root concentration, seems to function as an erosion-resistant capping.

THE PIPING

Pipes are common in the area, particularly in the isolated buttes and in association with the subsequent gullies. An indication of the density of the piping network is shown in figure 2 - the majority of the sites marked being water inlets. The piping is sometimes a single conduit from inlet to outlet, but more commonly forms a dendritic network of tubes. As with karst hydrologic systems, water input points exceed outputs by ratios of up to 30:1, though there are indications that this ratio may decline as the pipe network on a particular slope facet becomes more mature and hence integrated. The majority of the pipe outlets are at the level of intermittent stream channels but inlet points do not seem to relate to either topographic or lithological factors, except that removal of the Prairie sod seems essential for their evolution. Most pipes are too small to enter (5-30cm diameter) but several proved to be accessible and were examined in some detail. The typical morphology of a pipe system is:

1. An inlet shaft, 2-10m in depth.

Table 16.1 The geological succession in the Big Muddy Valley area

	EPOCH	FORMATION	THICKNESS Metres	CHARACTER
TERTIARY	Miocene	Wood Mountain	15+	Gravel, sandstone
	Oligocene	Cypress Hills	40+	Gravel, sand, conglomerate
	Eocene	Swift Current	15+	Gravel, sand, conglomerate
	UNCONFORMITY			
		Upper	30 - 105	Buff grey sands, coal, clay-shale
	PALEOCENE	Willowbunch	3 - 10	White, grey sandy clay, partly refractory clay
		Ravenscrag	125	Buff grey sands, shales, coal, clay
MESOZOIC		Lower Ravenscrag	6 - 60	Grey sands, shale, clay
	UNCONFORMITY			
	UPPER	Whitemud	4 - 23	White/grey sand clay clay - part refractory
		Eastend	6 - 30	Yellow, yellow/green very fine sand, silt, grey shale
	CRETACEOUS	Bearpaw	215+	Dark shales

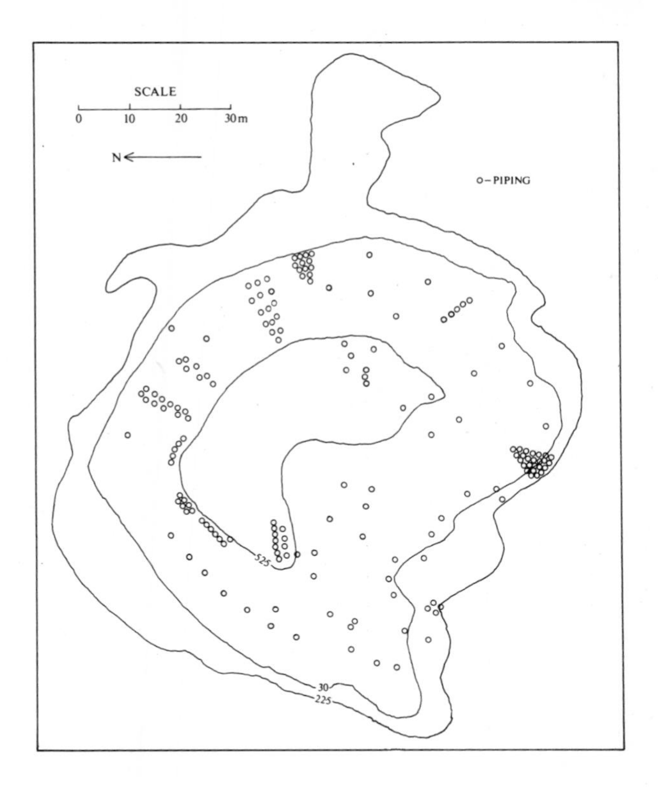

Figure 16.2 Location of piping features on Castle Butte (contour lines in metres above arbitrary local datum)

2. A conduit leading from the foot of the shaft, initially at 5-20°, then lessening in gradient toward the outlet point.

3. Roof breakdown and silting are common features as are vertical steps down in the passage.

Two surveyed pipe systems are described in detail: Pipe 1 is an outlet. Pipe 2 is developed in a bench on the valley to the east of Main Ridge and can be followed from entrance to exit. Pipe 1 is entered via a low crawl (figures 3 and 4). The passage (area 'd' in figure 4) has a well-defined vertical trench cut down from an originally ovoid passage (section d). This is blocked by liquid mud, apparently due to basal sapping by summer runoff. There is a bypass over the blockage (area 'c') and at the highest point of this loop is a major water inlet in the roof which was observed to function following heavy rain. Beyond the bypass is a chamber some 10m long, 2-3m wide and up to 5m high (area 'b'). The floor is composed of large blocks of siltstone from collapse, whilst the walls show extensive mud deposition features, often resembling calcite flowstone

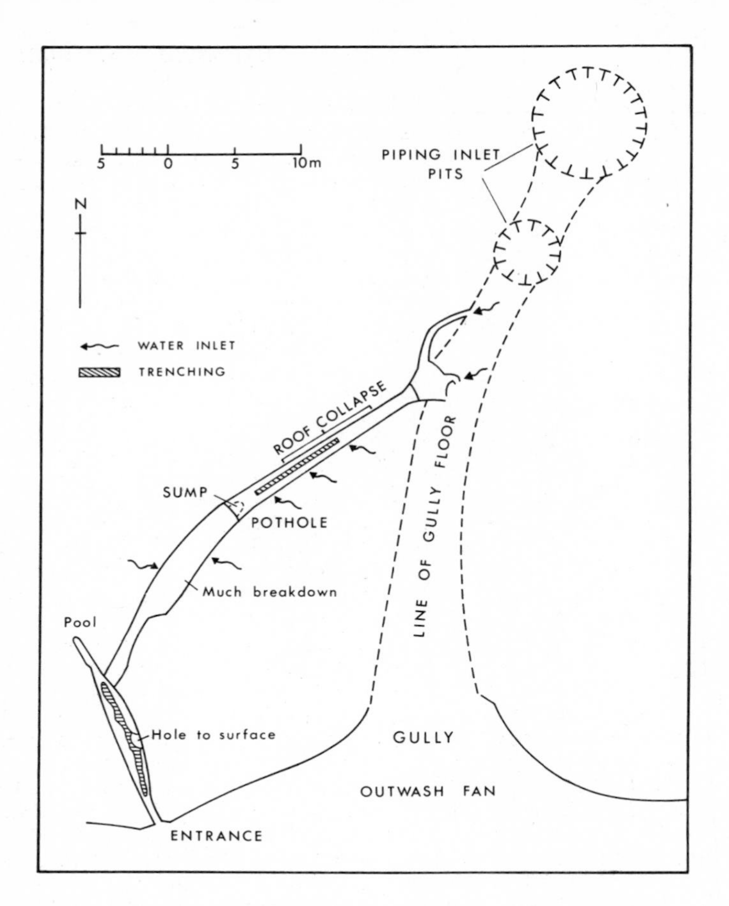

Figure 16.3 Plan survey of pipe 1

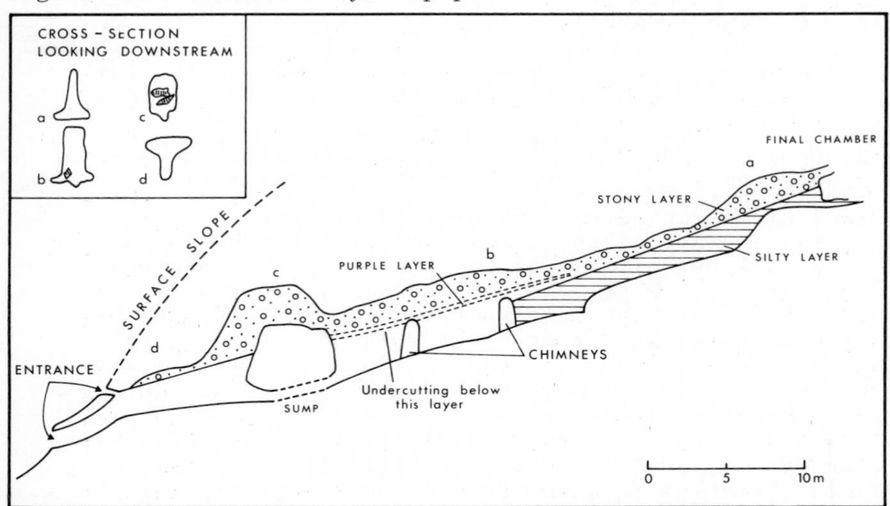

Figure 16.4 Long profile and passage cross-sections, pipe 1

in appearance. Mud stalagmites occur beneath points of heavy drip. In this area the passage walls are distinctively stratified. The lowest 3m is composed of hard silty sandstone capped by a purple clay layer rich in organic material; the passage walls are somewhat undercut below this layer. The top 2m consists of a stony clay-rich deposit, the individual cobbles rarely exceeding 10cm in diameter. Further upstream the purple layer often forms the passage roof. Throughout the cave the stony layer only seems to have been exposed by collapse. At the upstream end of the chamber a 2m vertical rise in the floor leads to a narrow, tall rift-like passage with an even floor gradient. At the foot of the drop is a small pothole with a low crawl to a mud sump; this is the usual water inlet to the chamber.

At its upper end the cave terminates in a small collapse chamber from which several tight sinuous tubes lead off. These seem to be very close to the surface (area 'a' in figure 4). This area lies under the floor of a major gully on the butte and is only some 5m distant from two large sinkholes at the head of the gully. These are the major sinks for the system, although the pipe system as a whole lies to the west of the gully.

Observations on the active hydrology of the system were limited owing to the infrequency of storms, the inaccessibility of the area and the nerves of the field workers. During one storm (14mm of rain in 2 hours), however, Pyranine dyestuff was placed in the two main sinkholes and was found to emerge from the upstream mud sump within 15 minutes. Total flow-through time for the system was 35 minutes. Flow rates are thus 200-300m/hour, of the same order of magnitude as flow rates for concentrated surface flow. In areas of well developed piping virtually all of the precipitation is converted into pipe flow and thence concentrated into the major outlet pipes. Thus, erosion is very localised compared with more widespread erosion on the surface from sheet flow and numerous shallow gullies. Observations within the pipe showed that flow continued at low level for two hours following the cessation of rain. One and a half hours after rain began, water flowed from the tubes beyond the final chamber. The stream was rarely more than 20cm deep in any part of the pipe but flow was rapid and turbulent. A sediment trap installed near the entrance of the pipe yielded negligible sediment, and analysis of the water near the entrance showed that the load was almost entirely in the silt/clay range. However, large cobbles up to 40cm in diameter are transported completely through the system by heavy flows. During the period of observation, no movement of the large cobbles occurred. No appreciable changes were observed between Fall and Spring suggesting that snowmelt is probably unimportant - the pipe systems are likely to remain blocked with snow during the melt. Thus, convective summer storms may be associated with movement of the large calibre bed load. Several small, discrete inlets in the roof of the pipe produced heavy drips for up to four hours after the rain ceased. The total dissolved solids in the water remained unchanged through the length of the pipe, averaging 180-210 ppm (expressed as $CaCO_3$). However, considerable solutional leaching is suggested by the markedly higher salt concentration in the sediments at the pipe entrance than those in the sediment transported through the pipe.

Pipe 2 (figure 5 and 6) lies parallel to a minor gully near the southern end of Main Ridge. The gully is of very low gradient (4-6°) and only active following heavy rain. The exit from the pipe is at gully level, but the sinks are in a large collapse depression on the flank of the Main Ridge.

Overall, the pipe is very similar in form to pipe 1, having two inlet passages with markedly trenched floors and some degree of lateral undercutting, converging on a large collapse chamber. From this chamber a single large passage leads to the exit (plate 2). The floor and the lower 1m of the passage walls are excavated in a hard silty sandstone, similar to that in pipe 1: 73% sand, 17% silt, 10% clay (Table 2). The sand is 90% quartzitic whilst the clay fraction is predominantly montmorillonite with lesser quantities of kaolinite and micaceous clay minerals. Above this layer is a 5cm horizon of clay and lignite (64% clay) that is black in colour. The clay is dominantly a poorly crystallized

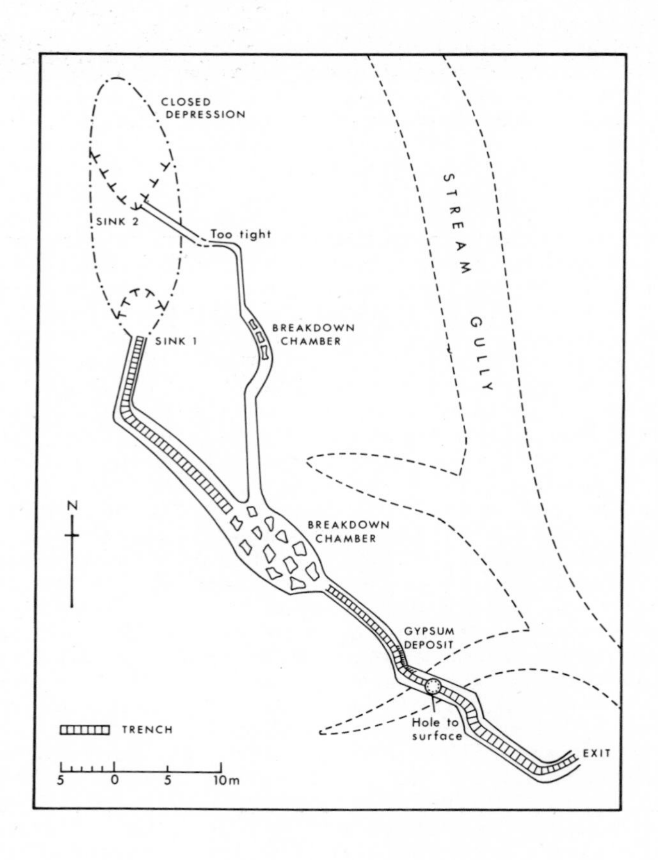

Figure 16.5 Plan survey of pipe 2

Plate 16.2 Passage 2m high, modified by collapse, pipe 2

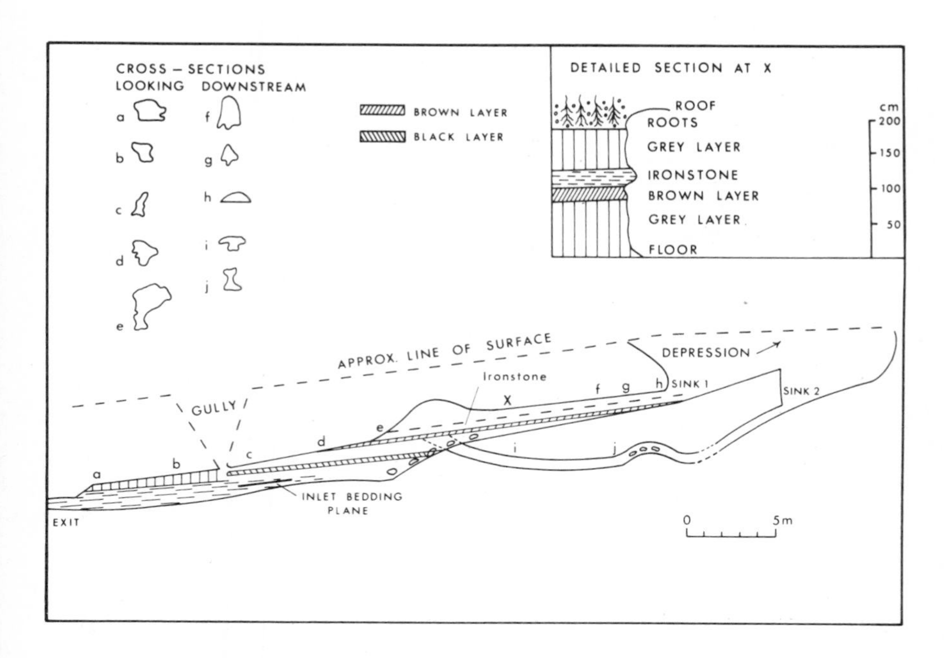

Figure 16.6 Long profile and passage cross-sections, pipe 2

Table 16.2 Data summary for sediment analyses from key sites in the Castle Butte area

Sample	Particle size					Mineral Frequency (Sand Fraction)					Heavy Minerals Occurrence					Salts ppm	
	clay %	silt %	sand %	light minerals %	heavy minerals %	quartz %	feldspar %	plagio-clase %	potash feldspar %	rock fragments %	ferruginous grains	biotite	garnet	zircon	hornblende	$CaCO_3$ ppm	total soluble salts ppm
Areas around pipe inlets	33	48	19	95	5	65	29	0	10	6	4	3	1	1	0	210	1740
Areas around pipe outlets	19	16	65	96	4	67	27	1	8	6	2	3	1	1	1	65	490
Discharged sediment from pipes	10	11	79	86	14	59	35	4	12	6	3	1	1	1	0	105	525
Black Layer Pipe (2)	64	34	2	99	1	86	12	1	4	2	3	3	0	1	0	170	1850
Side walls near exit Pipe (2)	16	19	65	99	1	78	21	1	1	1	2	3	0	2	1	45	750
Upper side walls, Main Chamber, Pipe (2)	10	27	63	98	2	91	8	0	2	1	1	2	0	1	2	400	2600
Pipe (2) Upper side walls near entrance	10	17	73	99	1	58	39	2	2	3	3	2	2	2	1	10	195

Key to heavy mineral occurrence: 0 = Nil 1 = Trace 2 = Common 3 = Abundant 4 = Very abundant

kaolinite. This layer first appears in the walls of the chamber and forms the roof of the passage for some distance below this point. Upstream of the chamber the grey horizon is overlain by a brown organic layer, again rich in clay, and above this again is an ironstone band some 3-20cm in thickness. Above the ironstone is a further layer of the compact silty sandstone (63% sand, 27% silt, 10% clay) and again the dominant clay mineral is montmorillonite. The roof in this area of collapse extends into the regolith. No observations on the hydrology of this system were made.

Origin and development of the pipes

Although surface drainage does occur producing typical badland erosional forms, it is estimated that over 50% of the drainage of the butte and gully slopes is via piping. All piping examined was developed within the consolidated or semi-consolidated bedrock of the Ravenscrag formation. In both pipe systems examined in detail, the initial development seems to have been concentrated at or just above a layer of wholly impermeable clay/lignite. Initial passage form was elliptical in cross-section and followed the bed down-dip to the exit.

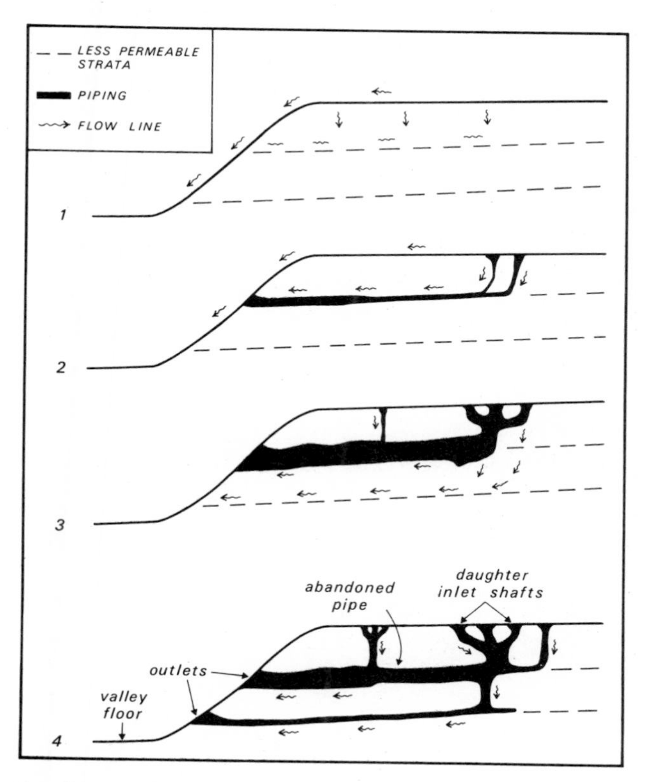

Figure 16.7 Suggested sequence of piping development, Big Muddy Valley

Later development included the breaching of this bed, the excavation of a trench in the siltstone beneath, some degree of roof and wall collapse, and eventually the development of a second tier of passages above the next organic layer down in the succession. Although the initial development may be partly solutional, the later stages would seem to be almost wholly corrosional. The large stones moved through mature systems indicate the mechanical erosive ability of the waters. The evidence from this area is that piping develops primarily in response to base control and that passages become progressively smaller and more poorly developed towards the sinks although these may often be enlarged by wall collapse. Mature pipes are well graded in their lower sections, but the profile is more irregular closer to the sink. There appears to be no general water table as such in these deposits, as their inter-granular permeability is very low. Even following heavy rain the wetting depth is only 25-30cm. Drainage into the pipes is via the sinkhole passages or through well-defined points of leakage in the roof.

A possible sequence of development for the piping of this area is indicated in figure 7:

1. The majority of runoff is overland, but a small proportion will percolate via cracks, secondary openings, root paths, etc. until it reaches the topmost organic/clay layer. Water movement is then by throughflow down-dip to the point of exit. Erosion is accomplished by limited solution of soluble salts and subsequent removal of fines in suspension until a small elliptical conduit is formed at the level of the organic horizon. At this stage the drainage is likely to be via many minor parallel systems with exits about 2-10cm in diameter.

2. As the passage enlarges greater flow through is possible, the sinks enlarge and the organic layer is breached to form a T-shaped passage in cross-section. A degree of lateral integration of piping is probable at this stage.

3. Roof and floor collapse occurs and 'daughter' shafts evolve in the area of the main sink. Runoff is now predominantly underground.

4. Pipes begin to develop above a lower aquiclude horizon, in some cases as an extension of the pre-existing passage and in others as a separate lower network. The system is now integrated laterally, and the whole slope is drained via one main conduit and outlet. The cycle is repeated until the system is graded to the lowest possible level of outlet; this may be a gully floor or the floor of the main valley. A 'closed depression' landscape develops in the sink area.

17 The occurrence of piping and gullying in the Penticton glacio-lacustrine silts, Okanagan Valley, B.C.

O. Slaymaker

Discussion of the characteristics and mode of origin of silts underlying the depositional terraces of the interior valleys of British Columbia has been active since Dawson's description of the 'white silts' in the north central British Columbia. (Dawson 1978; 1879; 1891; 1895; Daly 1915; Flint 1935; Meyer and Yenne, 1940; Nasmith, 1962; Fulton 1965; 1967; 1975 and Shaw, 1975). Post-glacial modification of these silts has received more recent attention (Cockfield and Buckham, 1946; Buckham and Cockfield, 1950; Wright and Kelley, 1959; Nyland and Miller, 1977; Buchanan and Evans, 1977 and Buchanan, 1977). It is thought that the silts at Penticton, which form the focus of this discussion, and those at Kamloops, discussed by Cockfield and Buckham (1946), accumulated under roughly similar conditions in late-glacial lakes. In the last thirty-years increasing interest has been shown by local and provincial government agencies in the problems associated with piping phenomena in these silts.

Piping is: 'a subsurface, primarily mechanical, process of erosion by which sediment is dislodged and entrained in water, generating tubular underground conduits' (after Mears, 1968). The primary landform which results from this process is a 'pipe', but an associated suite of secondary land forms has also been identified such as caves, sinkholes and other karst-like features. In addition, the formation and widening of gullies is facilitated. Recent land use changes on the terraces underlain by glaciolacustrine silts in interior British Columbia have apparently led to intensified piping development. British Columbia's Departments of Agriculture and Highways have published a number of reports assessing the nature of this natural hazard. Acknowledgement is made to Wright and Kelley (1959) for their recognition of the essential dimensions of the problem and to Buchanan and Evans (1977) for their valuable discussion. Two recently completed undergraduate field studies and the author's observations in May, 1979 and 1980 form the empirical data base for this report.

STUDY AREA

Okanagan Valley, incised into the British Columbia Interior Plateau, has a general north-south trend and has been deepened and widened by repeated Pleistocene glaciations (figure 1). The elevation of the water surface of Okanagan Lake is 340m, but bedrock under the lake is close to sea level (Macaulay *et al.*, 1972) and adjacent mountains rise to 1500m. The glacially eroded form of the valley has been modified during late and post-glacial time by deposition and by erosion of unconsolidated sediments (of which the glacio-lacustrine silts are one example) (Nasmith, 1962). The distribution of a number of these sediment types in the Penticton region is shown in figure 2. Of the units mapped, two are of particular importance in the ensuing discussion:- (1) the glacial lake sediments and (2) the outwash terraces and kettled outwash.

The climate of the Penticton region is summarized as having a mean monthly temperature of 8.4°C and a mean annual precipitation of 273mm. (1964-1973). According to irrigation district estimates approximately 429mm from irrigation water and 87mm from domestic sources are added to the groundwater each year (Nyland and Miller, 1977). Table 1 summarizes the annual water balance of the Penticton region, pointing out in particular the contrast between an irrigated area, with a moisture deficit of 365mm, and a developed area, with a moisture surplus of 151mm.

Table 17.1 Average annual water balance for Penticton region (1964-73)

	P.E.T.[1] (mm)	Input (mm)	Deficit (mm)	Surplus (mm)
	638			
P (mm)[2]		273	-365	
P + I[3]		702		+ 64
P + I + D[4]		789		+151

1. Potential evapotranspiration calculated by Nyland and Miller (1977) from data supplied by Environment Canada.

2. Precipitation calculated by Nyland and Miller (1977) from data supplied by Environment Canada.

3. Irrigation water estimate by West Bench Irrigation District.

4. Domestic water estimate by West Bench Irrigation District

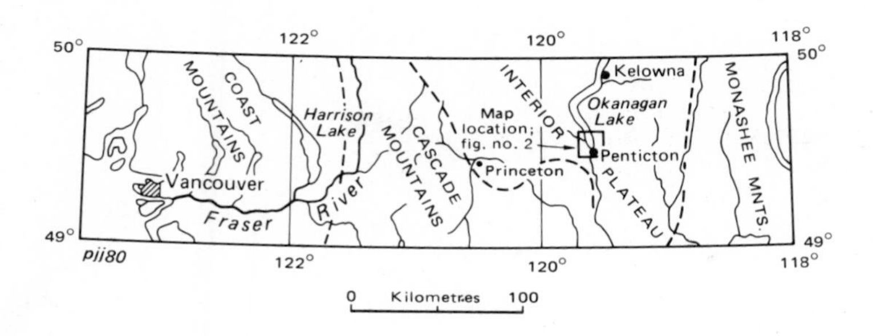

Figure 17.1 General location of study area

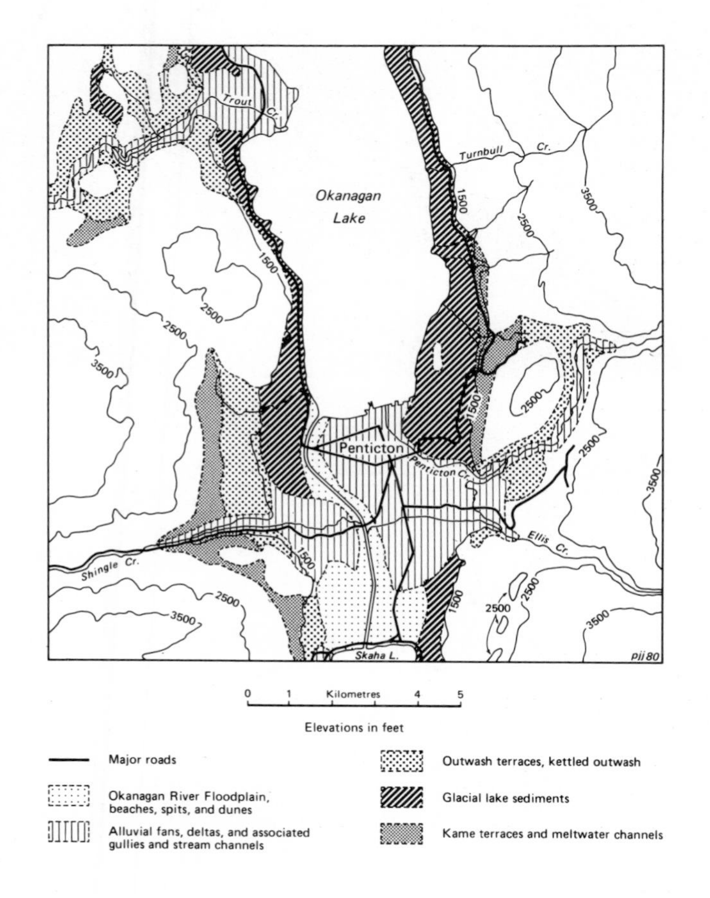

Figure 17.2 Distribution of major late-glacial and post-glacial sediments in the Penticton region (after Nasmith, 1962)

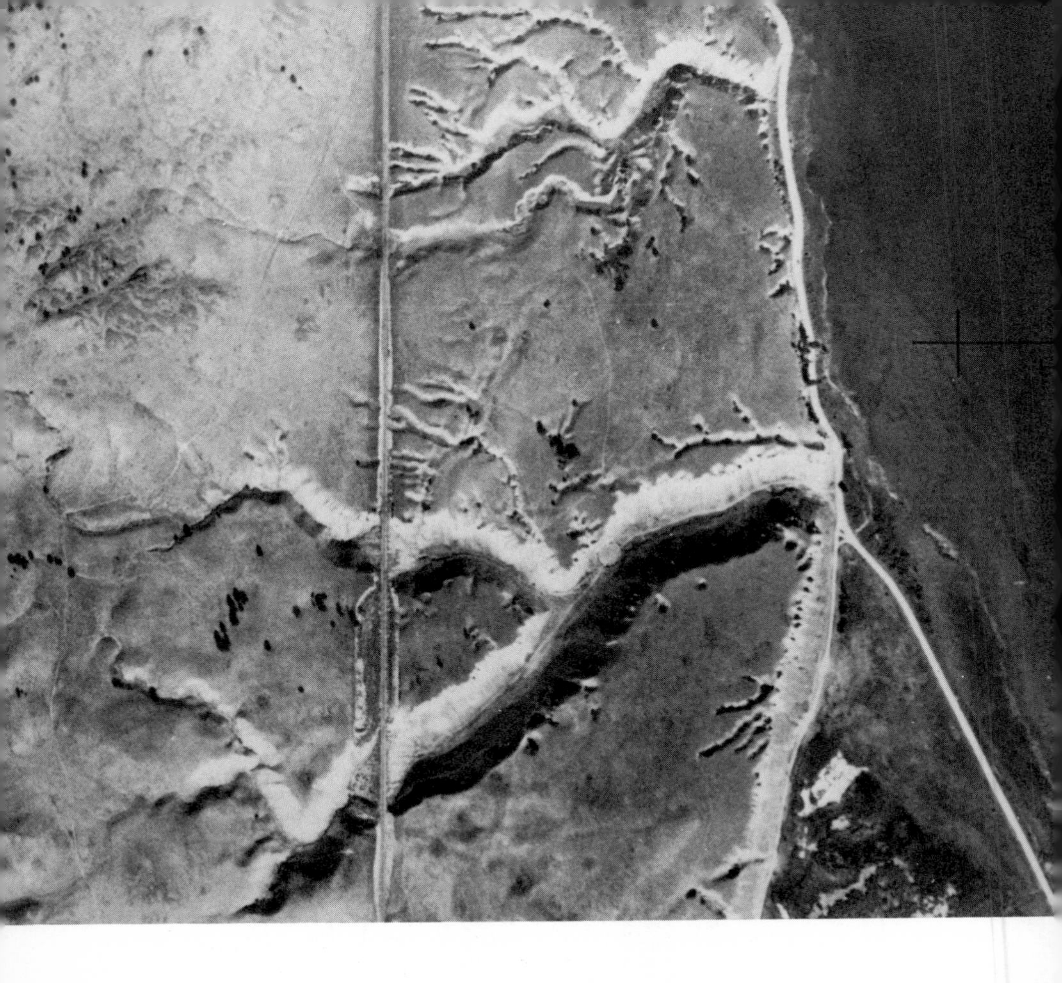

Plate 17.1 Dissected glaciolacustrine silts north-west of Penticton (detail from B.C. 105: 41; 1938 photography)

NATURE OF SEDIMENTS

The glacial lake sediments are of two kinds; both *in situ* silts and redistributed silts.

(1) The *in situ* silts have a maximum inferred thickness of 100m. They were deposited on top of downwasting ice so that when the ice melted, differential settlement occurred, leading to broad flexures and local normal faulting. Varving and intensive vertical jointing are characteristic. The general reasons for jointing are shrinkage and differential settling; in areas close to cliff faces stress release and wetting and drying cycles are probably important. (Nyland and Miller, 1977). These sediments are highly dissected by gully systems (plate 1 and Mollard, 1974).

(2) The redistributed glaciolacustrine silts result from mass wasting, surface erosion and piping in the *in situ* silts. They have similar grain size distribution but a very

different structure (non-varved) and are located in depressions, gully bottoms and below cliff faces.

(3) The glaciofluvial outwash deposits are generally at higher elevation than the silts and were deposited at the mouths of tributary meltwater channels where they flowed into glacial Lake Penticton. The outwash sediments are extensively kettled, are highly permeable and provide subsurface pathways for the lateral and downslope movement of water draining substantially larger drainage areas upslope (figure 2)

LANDFORMS

In the Penticton area, at least three scales of landform result from piping: (1) individual pipes, which can themselves be subdivided further according to scale, (2) sinkholes, (3) gullies, and gully systems.

Pipes

The distribution of pipes exposed in a 120m long cut 5-11m high, a part of which is shown in plate 2, was examined. The area of silts exposed in the vertical face is about 1600m^2. Vertical micropipes, of less than 5cm diameter, are localized along joints, which tend to be most prominent at apices of anticlines. Sloping or 'gradient' micropipes are

Plate 17.2 Railway cut through glaciolacustrine silts illustrating synclinal sedimentation units and the large vertical macropipe/sinkhole

difficult to identify because of the impossibility of checking on their lateral continuity without conducting individual tracer experiments. The total area of the section containing micropipes is estimated at 0.1m^2 (0.006% of the section).

Sloping or 'gradient' macropipes form along the junction between silt laminae of varying permeabilities. Approximately 0.9m^2 of the section (or 0.05%) consists of gradient macropipes. Vertical macropipes, of greater than 5cm diameter, are associated with synclinal troughs in the silt; these are difficult to distinguish from sinkholes. Approximately 4m^2 of the section (or 0.24%) is occupied by vertical macropipes and sinkholes. In total, then, approximately 0.3% of the silt section is occupied by 'macropores' of various kinds.

Sinkholes

Over one hundred sinkholes in the area of 5km^2 of silt west of Penticton can be identified from 1938 photography (plate 1). With an average planimetric area of 100m^2 per sinkhole, this gives at least 0.2% of the surface area as sinkholes. Such features have been well described by Cockfield and Buckham (1946) for the Kamloops region. The cross-section of a sinkhole, associated with the trough of a syncline in the varved silts, is shown in plate 2.

Gullies and gully systems

One of the most obvious features observable from airphoto analysis of the Penticton area is the contrast between the highly dissected glaciolacustrine silts and the kettled but undissected glaciofluvial sediments (plate 1; Mollard, 1974). There is little evidence of present extension of the major west-east sloping gullies - they are dissected to a maximum depth of 65m, but with mean dissection depths of 10-30m. The gullies have steep walls and their south or southeast facing upper parts are commonly vertical (plate 3). In these main gullies there is a nearly horizontal section across the gully floor, with a narrow vertical wall trench, sometimes discontinuous down gully, incised into the gully floor. In the above respects, there is a marked similarity with the gullies of the Kamloops silts (Cockfield and Buckham, 1946), though the depth of incision of the gully floor is greater in the Kamloops silts. The area of silts occupied by gully systems is approximately 0.3km^2 (or 6% of the planimetric area of lake sediments).

FORMATION OF PIPES

The *in situ* glaciolacustrine silts are varved and intensively vertically jointed. Overland flow is able to penetrate the silts via the vertical joints thereby forming vertical pipes. This process can be compared and contrasted with the formation of desiccation crack pipes (Bryan *et al.*, 1978). The variable saturated hydraulic conductivities of the varves encourages lateral seepage of water over the surface of

Plate 17.3 Down gully view looking towards the north-east. Gully wall asymmetry with south-facing bare vertical slopes prominently displayed

relatively impermeable layers in the silts and this, in turn, leads to the development of gradient pipes. Thirdly, the presence of a steep slope or a cliff face encourages large hydraulic gradients which are responsible for more gradient pipes. By these three processes there is initiated a subterranean 'plumbing system' in the silts which, given an unlimited moisture supply, would lead to the removal of the Penticton silt terraces. Jones (1971) has discussed the factors favouring the occurrence of pipes in humid Britain as (1) susceptibility to cracking in dry periods, high silt-clay content and high percentage of montmorillonite (2) periodic high intensity rainfall and devegetation (3) biotic break up of soil and a relatively impermeable basal horizon (4) an erodible layer above this base, high exchangeable sodium and high base exchange capacity or high soluble salts and (5) steep hydraulic gradient. Heede (1971) has emphasised the importance of steep hydraulic gradient, high exchangeable sodium percentage, low gypsum content and montmorillonite in piping initiation in semiarid Colorado.

Only the factor of steep hydraulic gradient is common to the Penticton silts and the discussion of Jones (1971) and Heede (1971). Nevertheless the jointing and varved nature of the Penticton silts allows water infiltration and lateral movement in ways similar to that induced by cracking and impermeable basal horizons elsewhere.

THE FORMATION OF SINKHOLES

The incidence of sinkhole formation was a major reason for the preparation of the reports by Wright and Kelley (1959) and Nyland and Miller (1977). As described by Wright and Kelley (1959), a hole about 4m in diameter and 6m deep was formed in December, 1958, in the front yard of one of the local residents. In addition, new incidences of sinkhole formation are being reported at points where outflow from piped water is concentrated, where drainage tiles have been disturbed, below down pipes without gravel protection and at points of surface runoff concentration below road developments. Dudley (1970) has described the major reasons for collapse phenomena in soils associated with the decrease in bulk volume when water is added. The two prime requirements are: (1) a loose soil structure and (2) a moisture content less than saturation (Dudley, 1970, p. 927). The addition of water to the soil is the triggering action. He showed from a review of the literature on hydrocompaction (collapsing soils) that maximum subsidence is associated with dry unit weight of 65-106 lbs ft^{-3}, clay content about 12% by weight, montmorillonite as the dominant clay mineral and a honeycomb structure of bulky (silt) grains held in place by finer grained (clay) material. These criteria can be compared with a selection of analytical data for two West Bench Penticton samples (Table 2). From this evidence, it seems probable that sinkhole formation in the Penticton silts is, at least in part, a hydrocompaction phenomenon, which should be compared with the slumping described by Bryan *et al.*, (1978).

GULLY FORMATION

The most probable interpretation of the major gullies is that they were formed during, and immediately following the drainage of glacial Lake Penticton, due to the release of large quantities of water stored within the glaciofluvial sediments upslope and movement of this water along preferred pathways within the silt following the lowering of local base level; this combined with the effective surface erosion on newly exposed silts without vegetation cover produced very rapid dissection by surface and subsurface erosion. The precise relationship between these suggestions and the process proposed by Rubey (1928) and applied by Buckham and Cockfield (1950) to the Kamloops gullies is not clear because (a) the writer has not observed the process and (b) the relative importance of surface and subsurface processes is difficult to ascertain. Fulton (1975) summarises the general sequence of events, but ignores details of the process.

GULLY SLOPE MODIFICATION

Although the major gully systems appear to be relict and the location of their heads at the junction between silts and the coarser glaciofluvial sediments was established at an

Table 17.2 Aspects of mineralogy, chemistry and structure of two Penticton West Bench silt samples (from Quigley, 1976)

	%<2μ	D_{50}	e_o	% collapse	γ dry (lbs. ft^{-3})	structure	dominant clay mineral	cations in liquid extract Ca (p.p.m)	Na	precipitates from supernatant liquid
IN SITU SILT A	10	5-10μ	1.2	2.6	78.5	dense varved	montmorillonite	33	80	calcium sulphate
REDISTRIBUTED SILT B	11.5	10-20μ	1.05	7.2	83	open (honeycomb)	montmorillonite	40	45	calcium sulphate

D_{50} = average particle size by electron microscopy; e_o = void ratio; % collapse = collapse under 4 tons ft^{-2} loading and flooding; γ dry = dry unit weight; structure = as seen under electron microscope; clay size fraction = under X-ray diffraction; cations in liquid extract = from soil: water mix of 1:5; precipitates from supernatant liquid = under X-ray diffraction.

early evolutionary stage, considerable detailed modification of the gullies occurs under present hydrologic conditions. Wright and Kelley (1959) have indicated the striking contrast between north and south facing slopes of the major gullies and the apparently rapid modification of the south-facing slopes due to the large proportion of these slopes without vegetation cover (plate 4). Absence of vegetation seems to result from extreme dryness in summer; in turn, it reduces the infiltrability of the sediments and surface rilling is encouraged. There is consequently a tendency for asymmetric development of lateral gullies by surface erosion on the northern side of the main gullies (plate 1). Subsurface processes of piping and sinkhole development are apparently equally active on north and south slopes.

MAGNITUDE-FREQUENCY CONSIDERATIONS

The general question in relation to the evolution of gullies, sinkholes and piping is that of the role of high magnitude low frequency events as compared with seasonal snowmelt effects and individual storm events of moderate magnitude. Wright and Kelley (1959) point out the possibility that the gullies have originated in one action by stored water in the glaciofluvial materials to the west 'which cut through the silts like a bursting dam'. Probably the piping and sinkhole formation, by contrast, represent the aggregate effect of low magnitude high frequency events. Where such land forms are developed in a linear or dendritic pattern, high magnitude low frequency events can take advantage of such erodible pathways.

Plate 17.4 Abrupt cliffs of south-facing gully walls

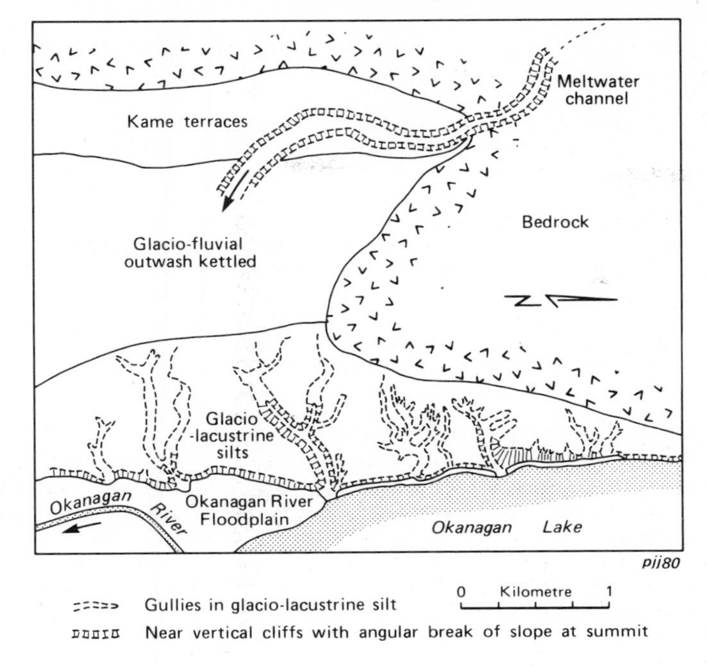

Figure 17.3 Relationship between dissected silts, kettled outwash and late-glacial meltwater channel north-west of Penticton

LAND USE AND PIPING

Intensification of land use in the area is intensifying the piping process through (1) increased moisture input and (2) increased localisation of runoff. As far as gullying is concerned, the regional control of moisture input from upslope seems to be the dominant factor. Gully development in the silts is most intensive in the areas located adjacent to and downslope from glaciofluvial outwash sediments (figure 3).

In the Penticton area, 'the severity of piping was concluded to be a function of the amount of seepage flow available and permeability of the silt deposit at a particular site. Control of excess moisture was concluded to be the best way to control piping' (Nyland and Miller, 1977). Continued development of West Bench, Penticton both in terms of private residences and, for example, golf courses will lead to an increased moisture surplus (Table 1) and piping and sinkhole formation will be intensified.

CONCLUSION

Piping is a major geomorphic process in Penticton glacio-lacustrine silts. 'Vertical' and 'gradient' pipes occur at the margins of benches and gullies. The distinctive pinnacles, columns and deep clefts associated with the abrupt cliffs of the silt formation originate from this piping. Sinkholes are also formed close to the bench margins. Vertical pipes and sinkholes are associated with hydrocompaction and are distributed almost randomly over the bench surface. Finally, it is suspected that piping is a factor in gully initiation and subsequent extension as pipes provide preferred pathways for subsurface water movement.

Acknowledgements

The assistance of Peter Chan, Diane Fowler, Sheila Loudon, Meena Oza and Greg Smith, undergraduate students attending the U.B.C. Physical Geography Field School, is gratefully acknowledged.

18 The role of piping in the development of badlands and gully systems in south-east Spain

A. Harvey

There are two general bases for our understanding of gully and badland development. The first, outlined by Horton (1945) and elaborated by Strahler (1958) which relates gully and badland morphology to surface erosion, provides the basis for an enormous body of literature on runoff and erosion processes, as well as on badland and gully morphology. (e.g. Schumm, 1956a).

The second, which has received much less attention involves subsurface erosion by pipeflow and the ultimate collapse of pipes to form gullies. In the growing body of literature dealing with pipes there is emphasis on pipeflow hydrology (Jones, 1971), the origin of pipes (Bryan *et al.*, 1978, Barandregt and Ongley 1977, 1979, Masannat, 1980), and some consideration of the influence of piping on gully development (Masannat, 1980, Brown, 1962, Heede, 1974, 1976, Parker, 1965). However, little of this considers the relative influences of piping and surface or near surface processes on gully and badland morphology. This paper examines piping in gullied areas in southeast Spain in the context of two questions: whether different types of pipe differentially affect gully morphology and whether the morphology of piped gullies differs from that of nonpiped gullies.

Gully systems may be classified into valley floor and valley side, or hillslope gullies (Brice, 1966). Valley floor gullies are essentially trenched channels of the 'arroyo' type (Cooke and Reeves, 1977, Leopold, 1978) and valley side gullies may either be discrete linear forms, or more extensive systems of the badland type. All three are common in southeast Spain, and examples have been examined in five areas (Figure 1a, Table 1). Each has been surveyed in the field and pipe systems and rill networks mapped. These maps have been used to provide morphometric and channel orientation data in an attempt to determine the influence of piping on gully morphology. In addition particle size analyses have been carried out on samples of surface materials, and infiltration tests undertaken to determine surface properties in piped and nonpiped sites.

Table 18.1 Description of Study Sites

Site	Geology*	Description	Piping
Altea	Triassic grey & purple Gysiferous Marls. (62)	Badlands.	None.
Tapia 1	Quaternary pink & pale silts (82) over Eocene grey marls.	Trench gullies, limited badland development. Major within-gully headcuts.	Numerous large, deep pipes.
Tapia 2	Quaternary pink & pale silts.	Trench gully. No within-gully headcuts but gullies start at large headcuts.	All types of piping including deep tension crack pipes.
Tapia 3	Quaternary pink silts (75) over Ecocene grey marls. (82)	Trench gully with tension crack influence. No headcuts within major gully, but present in tributary gullies. Includes badland area on Eocene marls (3a).	All types of piping present including deep tension crack pipes. Piping on 3a restricted to occasional small shallow pipes.
Sucina 1	Plio-Pleistocene red silts (91)	Small badland.	None.
Sucina 2	Plio-Pleistocene red silts.	Shallow linear gullies with some badland development. Headcuts present on gully floor.	Occasional shallow gully-floor pipes.
Sucina 3	Plio-Pleistocene red silts (90)	Mature badlands and adjacent trench gullies. No within-gully headcuts but trench gullies start at headcuts.	None on badlands. Some small pipes at the heads of trench gullies.
Vera 1	Pliocene soft yellow sandstone over grey gypsiferous marl & silt (96)	Full range of erosional forms from shallow discontinuous gullies to deep linear gullies and badlands (Vera 2). Headcuts present in discontinuous and linear gullies.	Numerous large, deep pipes (except on Vera 2).

Vera 2	Pliocene grey gypsiferous marl & silt. (95)	Badlands.	Occasional small shallow pipes.
Vera 3	Pliocene grey gypsiferous marl & silt.	Badlands and adjacent pediment.	Occasional small shallow pipes.
Vera 4	Pliocene grey gypsiferous marl & silt.	Valley floor trench gully, no within-gully headcuts, but gully starts at major headcuts. Adjacent badlands and pediment.	No piping on main channel, occasional pipes at tributary junctions, and on pediment.
Tabernas	Upper Miocene dark grey marls & shales (94)	Mature badlands	None.

* Geological information derived from IGME (Intituto Geologico y Minera de Espana) 1:50 000 maps and accompanying memoirs. Old series 848 Altea, New Series 934 Murcia, 1014 Vera, 1030 Tabernas.

Figures in brackets are% of total material finer than 4.25 for surface layers.

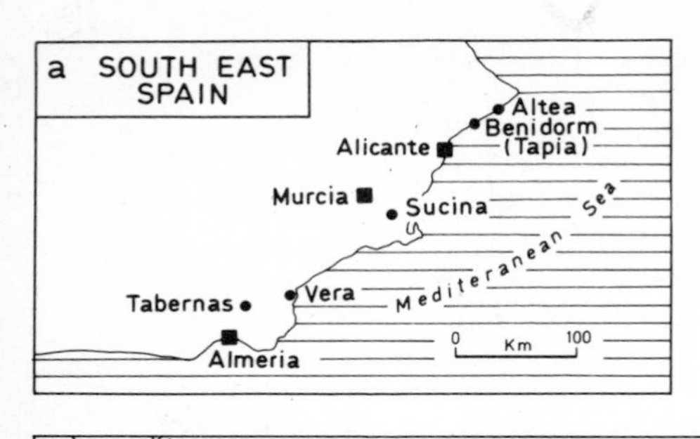

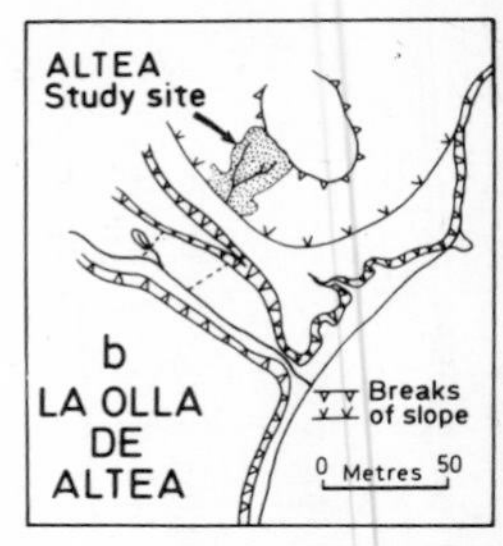

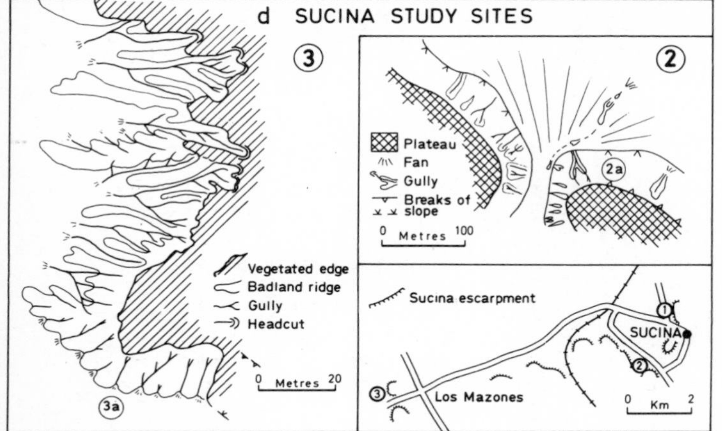

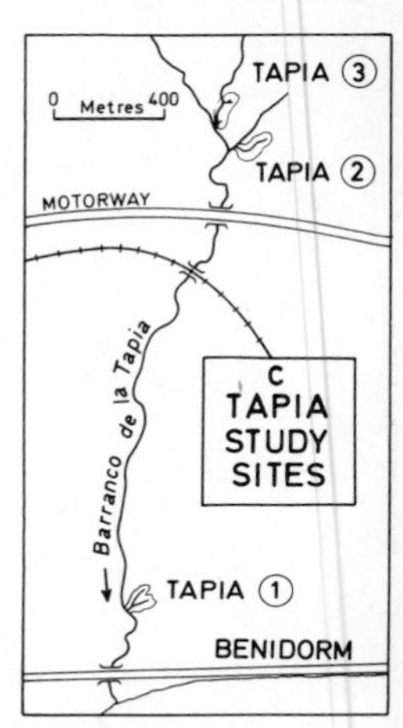

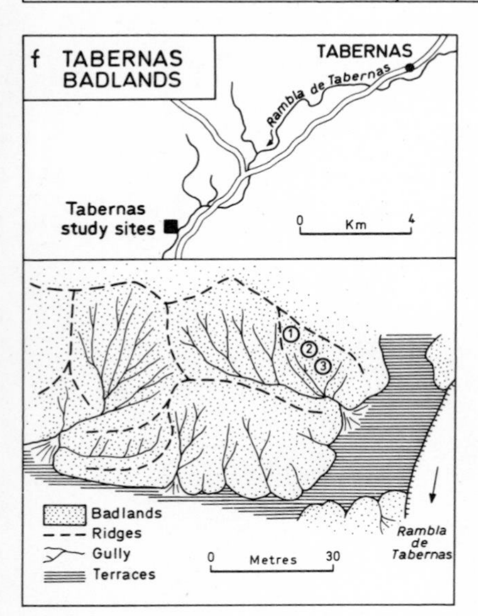

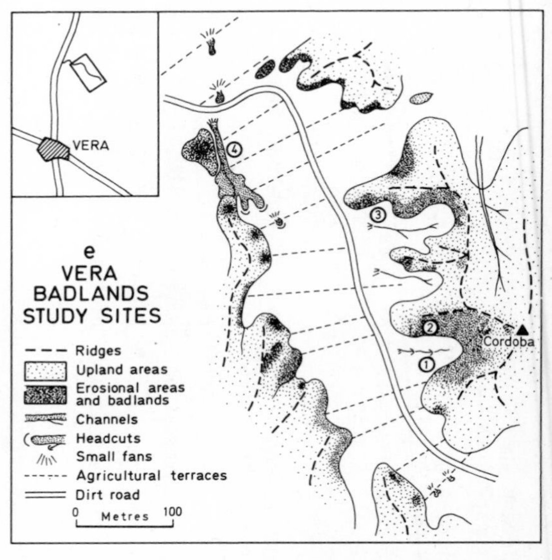

Figure 18.1 Location of the study sites. a) locations within southeast Spain b) Altea; c) Tapia; d) Sucina, sketch map of Sucina 2, triangulation survey of Sucina 3; e) Vera; f) Tabernas. For detailed maps of Tapia sites see Fig. 2, of Vera sites see Fig. 3 and of Altea, Sucina 1 and Tabernas see Fig. 4.

TYPES OF PIPING

Several types of piping have been observed, with apparently differing influences on gully morphology.

a) Shallow pipes, irregular in section but never more than a few cm in diameter occur on badland slopes. (e.g. at Vera, Tabernas and Tapia, Figure 1a) They appear to be related to the presence of a weak disaggregated layer a few cm below a light surface crust. Although they may influence surface erosion processes they appear to have no major differential influence on gully or badland morphology and where no other pipes occur morphology is dominated by surface or near surface rather than piping processes.

b) Deep pipes, from 50 cm to several metres below the surface occur in several forms.

i) Pipes may develop by the removal of weak material from below lightly crusted surface layers but on a larger scale than in the shallow pipes. They often occur in or adjacent to deep trench gullies and may be responsible for subsurface capture thereby influencing network development. This is evident at Tapia 1 (figure 2) and in a trench gully 500m north of the Vera sites, where large cavernous pipes up to a metre in diameter occur in weak materials roofed by lightly crusted surface layers.

ii) Deep pipes may also develop in relation to differential porosity, cementation or solubility within the material. Cemented gravel layers in the Quaternary fills at Tapia 1 (Harvey 1978) have led to the formation of large gully floor pipes by the erosion of underlying silts. Vertical gypsum veins and horizontal gypsum sheets within the Pliocene silts and marls at Vera 1 have strongly influenced pipe development, where gully floor pipes are common within the discontinuous and linear gullies (plate 1d). They vary in shape and size but are usually from 10 to 40 cm in diameter, and generally occur adjacent to gypsum sheets or veins, sometimes one above another. They strongly influence gully development as their collapse may create within-gully headcuts or cause the fusion of discontinuous gullies.

iii) Deep pipes may also form in relation to tension cracks in thick silt formations. The steep slopes on Quaternary fill along the Barranco de la Tapia resulting from entrenchment of the barranco are characterised by large tension cracks. These cracks focus pipe development whose collapse produces a gully network with strong parallelism along the major valley axis. This is well illustrated at Tapia 2 and 3 (figure 2). Smaller tension cracks may form around trench gully heads as at Sucina 3 (figure 1d) but subsequent pipe development has little effect on gully morphology, merely accentuating headcut recession.

c) A final type of piping could be described as bridge piping. Entrenchment or lateral undercutting producing

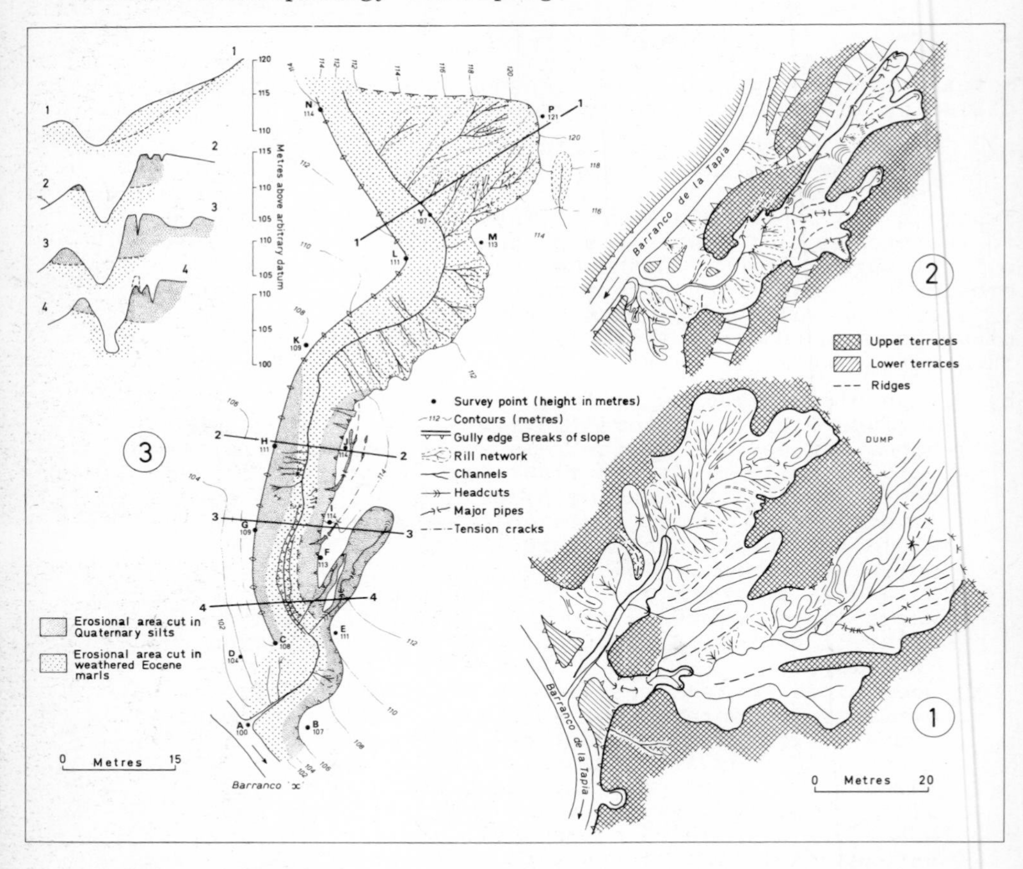

Figure 18.2 The Tapia study sites. Tapia 1 and 2 based on traverse surveys, Tapia 3 based on triangulation survey

an overhang which on collapse may create a bridge over the gully floor. Bridge pipes exert a minor influence on gully form if their subsequent collapse causes ponding or diversion of the channel against the opposite wall. Bridge pipes due to lateral undercutting are common in Quaternary silts at Tapia 1 and 2 and due to entrenchment have been observed at Tabernas but not within the study site.

THE STUDY AREAS

Southeast Spain has a semi arid climate with mean annual precipitation ranging from c350 mm at Altea to c170 mm at Vera and Tabernas (Geiger, 1970). Much of this falls as storm rainfall in autumn and spring with an intervening drought in midwinter and a long dry summer. Winters are mild (January mean temp. c 11°C) and summers hot (July mean temp. c 27°C) (Geiger, 1970). In the north the vegetation of the non-farm land is open Aleppo pine woodland inter-

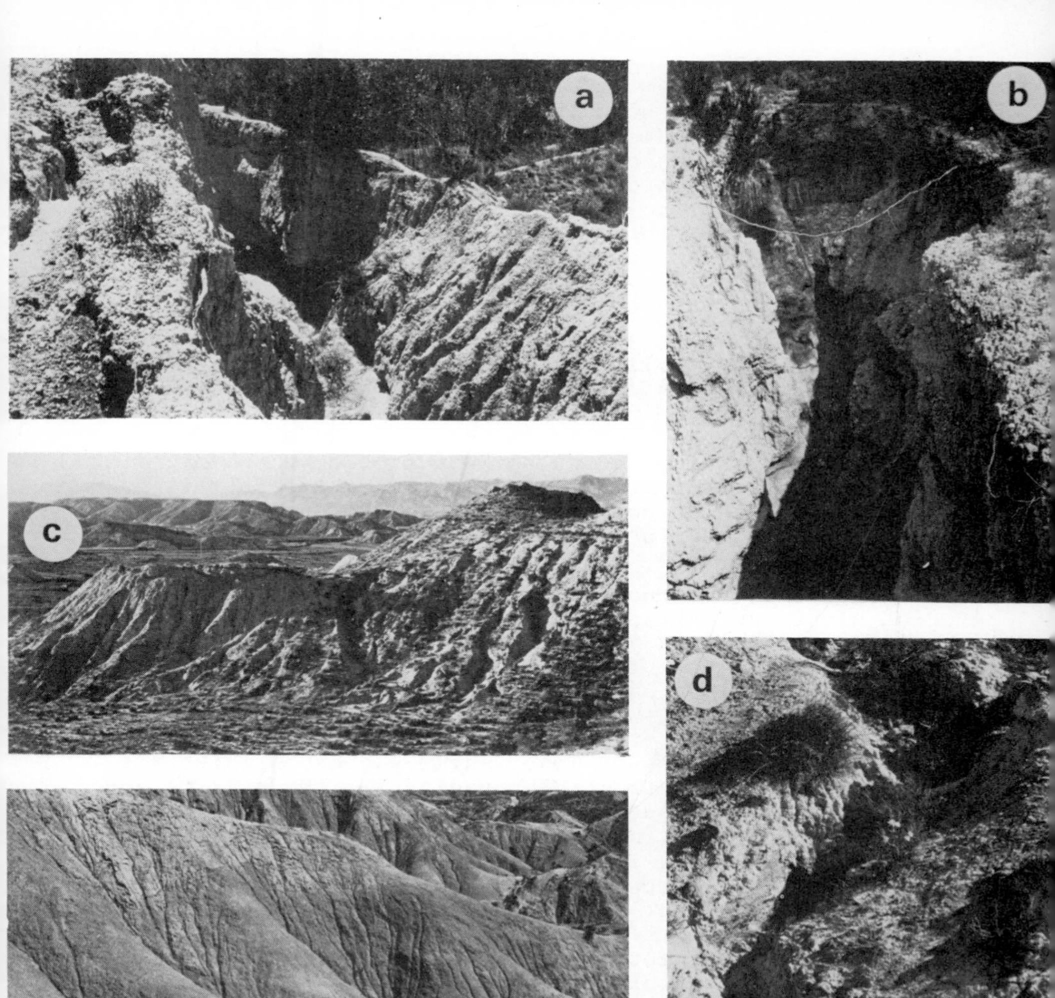

Plate 18.1 a) Tension cracks on the East wall of Tapia 3.
b) Deep headcut at the head of Tapia 2 tributary. Note root suggesting former pipe.
c) Vera badlands, view across Vera 1 to Vera 2 linear piped gullies (Vera 1b) on right, badlands (Vera 2) on left.
d) Gully floor pipes at Vera 1b.
e) Tabernas badlands.

spersed with garrigue scrub and further south is dominated by asphodel steppe and sparse open garrigue (Birot, 1964). The long history of uncontrolled grazing by sheep and goats has led to degradation of the vegetation cover and widespread soil erosion (Bennett, 1960). Gullies and badlands are common on the softer rocks especially on Triassic and lower Tertiary marls in the north and on upper Tertiary marls and silts south of Alicante. Quaternary valley fills throughout the area are also subject to gully development.

Altea (figure 4). The Altea site comprises a gully system cut into diapiric gypsiferous Triassic marls capped by indurated Quaternary fan deposits. It is a small badland with rilled surfaces and rounded interior divides. The surface is desiccation cracked and loose to depths of 3-5 cm but there is little caking and no obvious pipe development. Erosion processes appear to be surface dominated but with the poor rill development and low drainage density (table 3) surface mass movement by creep or mudsliding may be more important than surface wash.

Tapia (figure 2). Three trench gully systems have been studied along the Barranco de la Tapia, each cut in Quaternary fill, predominantly of silt but with occasional gravels. Tapia 1 and 3 cut through to the underlying Eocene marls on which at Tapia 3 is a badland area (Tapia 3a).

Tapia 1 includes two trench gullies which seem to have developed by headward extension. Above mid-gully headcuts vertical walls give way to lower angle slopes on which rill systems are beginning to develop badlands. The gullies head in vertical walled headcuts except for part of the southern gully where rills continue from the adjacent building excavation dump. Piping is common especially in the southern gully and includes large cemented gravel protected pipes below the major headcut and large surface crust protected pipes in the upper part of the gully. Two of the latter have effected gully capture and modified the network. Small gully floor pipes and bridge pipes are also common. The northern gully shows fewer large pipes and may represent a more advanced stage of development with larger areas of incipient badlands.

Tapia 2 is a trench gully cut in Quaternary fill, whose main axis parallels that of the barranco. At the lower end of the gully are rilled side slopes but further up-gully the walls are vertical and the gully heads in deep headcuts (plate 1b). There are numerous gully floor and bridge pipes but no major within-gully headcuts. Tension crack piping is important and has produced tributary gullies parallel to the main gully and it is possible that the parallelism of the whole system reflects tension crack formation. The gully dissects abandoned agricultural terraces but these seem to have had little influence on the network.

Tapia 3 is a large trench gully cut through Quaternary fill into underlying Eocene marl and showing many of the features apparent in Tapia 2. It is cut in the steep valley side above Barranco 'x', a tributary of the Barranco de la Tapia, and runs parallel to the valley side, suggesting an enormous tension crack induced gully. Its cross profile (4 on figure 2) suggests that its formation may have been

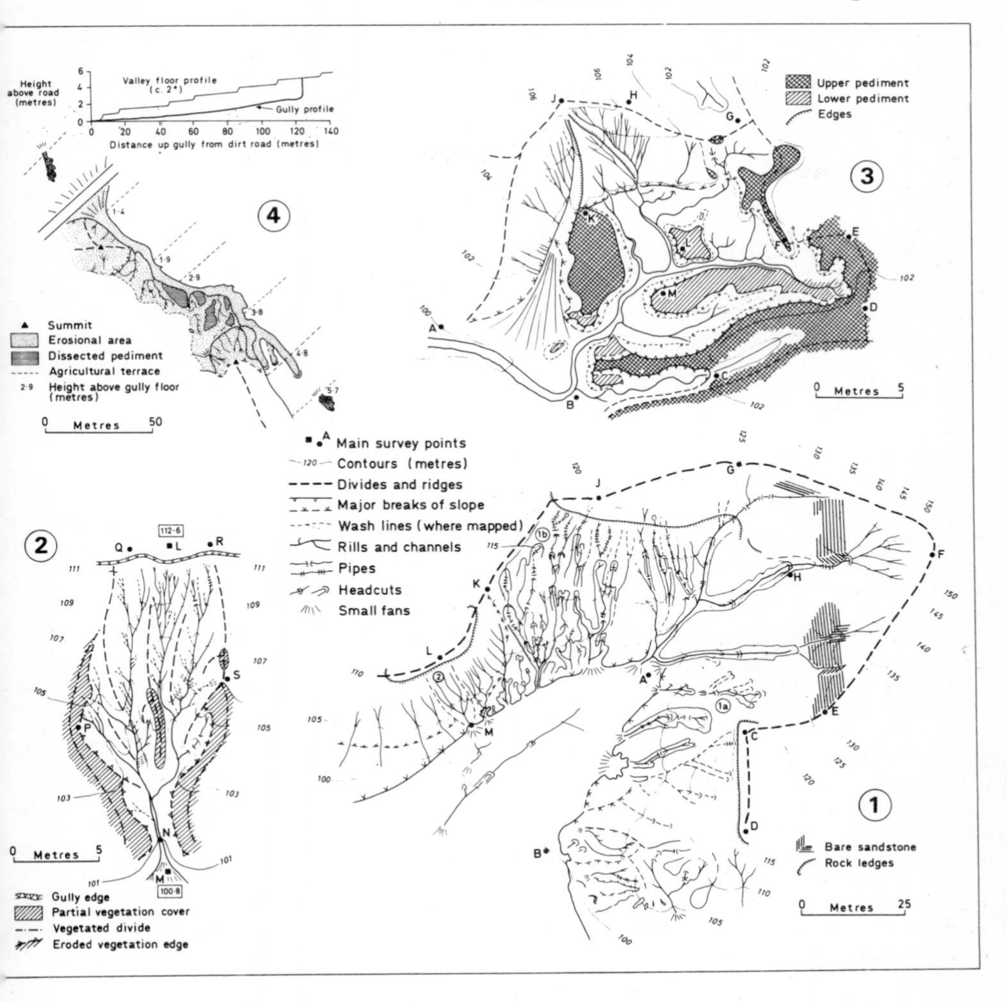

Figure 18.3 The Vera study sites. Vera 1, 2 and 3 based on triangulation surveys, Vera 4 based on traverse survey and measured profile

due to the collapse of a very large pipe 1.5-2 m in diameter. The main channel heads outside the gully as a valley floor channel which was apparently diverted against the valley side by the construction of agricultural terraces. It then leaves the valley to enter the gully system (near survey point N, figure 2). Whether this was an artificial diversion or a natural, perhaps pipe-induced capture is uncertain. Within the gully system tension crack piping is very apparent. Not only does the main channel appear to be tension crack-induced but on the east wall are large cracks (plate 1a) feeding into pipe systems that issue on the wall of the main gully. The major tributary gullies also appear to be tension crack pipe-induced. At the head of the gully system is a

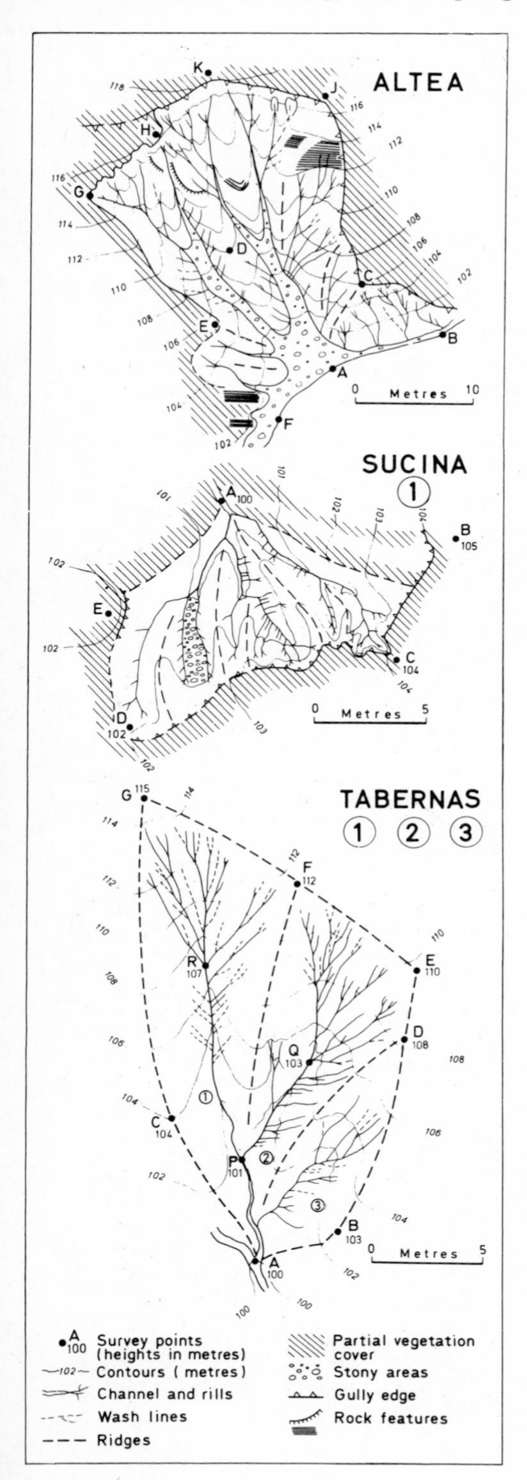

Figure 18.4 The simple badland sites at Altea, Sucina 1 and Tabernas. Based on triangulation surveys

badland area cut in Eocene marl, part of which has been mapped in detail (Tapia 3a). It shows some shallow piping and bridge piping, neither of which appear to influence the gross morphology.

Sucina (figures 1d and 4). The three Sucina sites all occur as scarp face gullies cut into Plio-Pleistocene silts exposed on the Sucina escarpment below a massive indurated caliche.

Sucina 1 (figure 4) is a small midslope badland with a well developed rill network and smooth rounded divides. It is clearly a recently developed badland area showing all the characteristics of surface erosion although as at Altea, the thin friable desiccation cracked surface layer may be subject to creep or mudsliding as well as to surface wash. The major rills are clear and obviously carry considerable runoff.

Sucina 2 (figure 1d) represents another form of gully initiation on the Sucina escarpment. Here linear gullies and small badlands occur together. The main gully on the south face of the scarp embayment is a discontinous linear system with headcuts in mid gully and incipient badland development on the side slopes. Although small gully bottom pipes are present they do not appear to exert a major influence on morphological development. It seems likely that as headcuts migrate gully side slope angles reduce and rill and badland development proceeds.

Sucina 3 (figure 1d) represents a more advanced stage in gully development on the escarpment. Here the whole of the scarp slope is in badlands with well developed rill systems and smooth bare rounded divides. At the northern end of the site where the relief is less, major gullies cut back to form headcuts. Only here is there any piping, with small headcut tension crack pipes, but these have little influence on the overall morphology.

Vera (figure 3). The badlands at Vera, cut in Pliocene gypsum-veined marls and silts, include a wide range of gully type. Vera 1 is a large valley which exhibits three distinct types of erosional development. The south slope (Vera 1a) is well vegetated but has several discontinuous gully systems with minor headcuts, and some minor gully floor piping. The east slope (Vera 1b) is dissected by deep subparallel linear gullies but some of the original vegetated surface remains (plate 1c). There are numerous within-gully headcuts and some of the gullies are discontinuous. They are riddled with pipe systems (plate 1d) and some of the headcut formation and gully entrenchment can be related to the collapse of pipe systems. The pipes themselves are influenced by the presence of gypsum sheets and veins. On the north slope virtually all of the original vegetated slope has been removed and badland gullies are cutting directly into bedrock. There are no headcuts and only superficial pipes. The badland surfaces are rilled and the divides smooth but in places the surface layer is desiccation cracked and caked and there is local evidence of superficial mudsliding indicating that surface mass movement may be as important as surface wash. Vera 2 is part of this badland area (plate 1c).

Table 18.2 Infiltration Characteristics

	30 min. Infiltration (mm)	20-30 min infiltration rate (mm/min.)	Wetted depth after 30 min.
Bare Sites			
Altea	67	1.0	25
Tapia 1 (Quat. younger fill)	46	1.5	37
Tapia 3 (Quat. older fill)	113	4.1	55
Tapia 3 (Eocene)	84	2.7	45
Sucina 1a	91	1.9	45
Sucina 1b	126	3.8	70
Sucina 3	175	5.1	150
Vera 1	39	0.8	35
Vera 2	32	0.5	30
Tabernas	32	0.5	25
Vegetated Sites			
Altea	10	0.4	18
Tapia 1 (Quat. younger fill)	126	3.6	46
Tapia 3 (Quat. older fill)	84	1.6	58
Tapia 3 (Eocene)	73	1.8	52
Sucina 1	54	2.4	50
Vera 1	68	1.9	50
Tabernas	112	3.7	60

Vera 3 is also a badland area, but one where slope recession has produced a frontal pediment (Smith, 1958). The pediment is multiple with the present rill network trenched into the youngest pediment surface. On neither backslope nor pediment are there more than shallow pipes.

Vera 4 is a trench gully with adjacent badlands and frontal pediment. The trench gully cuts into bedrock through only a thin veneer of alluvium. It starts in dramatic headcuts but has no within-gully headcuts. Piping is common on the pediments and where tributary rills join the trench gully. Some of this appears to be tension crack piping related to the incision of the trench gully, but the gully itself appears to have developed entirely by headcut recession with little or no piping influence. Erosion of the pediment and possibly of the badland slopes is at least in part influenced by piping.

Tabernas (figure 4). The Tabernas badlands are cut in a variety of sediments of Upper Miocene age, mostly marls, silts and shales, on some of there is minor superficial and

Table 18.3 Morphometric Properties

	Area (m^2)	relief (m)	mean slope^o	max. order (iii)	Bifurcation ratios (iv) Rb_1	Rb_2	Length ratios (v) Rl_1	Rl_2	Drainage density(vi)	Pipe (vii) index
Altea (i)	257	18.1	33	4*	3.24	4.25	2.71	2.44	0.76	0
Tapia 3 (ii)	95	14.0	33	4	3.31	3.25	2.18	1.63	1.01	0.6
Sucina 1	83	4.1	22	4*	4.67	6.00	3.75	5.90	0.88	0
Sucina 2a	715	22.0	26	3*	6.00	(3.00)	5.01	(4.46)	0.19	0.9
Sucina 3a	40	5.5	24	4*	3.36	7.00	2.42	2.44	1.33	0
Vera 1a	348	27.0	34	3*	5.50	(2.00)	4.46	(1.54)	0.19	3.4
Vera 1b	532	16.5	19	5*	2.67	2.40	3.83	1.70	0.40	5.9
Vera 2	116	11.6	34	5*	4.64	5.00	4.38	2.85	1.21	0.9
Vera 3	308	6.5	18	5*	3.30	5.40	3.17	2.07	0.63	0.7
Tabernas	157	15.2	36	5	3.74	3.44	2.55	1.96	1.29	0

(i) 4th order (south basin only. (ii) 4th order (south) basin. (iii) * indicates open ended network.

(iv) $Rb_1 = n1/_{n2}$, $Rb_2 = n2/_{n3}$ where n is the number of segments of given order.

(v) $Rl_1 = L2/_{L1}$, $Rl_2 = L3/_{L2}$ where L is cumulative mean segment length at the given order. (Figures in brackets are ratios for open ended 3rd order networks)

(vi) $m/_{m^2}$

(vii) length of pipes as % network length.

bridge piping. The dark shales on which the study site is located show no evidence of piping and appear to represent typical badlands developed entirely by surface processes. The divides are smooth and the rill networks clear and well developed (plate 1e). However even here the thin desiccation cracked and caked surface layer suggests that mass movement may be significant in supplying sediment to the rill systems but the clarity of the rill systems themselves obviously reflects the importance of rillflow.

DISCUSSION

The morphological evidence indicates that pipe erosion is important in the Tapia trench gullies and Vera badlands. In the other areas, Altea, Sucina and Tabernas and also at Tapia 3a, surface or near surface processes are dominant.

Surface Properties

Particle size analyses of surface samples all show a predominance of silt grades with mean sizes between Ø5.1 and Ø6.9 and silt clay percentages between 61% and 97% (table 1). Although the distributions differ, piped and nonpiped sites cannot be differentiated. Surface materials are coarsest at Altea and finest at Vera and Tabernas. Shear van tests were also carried out on both surface and subsurface materials but the results were inconclusive.

Thirty minute infiltration tests were carried out both on bare surfaces and adjacent vegetated surfaces (table 2, figure 5). The cylinder infiltrometer method (Hills, 1970) was used but because of difficulties in maintaining operational consistency the outer ring was dispensed with and as an alternative the surrounding soil was maintained in a constantly wet condition. The test sites were located on level surfaces on spur tops and the tests were carried out under very dry conditions. The infiltration rates measured therefore do not reflect actual infiltration behaviour during storm conditions but do allow comparisons between sites. As a whole the results do not enable clear differentiation between piped and other sites but do throw light on the nature of surface processes, in terms of comparisons both between bare sites and between bare and vegetated sites.

Infiltration rates at bare sites were low at Vera, Tabernas and Altea and high and variable at Tapia and Sucina. At Tapia 3 infiltration into the Quaternary fill was greater than into weathered Eocene marl. In comparing bare and vegetated sites the expected pattern of lower bar site infiltration is clear only at Tabernas and Vera. Elsewhere the position is more complex, in some cases with infiltration rates into lichen crusted vegetated surfaces being less than into loose friable bare surfaces. In general the measured infiltration rates are high, and even allowing for the differences between test conditions and natural infiltration under storm rainfall, it seems unlikely, except perhaps at Vera and Tabernas, that much surface runoff is generated by simple rainfall excess on the slopes away from the rill floors. This accords with results reported from eroding

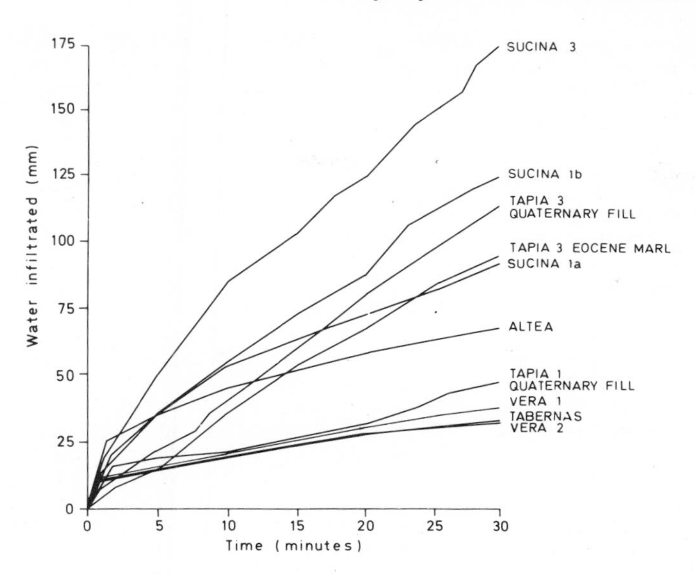

Figure 18.5 Infiltration tests

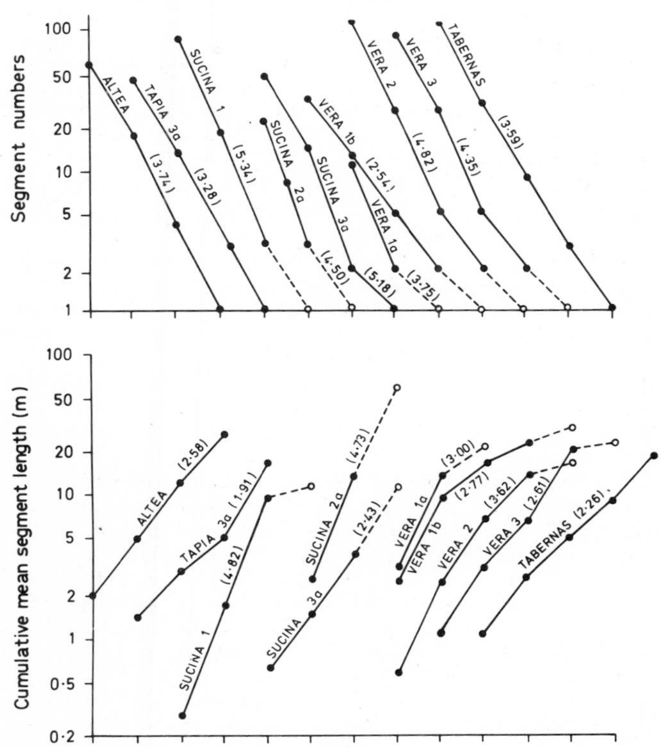

Figure 18.6 Morphometric properties of selected study sites. Figures in brackets are mean of first and second order bifurcation and length ratios respectively. Open symbols are for open-ended networks.

areas elsewhere in Spain (Thornes, 1976, Scoging and Thornes, 1979). Rillflow is important as is evidenced by the clarity and continuity of the rill networks on the badland sites. Supplies of rillflow may move through the surface material at depths of up to c30 mm, rather than over the surface itself, at least until after thorough wetting. In places the surface is caked and sediment appears to be washed from the weaker layers just below. At Altea there is evidence of surface creep and at Sucina 1 and Vera of mudsliding of thin surface layers.

Surface properties do not clearly differentiate piped from non piped sites. What does appear important is the nature of the material at depth, particularly the presence of differential porosity, solubility or strength together with surface features that allow the concentrated penetration of surface water, such as deep tension cracks or desiccation cracks. In southeast Spain the tendency for surface crusting is undoubtedly conducive to pipe formation. Over long periods of time crusting produces massive indurated caliche crusts (Harvey, 1978) but over shorter periods the vertical movement of soil moisture by capillarity and intense surface desiccation produces a lightly crusted layer over less consolidated subsurface layers. In the Quaternary fills at the Tapia sites this factor associated with differential cementation at depth and locally with tension cracks is conducive to piping. At Vera lightly crusted surfaces and the presence of gypsum sheets and veins within the marls leads to piping. The poor development of piping elsewhere appears to be due to the uniformity and compactness of the other deposits.

Piping and Gully Development

Although there has been a long history of Holocene hillslope erosion in the western Mediterranean (Vita Finzi, 1976, Butzer, 1961), erosion does appear to have increased recently. The Tapia gullies have developed since the abandonment of valley-side agricultural terraces. The active gullies at Vera are cut into relatively stable vegetated hillslopes and the trench gully at Vera 4 cuts directly into bedrock without any evidence of previous trenching. The study sites seem to represent not only differing patterns and rates of erosion but differing ages and stages of gully development within three general groups.

The simple badland sites are dominated by surface or near surface processes and show little piping. Sucina 1 and 2 represent early stages of gullying on otherwise stable vegetated hillslopes with more advanced badland development at Altea, Tapia 3a, Sucina 3 and particularly at Tabernas.

The Tapia trench gullies show deep pipe development both of crust protected pipes (Tapia 1) and tension crack pipes (Tapia 2 and 3). Subsurface flows enter through cracks, and concentrate in pipes which may be enlarged to a considerable size before they collapse to form vertical walled trench gullies. Piping is most strongly developed where the trenches are most recent, in the steep-walled head areas or below major within-gully headcuts at Tapia 1. As the side slope angles reduce rilling and badland development begin and piping becomes less important.

Vera badlands show important pipe development in relation to a lightly crusted surface and the presence of subsurface gypsum veins. The contrasts between Vera 1a, 1b and 2 reflect differences in aspect controlled erosion rates (Yair *et al.*, 1980). North facing slopes have a denser vegetation cover and a thicker and more resistant lichen crust than south facing slopes, even where both slopes are ungullied. Since the lichen crusted surface is largely intact on the north facing slope (1a), partly intact between gullies on the intermediate slope (1b) and has been largely removed by gullying on the south facing slope (Vera 2) it seems likely that these gullies represent an erosional sequence, with a progressive change in the role of piping (Table 3). Initially widely spaced discontinuous gullies (Vera 1a) fuse, trench and develop associated pipe systems (Vera 1b). Gully and pipe erosion continue, most of the pipes collapse and the lichen crusted surface is largely removed by which time (Vera 2a and 3) badlands have developed and surface or near surface processes now dominate. Further major piping appears to be prevented by the loss of the crusted surface and by the rapidity with which weathered material is removed.

Morphometric Implications

Two questions may be asked in relation to the morphometric implications. First, do the three conditions suggested above show distinctive morphometric or network characteristics that may be related to piping? Second, on the assumption that the progressive development of piped areas may lead to the development of poorly piped or non piped badlands, do such badlands differ from those developed through surface processes?

Conventional network analysis using the Strahler (1957) method has been carried out on examples of each type of network except the trench gullies where comparison is impossible. (figure 6, tables 3). Drainage densities vary considerably and although the low values at Vera 1a and 1b are associated with important piping this, as at Sucina 2, is due primarily to the limited development of gullying and the inclusion of ungullied areas within these drainage areas. Tabernas and Vera 2 with low infiltration rates, high silt clay content and steep slopes both show similar and high drainage densities. Likewise network characteristics reflect maturity of development rather than piping conditions, the more mature networks tending towards lower bifurcation ratios and lower length ratios.

Figure 7 shows 10° orientation rose diagrams for example networks. Tapia 2 and 3, the two gullies strongly influenced by tension crack piping, show multi-peaked distributions that reflect crack orientations at variance with other channel segment alignments. Tapia 1, without major tension cracks has a simpler orientation distribution. Vera 1b, with the maximum piping at Vera, shows orientation peaks both with slope and gypsum vein alignment. The simple badlands show strong main channel alignment with low order channel orientations grouped, often bimodally, on either side. No distinction can be made between those badlands that appear to have developed through piping (Vera 2 and 3 backslope) and the others.

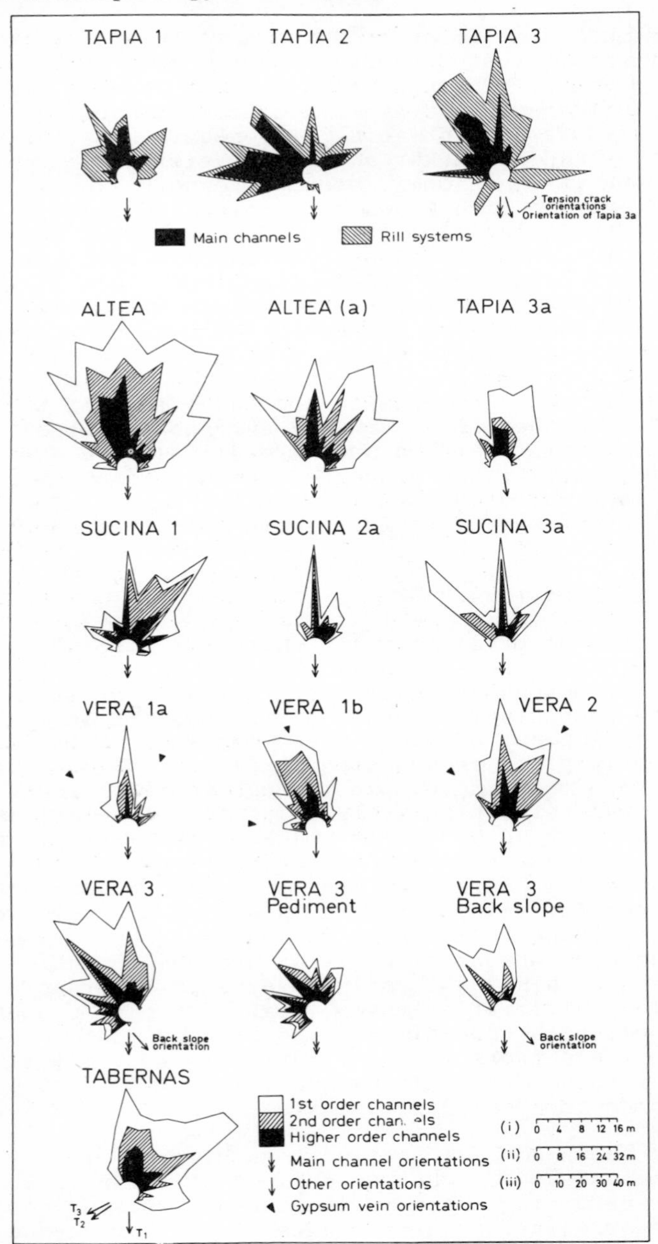

Figure 18.7 10° channel orientation rose diagrams for selected study sites. Tapia 1, 2 and 3 do not show orientations of minor rills and wash lines. Scales: (i) Sucina 1, (iii) Tapia 1 and 2, (ii) all other sites

The main morphometric characteristics related to piping are expressed in terms of channel alignment, particularly in relation to tension cracks and to subsurface veining but where erosional development has progressed to form a badland topography no morphometric distinctions can be made between badlands where pipe collapse appears to have been important and areas developed by surface processes.

CONCLUSIONS

Piping has an important role in the erosional development of some gully systems and badland areas in southeast Spain, especially where soil conditions produce a light surface crusting over a less consolidated layer in thick poorly consolidated alluvial fills. Secondly it develops where similar soil conditions occur over relatively weak bedrock but one which includes differential strength, or porosity such as created by the presence of gypsum veins. Thirdly it occurs where deep trenching into weak materials has produced near vertical walls conducive to tension crack formation.

The most obvious morphometric influence of piping is on channel alignment, particularly in the tension crack, trench gully situation. Elsewhere even in areas prone to piping its influence may show progressive change, decreasing in importance as badland topography develops.

Acknowledgements

I am grateful to the University of Liverpool research fund for a grant towards the cost of the field work, to the staff of the drawing office, particularly to Sandra Mather, and of the photographic section of the Department of Geography, University of Liverpool for producing the diagrams. I am also grateful to Dr. J.M. Hooke for assistance with some of the survey work.

19 Throughflow and pipe monitoring in the humid temperate environment

M. Anderson and T.P. Burt

The role of variable or partial source areas in controlling the hillslope soil and water discharge to streamflow is now well established (Hewlett and Nutter, 1970). Whilst such contributing zones are manifest in numerous field study catchments in the humid temperate region, it is difficult to make useful empirical generalisations regarding streamflow generating source areas due to the inherent variability of such parameters as infiltration rates, suction-moisture characteristics and other soil properties. Nevertheless research to date has illustrated the spectrum of streamflow hydrographs which we can associate with different soil and basin types. Figure 1 illustrates the range of response with (a) gentle slopes of low permeability soils yielding return flow with the consequential near immediate streamflow peak generated by overland flow, (b) somewhat steeper slopes with more permeable soils generating a significant delayed throughflow pulse and (c) on steep slopes (30°) with highly permeable soils where the throughflow pulse can dominate the hydrograph in yielding a peak flow in excess of the immediate overland flow and direct precipitation induced peak. As figure 1 indicates there is now ample field evidence available to support this spectrum of responses and coarse assignation of process dominance to response is possible.

Currently, however, research in contributing area modelling faces something of a dilemma. To refine the 'coarseness' of process association it would be seen by some workers as necessary to treat small plot studies in detail with short time increments to establish a physically-based distributed model of slope segment response to precipitation input. Whilst acknowledging this is one possible research strategy we argue here that it is not optimal in variable source area monitoring and modelling. Rather we stress the need to maintain a spatial sampling zone commensurate with the spatial extent of the contributing area and the processes associated with its maintenance. Two principal problems ensue from adherence to the former strategy:

(i) the very great spatial variation in soil parameters such as hydraulic conductivity (e.g. anisotropy, and variations in anisotropy) can result in severe calibration problems as spatial generalisation of the model is enhanced from the plot study,

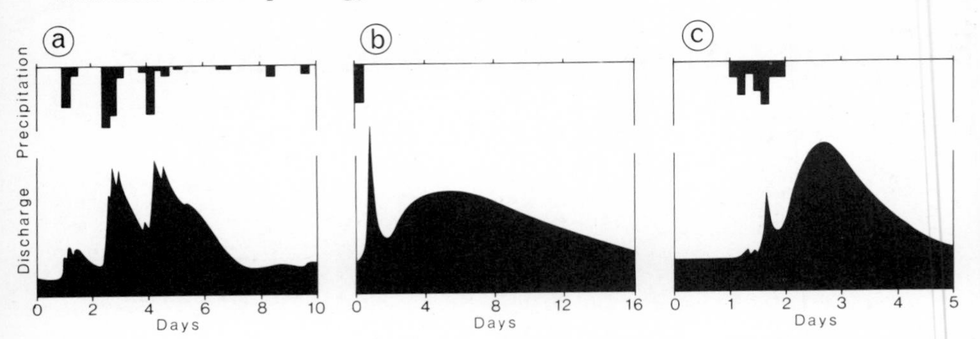

Figure 19.1 Hydrograph form variation
(a) overland flow generated - shallow, clay slopes, (e.g. Anderson and Kneale, 1980a)
(b) overland flow generated with throughflow component - somewhat steeper, more permeable soils, (e.g. Weyman, 1970).
(c) throughflow dominated hydrograph - steep (30°) and very permeable soils, (e.g. Anderson and Burt, 1977).

(ii) concentration on small plot study cannot guarantee to isolate, the perhaps dominant, processes occurring at a larger scale. Anderson and Burt (1980), for example, argue this in the context of soil water convergence induced by topographic control necessitating relatively widespread instrumentation to identify its existence and dominance (figure 1c) in the hydrograph.

Accordingly in this paper we examine the identification of throughflow and pipeflow generation from field research we have undertaken *apropos* the strategy of improving field monitoring methods in line with dominant process investigations at scales on which those processes, such as topographic convergence, are found to occur. In addition this stresses the importance of 'three dimensional' aspects of variable source area monitoring which have hitherto been explored only by implication (Weyman, 1970) or qualitatively (Dunne *et al.*, 1975).

GENERAL PROBLEMS OF IDENTIFYING CONTROLS ON THROUGHFLOW GENERATION AND PIPEFLOW

Numerous instrumented catchment studies over the last twenty years or so have done little more than to illustrate the range of hillslope hydrologic processes that occur. Generalisation of controls on throughflow is difficult but must be attempted. There is ample empirical evidence to demonstrate the generation of throughflow by (a) reduction in permeability with depth (Whipkey, 1965) and (b) topographic hollows inducing soil water convergence (Anderson and Burt, 1977 - discussed in more detail below); where soils are deep and permeable, there is evidence that throughflow is frequently a dominant process. Moreover in the context of the controls

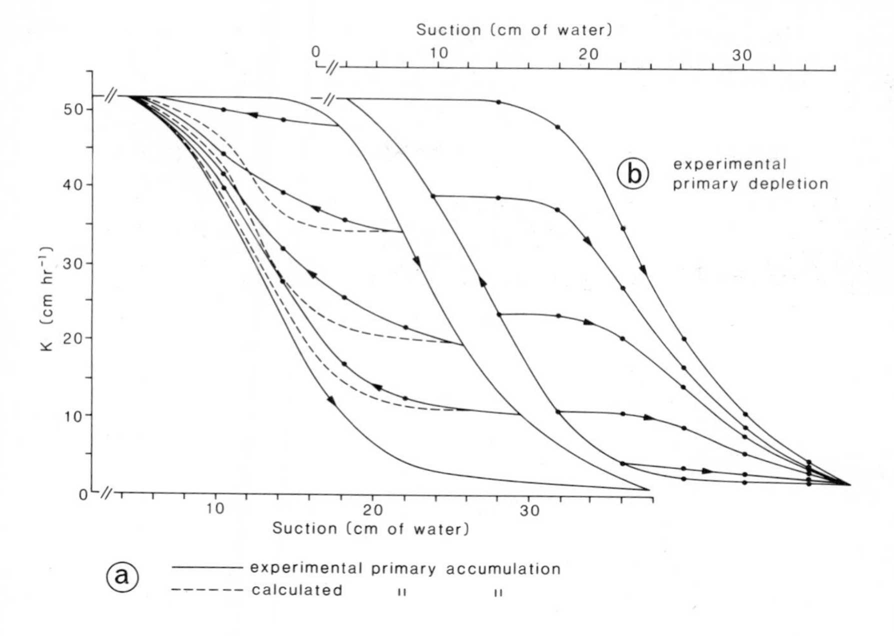

Figure 19.2 Hysteresis in the unsaturated hydraulic conductivity-moisture content curve
(a) set of experimental primary accumulation curves (solid lines) and calculated primary accumulation curves (broken lines)
(b) set of experimental primary depletion curves (after Poulovassilis and Tzimas, 1974)

on permeability itself, not only is hysteresis in the suction-moisture curve of significance, but there is therefore a significant hysteresis in the unsaturated hydraulic conductivity-tension curve (figure 2 illustrates the importance of such a curve).

The salient point is that whilst nearly all the controls on throughflow can be specified from theoretical, laboratory and field observations *singly*, the most important research question relates to the prediction of the occurrence and volume of throughflow in a *combinatorial* field situation, where many but not necessarily all controls operate simultaneously at different scales (macro-spatial in terms of hollow convergence to micro-temporal in terms of unsaturated hydraulic conductivity hysteresis). Current field research must increasingly concern isolation of the *dominant* control or controls in space and time rather than mere replication of detailed site to site instrumentation, if predictive models are to be realistically calibrated. The Freeze (1972), Beven (1976), Kirkby *et al.*, (1976) approaches to modelling hillslope hydrographs illustrate just this requirement with distributed physically based models dealing in discretised time and space domains. From field throughflow control-orientated research we must be able to specify the nature of these domains increasingly closely, as well as the nature of the representative parameter values - an element which, as Freeze (1978) observes, is lacking in current field approaches. It is evident from the foregoing discussion that this will only be feasible *if* the field experimental design is established with this as the principal criterion of operation. We have already shown above and discussed in detail elsewhere (Anderson and Burt, 1978a), that insufficient attention has been attached to field equipment design and installation to estimate throughflow process controls with adequate precision. Below we argue the need for an equally sophisticated and rigorous approach to control process identification in tackling topographic soil water convergence and throughflow generation. The review by Dunne (1978) of field studies of hillslope hydrology processes as well as specific investigations is seen as supportive of our view that accurate throughflow parameterisation for modelling, as well as for process identification, can be achieved only by instrumentation capable of yielding continuous time soil water potential data over appropriate spatial units such as hillslope hollows and spurs. Sectional, two-dimensional studies (Weyman, 1970), may illustrate and calibrate certain points of relatively important detail, but are not sufficient to guarantee isolation of dominant controls if these are three dimensional in nature.

At all scales of flow within the soil the elements of the pore size distribution which contribute most significantly to soil water discharge are of principal concern. Beven (1979), for example, has formulated an initial model for soil water flow in a combined micropore-macropore system. Although macropores may provide only a very small proportion of the total pore space of the soil, when flowing under saturated conditions they may allow flow velocities several orders of magnitude higher than in the micropore system. Figure 3 illustrates predicted macropore discharge rates as a function of macropore water content for a hypothetical porous medium with microporosity of 0.49 and a macroporosity of 0.01. Of course, relatively large discharge *variations* at this scale can be meaningfully averaged at the hillslope scale. However, increasing the scale of 'macropores' to that of pipes, whilst analogous to that described, presents a much greater field identification and modelling problem. Although discharge variations as controlled by the soil water regime may be similar, (figure 3), the *absolute* discharge yield can now dominate the type of response. An additional complication is that unlike macropores and small cracks, pipes are not present on all slopes or in all catchments. The principal initial problem is thus pipe identification and mapping, for, as Newson (1976) observes, the hydrological significance of pipes may be reduced because pipes rarely discharge directly into the surface streams but rather play a most important role in controlling the partial contributing area as they terminate in the valley bottoms or hillslope basal segments. Whilst the controls upon pipe formation have been reviewed by many workers (Jones, 1976), the detailed hydrological response has been relatively neglected (Atkinson, 1978).

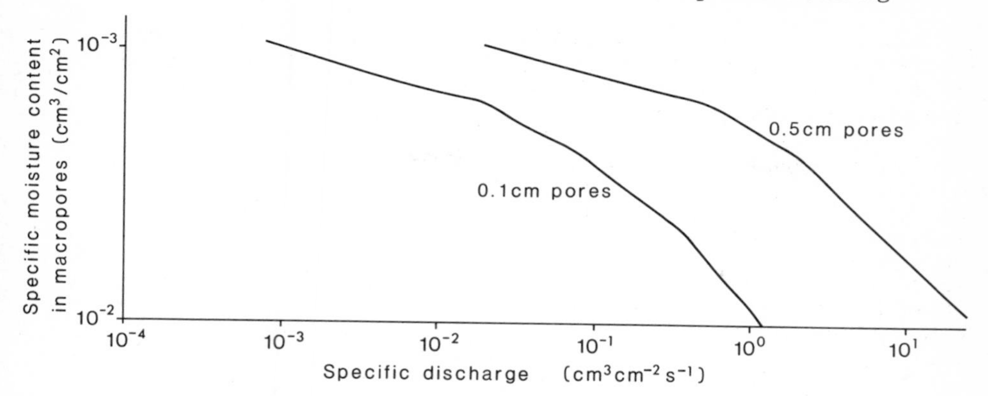

Figure 19.3 Predicted macropore discharge change with macropore water content with
(a) 0.5 cm pores and
(b) 0.1 cm pores with a background micropore flow at 0.01 cm min^{-1}. (after Beven 1979)

Thus pipes are seen here as a special end member of the spectrum of 'micropore/macropore' pore size distribution which in relation to their absolute discharge capacity deserve much greater attention than has hitherto been acknowledged.

MONITORING STRATEGIES FOR THROUGHFLOW AND PIPEFLOW

Because of the spatial variations thought to exist in the parameters controlling hillslope discharge, identifiable through Darcy's Law

$$V = -K\nabla\phi$$

where V = water flux
K = hydraulic conductivity
ϕ = total hydraulic potential

which controls the flow of water in both the saturated and unsaturated cases (in all but the very highest velocities), the emphasis has been upon the direct measurement of throughflow. This has been undertaken (in the majority of field investigations) in preference to the intermediate steps of measuring K or ϕ. However there are considerable problems in correctly inferring the active processes from direct measurement techniques. Trough installation can distort the local soil water regime so that in an extreme case saturated throughflow is observed on a slope where unsaturated flow is occurring. Secondly, improper installation of throughflow troughs on slopes where permeability decreases markedly with depth can lead to incorrect association of horizon discharges. Anderson and Burt (1978a) report that laboratory testing of trough installations shows that such discharge errors can exceed an order of magnitude. Certain recent studies have, however, sought to monitor throughflow and soil water movement using tensiometers. Weyman (1974),

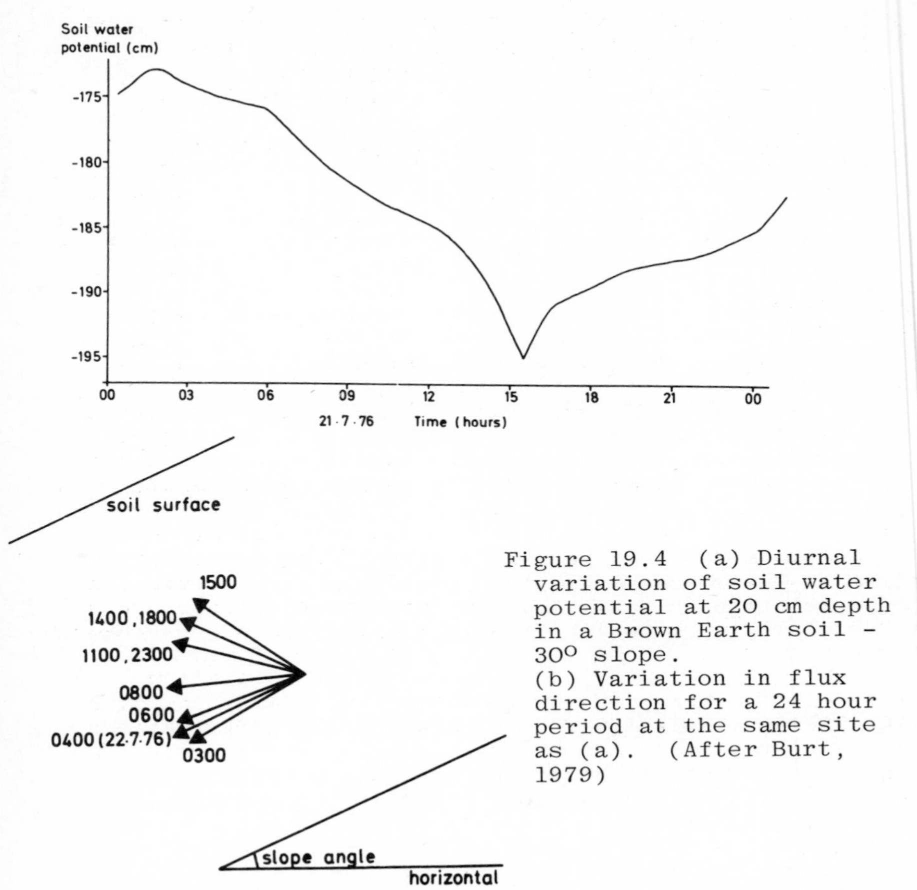

Figure 19.4 (a) Diurnal variation of soil water potential at 20 cm depth in a Brown Earth soil - 30° slope.
(b) Variation in flux direction for a 24 hour period at the same site as (a). (After Burt, 1979)

Harr (1977), and Nortcliff *et al.*, (1979) used manual tensiometers read when an operator is available, to isolate hillslope soil water processes. Automatic and continuous reading of soil water potentials, however, is desirable, and has been undertaken since the early 1970s (e.g. Wilson and Ligon, 1973). In steep, well draining soils continuity of recording and reading coupled with the accuracy of a transducer-based system for the tensiometers are necessary. Williams (1978) has outlined the general high accuracy of such a system, whilst Burt (1979) has provided field evidence of the rapidity of soil water potential change and flux direction demonstrating clearly how the controls on soil water movement change dynamically (figure -4). Whilst such changes are expected in permeable material, in low permeability clay Anderson and Kneale (1980b) report *equally* fast but shorter duration changes in soil water potentials occurring on 26° clay slopes; changes facilitated by high permeability reswelled lenses in the clay after summer cracking. The salient argument here is that high resolution field monitoring must be undertaken - it is insufficient to rely too heavily upon inductive instrumentation schemes since these cannot be guaranteed to yield dominant soil water processes. In this field monitoring has hitherto relied too heavily upon an insufficiently rigorous sampling approach and too much attention has been devoted to two-dimensional, non-continuous soil plot based studies.

The relative absence of pipe discharge monitoring has already been noted. Gilman (1971) and Pond (1971) describe a tank syphon system which is capable of recording the very flashy response typical of pipeflow. Other workers have used V notch weirs (Knapp, 1973) and direct cylinder measurement. Anderson and Burt (1978b) utilised a more precise electrical stage recorder system to continuously monitor discharge emerging from a pipe at the base of a topographic hollow. This system facilitates simultaneous input of a number of channels so that exact synchroneity between recording weirs on both the stream and pipe outflows can be obtained. Wilson (1977) reports on the use of fluorescent dyes to estimate average travel times in pipe networks on the Brecon Beacons, but discontinuous pipe networks make at best for somewhat fragmented spatial estimates which are difficult to treat analytically.

To stress the need for appropriate instrumentation strategies, that we have described as a prerequisite for the correct isolation of the controls on throughflow generation, the following two sections provide for a discussion of such controls in two contrasting catchment types; one where throughflow dominates the storm peak flow generation (figure 1c) and one where throughflow dominates baseflow generation (figure 1a). We also provide results from detailed pipeflow monitoring and argue for the continuation of more detailed field investigations.

THROUGHFLOW DOMINATING THE STORM PEAK FLOW GENERATION

Plate 1 illustrates a typical catchment with steep slopes (25°) and permeable soils ($K = 1 \times 10^{-6}$m sec^{-1}). The typical hydrograph form from the catchment is dominated by the delayed throughflow pulse (figure 5). Such responses were observed following all significant precipitation events, except those relating to the drought of 1976 when the soil moisture deficit was sufficiently large to eliminate the throughflow peak. The production of saturated overland flow is limited to marshy headwater zones (in common with similar studies; Dunne, 1978), for in general the soils are too deep and permeable, and the slopes too steep to allow more widespread production of surface saturation. Following the period of storm runoff, the delayed peak in stream discharge is provided entirely by throughflow, generated by the convergence of soil water in the hillslope hollows distinguished in plate 1. An automatically recording tensiometer system employing a Scanivalve fluid switch and pressure transducer (Burt, 1978) was installed in a hillslope spur and hollow, with 44 tensiometers on a grid basis covering 2500m^2. (Anderson and Burt, 1977). Thus the deepening of the saturated wedge at the base of the hillslope hollow could be continuously monitored together with the degree of convergence

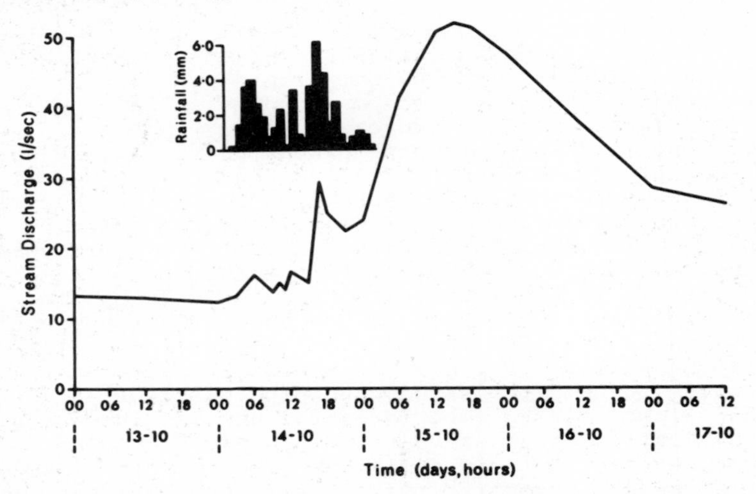

Figure 19.5 Precipitation and discharge for a storm in Bicknoller Combe. (After Anderson and Burt, 1978c)

exhibited by the soil water flow paths upslope. This deepening is responsible for the delayed peak in stream discharge noted in figure 5. Volumetrically the delayed throughflow response dominates the total storm runoff response: using simple hydrograph separation methods it is calculated that the runoff percentage provided by the 'precipitation' peak was 0.78%, whilst the delayed throughflow response represented 20.57% of the precipitation input. The throughflow peak for the event illustrated is almost double that of 'precipitation' peak.

Figure 6 shows maps of soil water potential at 60 cm depth for the instrumented hollow and spur during the event shown in figure 5. The maps show that during and soon after the precipitation there is widespread saturation in the hollow. The slope is then observed to drain progressively except in the lower hollow where convergent flow causes the redevelopment of an enlarged saturated wedge, which provides the delayed peak in stream discharge. Once the convergent flow ceases to provide a significant input to the hollow base, the saturated wedge quickly contracts to its present extent. Figure 7 shows a downslope section for the spur zone providing plots of soil water potential. This indicates that even in the largest throughflow events the saturated wedge on the spurs is limited to a relatively small temporary development at the slope base. By contrast the hollow wedge at this time is greatly expanded and is very close to the soil surface.

Not only was throughflow generated by soil water convergence due to topographic controls, found to occur in all similar hollows throughout the catchment, but the nature and degree of convergence remained constant irrespective of the soil moisture state; in addition a saturated wedge was maintained throughout the entire range of soil water states occurring (Burt and Anderson, 1980). This condition, impor-

Plate 19.1 View of Bicknoller Combe, Somerset

tant from the modelling standpoint, is due to the dominance of elevation potential in the total potential. A most important contrast to this is illustrated below, where shallow slope throughflow generation is described.

Throughflow dominating baseflow generation

On gentle slopes (approximately 6° - plate 2) with generally low permeability ($K = 1 \times 10^{-8}$m sec^{-1}) previous evidence suggests the relative 'unimportance' of throughflow. Unquestionably this view stems from the emphasis accorded to the discharge peak as the principal component to be predicted. Figure 8 illustrates the hydrograph form from the catchment shown in plate 2 in which overland flow dominates the discharge peak. This contrasts with the throughflow dominated peak of figure 5, and this contrast is further augmented by soil water potential plots (80 cm depth) - figure 9). In this instance saturation at the hollow base is maintained throughout the period of recession, and soil water convergence is seen to initially accord with the hillslope hollow, but then through the recession to shift across

Plate 19.2 View of Winford catchment, Avon

onto the downstream spur. Calculations based upon Darcian models confirm that throughflow is the generating mechanism for baseflow (Anderson and Kneale, 1980a). Thus in low angled topography, hillslope hollows do not necessarily remain the principal source areas at all stages of drainage and the hillslope contributing areas of saturation are maintained for comparatively long periods - in summer storms the hillslope saturated areas generally do not decline to pre-storm levels even one month after the precipitation event.

It is emphasised here that to isolate the baseflow generating mechanism, even in relatively impermeable material, a continuously recording, high resolution tensiometer array is needed. For gently sloping topography the zone of soil water convergence is seen to move rapidly in the shallow, low permeability material. These findings illustrate that even in gentle topography, elevation potential plays a significant role as convergence still occurs to the saturated wedge, but that the soil water potential has greater influence over the total potential (and hence the soil water flow paths) than in the steep topography illustrated in plate 1 and figures 5-7.

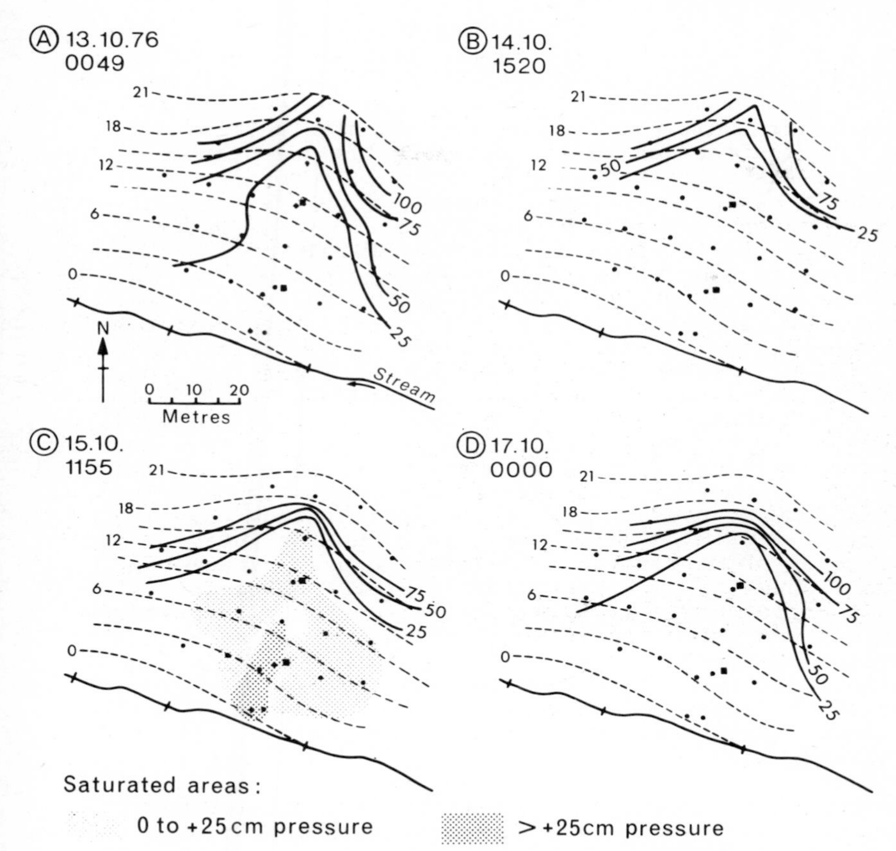

Figure 19.6 Soil water potentials at 60 cm depth for the instrumented hillslope area of the catchment (plate 1) at selected times during the storm event showin in figure 5. (After Anderson and Burt, 1978c)

THE ROLE OF PIPES

The importance of a natural pipe network in catchments is emphasised by Dalrymple *et al.*, (1968) and Jones (1971, 1976). Pipes can transfer water rapidly (Wilson, 1977, reports velocities of 1m sec^{-1}) increasing the variable source area considerably by effectively lengthening the stream network. Generalisation, and hence modelling, of pipe flow contributions has not been satisfactorily undertaken as far as the authors are aware. The principal reason here, already outlined above, is the very site specific presence/absence, continuous/discontinuous configuration of the network. The variable capacity and the non-linearities induced by the variable soil water content of the hillslope zone to which the network finally contributes, if not directly to the stream, are additional complications. With these qualifications which have hitherto restricted the inclusion of such phenomena into physically based distributed models, certain more qualitative comparisons of varying pipe systems can be made, together with some pointers on aspects of pipe flow which need future attention in studies of hillslope hydrology. Upland areas of Britain that have a well developed peat cover frequently exhibit pipe development. Pipe discharge records are rare in the literature, but figure 10 illustrates for the Shiny Brook catchment on the Southern Pennine hills of west Yorkshire (a) a quickflow pipe response in which the pipe response mirrors that of the overland flow generated discharge peak, and (b) a pipe with a response lagged two hours with respect to the stream discharge.

Figure 11 shows the position of the water table and equipotential lines for a slope in the undissected cotton-grass moorland subcatchment. The water table is typically very close to the soil surface, so that even in very low intensity rainfall events, saturation overland flow will be produced. In higher intensity events, Hortonian overland flow is common, because of the low infiltration capacity of the peat ($1x10^{-8}$m sec^{-1}). The flow directions suggest that drainage is primarily vertical into the peat except near the stream channel where a significant lateral component exists in the surface horizons. In fact, given the strong anisotropy of peat soils, it is apparent that greater volumes of lateral flow do occur than is suggested by the flow lines. Even so, this lateral throughflow only becomes significant close to the stream banks where a greater hydraulic gradient is produced due to the elevation of the peat relative to the stream bed.

This strong lateral flow close to the streams seems to be responsible for the development and the maintenance of one of the two types of pipes present at Shiny Brook. Periods of desiccation are common enough, even in an area of such high rainfall, to allow a network of shrinkage cracks to be created (Burt, 1980). Some of these are exploited by throughflow during subsequent wetter periods when the water table recovers to its normal position. The pipes which result, either within or at the base of the peat, produce outflow discharge for relatively long periods, being fed by the saturated zone within the peat. A typical pipeflow hydrograph for one of these pipes is given in figure 10(b). The other type of pipe at the Shiny Brook catchment is a near-surface phenomenon, possibly formed by erosion by surface water flowing between the cotton-grass mounds. Subsequent growth of the peat could then 'roof-in' these emphemeral channels. Flow through desciccation cracks may also contribute to the formation of these surface pipes. Such pipes mainly flow during periods of overland flow generation, and serve to conduct surface water from areas upslope to the stream channel, thus conforming to the type of pipe described by Jones (1971). These may provide contributing areas, which are not contiguous to the stream, with a means of rapid drainage into the stream system. A typical hydrograph for a near-surface pipe is shown on figure 10(a). The rapid response of the pipe system shows that it is fed largely by overland flow.

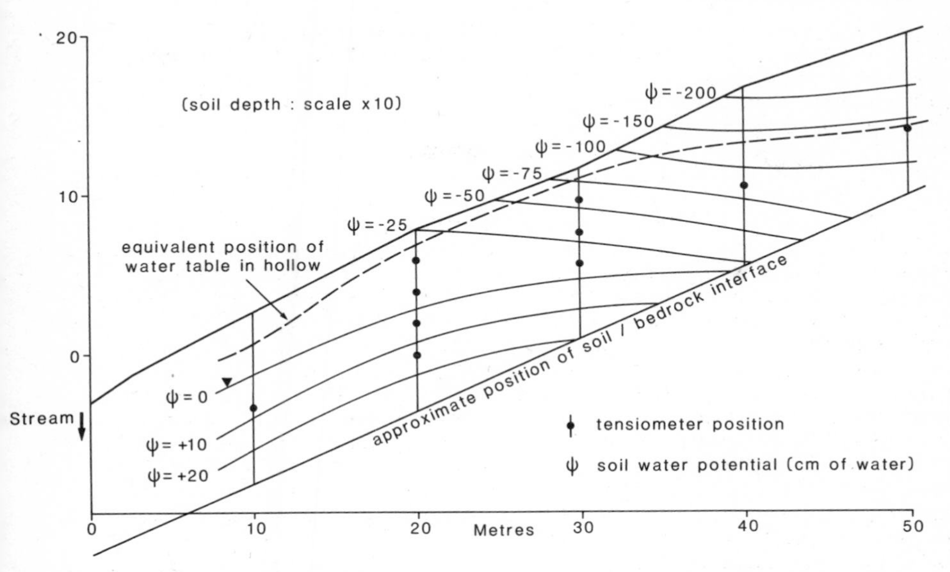

Figure 19.7 Contrasting soil water conditions on the spur and hollow at Bicknoller Combe for the storm shown in figure 5

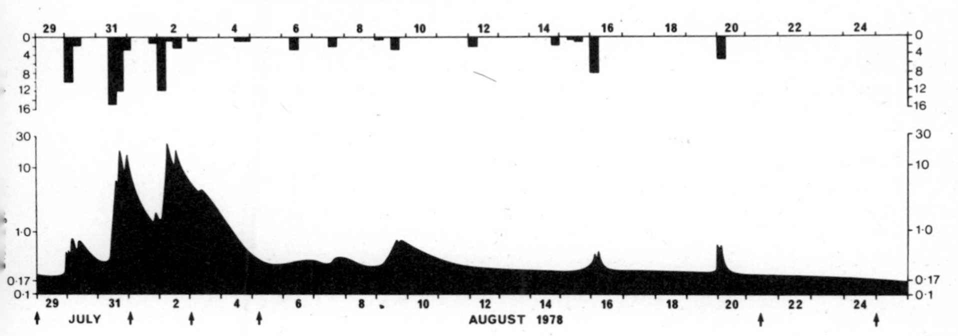

Figure 19.8 Typical hydrograph of Winford catchment

Both types of pipe are common in both erosional areas of Shiny Brook. In the open moorland, the pipes tend to remain small (below 5 cm. diameter); this may be related to the high soil moisture conditions which prevail in the uneroded peat, thus not allowing significant pipe erosion to develop. In the eroded peat, the lower pipes (fed by the saturated wedge) remain very small (below 1 cm diameter). However, the near-surface pipes which drain saturated areas not contiguous to the stream, appear to be much more liable to erosional development, perhaps since the surrounding peat is often unsaturated. Several pipes have been eroded sufficiently to cause collapse during individual storm events, and where the pipes remain intact diameters up to 50 cm have been observed. The headward erosion of such pipes may

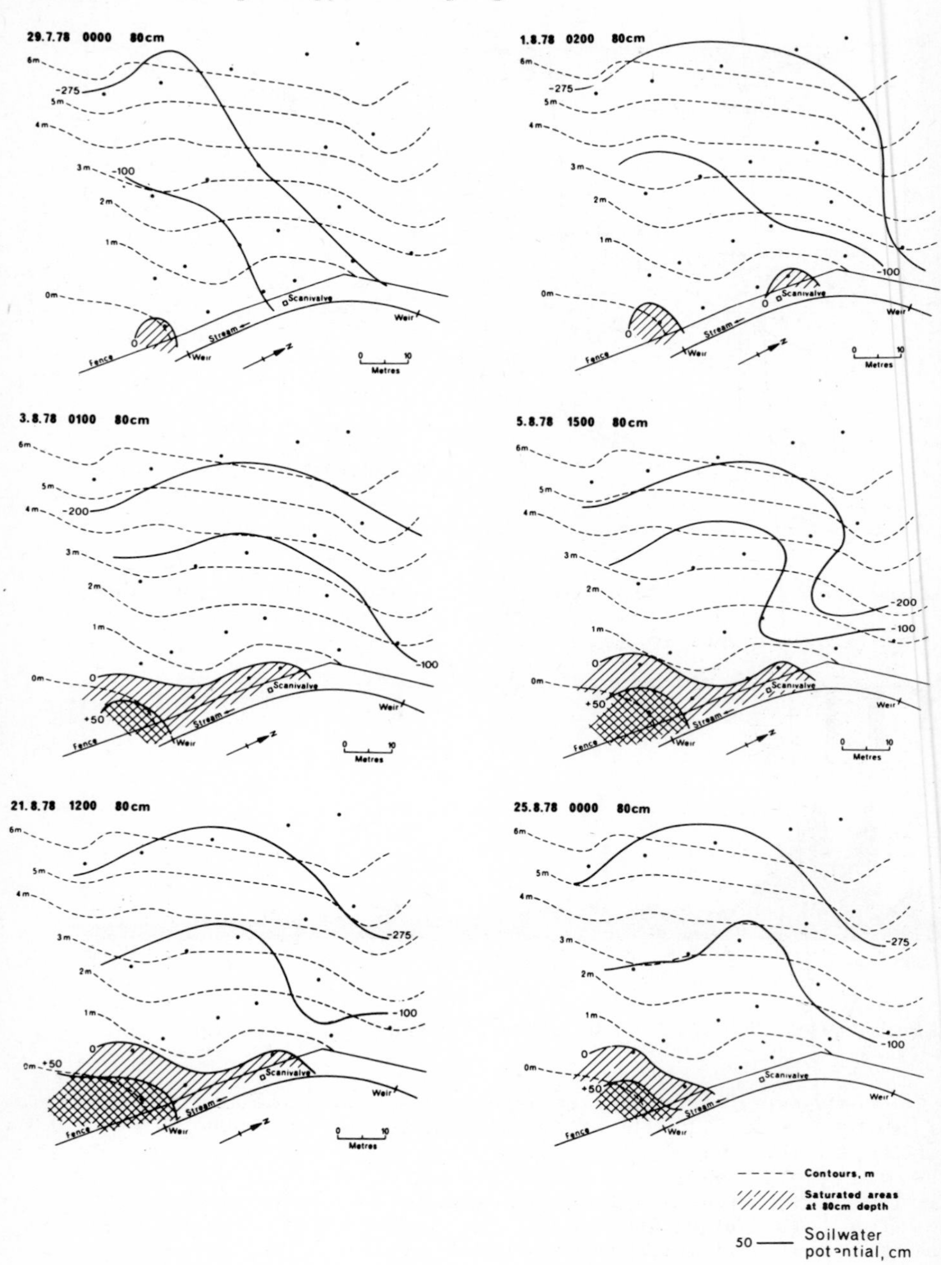

Figure 19.9 Soil water potentials for selected times during the precipitation event shown in figure 11

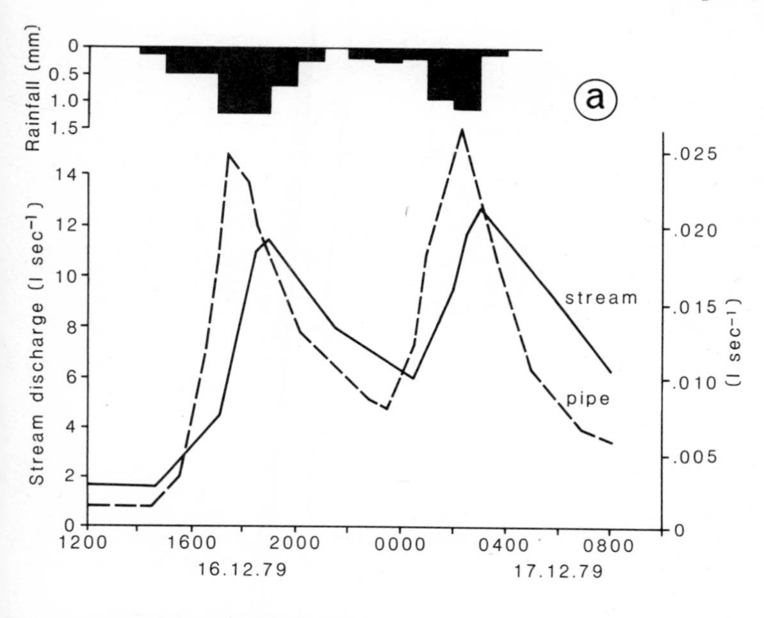

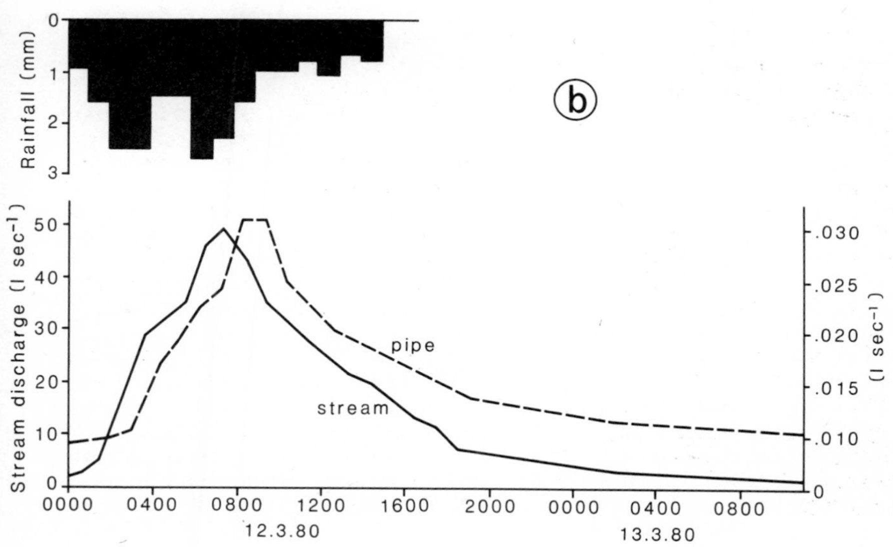

Figure 19.10 Pipe discharge responses in Shiny Brook catchment
(a) quickflow pipe
(b) delayed pipe

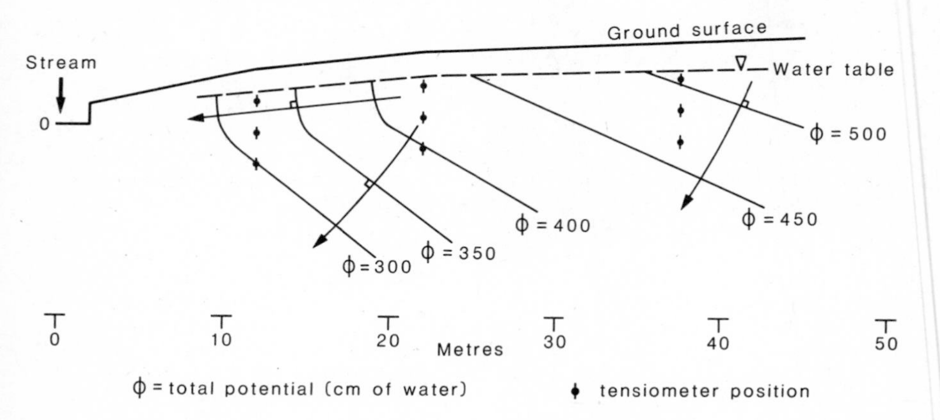

Figure 19.11 Total potential relationships on hillslope of Shiny Brook catchment for 6 March 1980

well be important to the continued extension of the stream network in the heavily dissected interfluve areas. Generally, however, the importance of pipeflow to the hydrology and geomorphology of Shiny Brook is limited, with overland flow erosion and channel incision providing the main erosive agents. In other nearby catchments the importance of piping may be greater due either to generally increased slope angles or to a change in the bed material beneath the peat from clay to weathered sandstone, which may allow larger pipes to develop at the base of the peat. Where the pipes remain small, the presence or absence of piping may not be particularly important therefore with respect to modelling the subsurface flow. However, where larger pipes do develop, the specific effects of such features must be considered with respect to both runoff and erosion.

The storm runoff at Shiny Brook is therefore dominated by overland flow. Throughflow dominates the recession flow, maintaining a low but relatively constant outflow over long periods. The presence of pipes contributes to both stormflow and recession flow. The near-surface pipes may be particularly important in modifying the storm hydrograph by routing overland flow fron non-contiguous contributing areas. Such a rapid transport process may help to produce the very 'peaked' storm hydrographs which are typical of Shiny Brook, but which are unusual given the low slope angles involved.

On a more general level we argue that the hitherto somewhat fragmentary approach to pipe hydrology in humid temperate areas must be augmented by a closer examination of the soil water potentials of the pipe areas. Only then can the non-linearities in pipe discharge response (analogous to those of the macropore simulation in figure 3) be explained from site to site and the dominant factors isolated so that more relevant generalisations can be made.

DISCUSSION

There are two principal demands for the undertaking of field based research - to provide representative parameter values for the calibration of physically based models and to explore the operation of processes in the field. The instrumentation experimental design for both these strategies need not be similar. In the case of throughflow generation on hillslopes we have been concerned with establishing dominant process mechanisms. It is clear that the relatively sophisticated soil water monitoring programme undertaken in two contrasting sites (plates 1 and 2) has facilitated the isolation of the exact nature of topographic control of soil water convergence and throughflow generation on hillslopes. We have stressed the need for rigour both in field instrument installation (Anderson and Burt, 1978a) and in adequate spatial and temporal parameter monitoring (figure 4). By contrast in the context of pipeflow we have stressed the lack of an equally sophisticated field instrumentation approach, which has resulted in the hitherto somewhat qualitative and fragmentary approach to the inclusion of pipe networks in hillslope hydrology studies.

20 Experimental studies of pipe hydrology

J.A.A. Jones

It is now a decade since soil pipes were first seriously considered as an element of the drainage network in humid temperate landscapes such as Britain. Considerable advances have been made in our understanding of their nature and function during this period, although current experimental studies are revealing that there is still much to be learnt. The feelings expressed by Carson and Kirkby (1972) that 'piping is very important for sub-surface flow and debris removal, but its quantitative contribution cannot yet be estimated' have since been largely substantiated and quantified in a number of experimental basins, notably in Wales in the Upper Wye catchment by workers at the Institute of Hydrology (Gilman and Newson, 1980) and in the headwaters of the River Rheidol (Maesnant) by Jones (1978) and Jones and Crane (1981), and in the University of Bristol East Twins Catchment in the Mendips (e.g. Stagg, 1974; Finlayson, 1977). As a result, more positive albeit brief statements have appeared in recent British texts, for example, Weyman (1975), Ward (1975), and Rodda *et al.* (1976), the latter suggesting that minor floods may be generated by flow from permanent and intermittent pipes. Chorley (1978), Atkinson (1978), Weyman (1973) and Whipkey and Kirkby (1978) all considered the role of soil pipes in hillslope hydrology, and brief references to this role have already filtered through into school literature (Weyman and Weyman, 1977; Weyman, 1978; Smith and Stopp, 1979).

Nevertheless, the quantitative foundation for most of the generalisations made to date remains quite thin. It is also largely limited to the British environment, by no means a world norm, although work in semi-arid and arid environments in Canada and Israel by Bryan, Yair and associates (Bryan *et al.*, 1978; Yair *et al.*, 1980) is suggesting not only that piping may be hydrologically important there, but that it is a major element in subsurface flow contributions to runoff in environments often considered the last preserve of the Hortonian overland flow model (cp. Ward, 1975; Chorley, 1978. Morgan's (1972) observations in Malaysia also appear to merit an experimental investigation of pipeflow in humid tropical environments. However, the hydrological role of soil piping is still generally ignored worldwide. Lvovitch (1980) makes no mention of it in his review of the hydrological role of the soil; nor, indeed, do recent American texts (e.g. Viessman *et al.*, 1977).

Closely related to the hydrological role of soil pipe as a source of discharge is their hydrogeomorphological role in the development and morphology of drainage networks. Dunne (1980) has recently drawn much needed attention to this, although his analysis of the mechanics of pipe initiation has been perhaps over-influenced by the classic engineering case of piping development beneath the phreatic level in earth dams, as presented in Kälin's (1977) paper. There is, in fact, abundant evidence of pipe development and indeed of pipe collapse and channel development within a vadose context on natural hillslopes, following and extending structural voids such as desiccation cracks (e.g. Barendregt and Ongley, 1977), mass movement cracks (e.g. Jones, 1978) and biotic holes (e.g. Cockfield and Buckham, 1946). Indeed, a number of civil engineers have recently noted the importance of this form of *supraphreatic* piping even in artificial earthworks (e.g. Sherard, Decker and Ryker, 1972). Jones (1979) has suggested that the classic engineering process of 'boiling' beneath the phreatic surface, originally outlined by Terzaghi back in 1922 (cp. Terzaghi and Peck, 1966), may occur more commonly on the lower slopes, the alluvial toeslope and colluvial footslope of Conacher and Dalrymple (1977), while the other processes dominate higher up the hillslope. Desiccation cracking in particular has been seen as a major source of pipe development on upper slopes even in the wet environment of Plynlimon in Wales, with an annual precipitation of nearly 2200 mm, by Gilman and Newson (1980). The majority of these pipes appear to be ephemeral and not directly connected to the stream network. On the other hand, Jones (1978) adduced evidence to suggest that no more than about 10% of the pipe outlets in the banks of the Burbage Brook, Derbyshire, were due to cracking and that most appeared to be due to 'boiling'.

It is clearly of paramount importance that researchers and reviewers alike recognise the great variety of pipe forms, in terms of both initiating process and hydrological function. The wide variety in form and response means that many of the bold generalisations made hitherto can be very misleading. The operational significance of this variety is discussed in the following section. Meanwhile, Table 1 is an attempt to bring order into this range of forms, although it must be admitted that this is a gross, heuristic simplification.

THE MAESNANT EXPERIMENTS

The basin and the pipe networks

The Maesnant catchment lies on the western, windward slopes of Plynlimon, which rises to 752 m O.D. in the Cambrian Mountains, Wales. The basin covers 0.54 km^2 of mountain moorland, dominated by heath, grassland and bog communities, with an outfall at the Welsh Water Authority combined V-notch and rectangular weir at 465 m O.D. Measured annual rainfall lies in the range 1700-2100 mm with a slight tendency to a wetter period in early winter, and the stream has a median discharge of 0.015 $m^3 s^{-1}$. The bedrock is Ordovician greywacké, mudstone and grits (James, 1971), covered in much of the area by peat ($\leqslant$ 1m) and by solifluction materials ($\leqslant$ 5m) which form the river terraces (Watson 1970). Soils are generally poorly developed, ranging from mountain rankers of the Powys Series on the upper slopes to peaty gleyed podzols of the Hiraethog series, peaty gley soils of the Ynys series and undifferentiated peat soils on the lower slopes and river terraces (Rudeforth, 1970). Piping appears in all soils except the mountain rankers.

In 1974 a pilot experiment was planned to obtain the first continuous record of pipe discharge. One pipe was chosen for detailed study within a strict budget, with little prior knowledge of expected flow levels or the size of the tributary network. The experiment was successful for a 6 month period from January 1975 until the final breakdown of the chart logger and over 20 storm hydrographs were obtained for the pipe and the adjacent stream. The pipe appeared to have a seasonal regime, drying up in summer, with a mean discharge of 0.17 $l\ s^{-1}$ and a recorded maximum of 0.38 $l\ s^{-1}$. Exploratory analyses of the data, summarised elsewhere (Jones, 1978), suggested a number of interesting features, many of which have been corroborated or expanded upon by subsequent work discussed below.

The current Maesnant experiment began in January 1979 with funding from the Natural Environment Research Council. It aims to determine the quantitative importance of pipeflow to stream discharge and to investigate the factors responsible for variations in pipe discharge in the basin. A second weir has been built 730 m upstream from the outfall weir isolating a reach between upper and lower weirs over which the increase in stream discharge can be related to measured inputs from pipeflow, diffuse seepage and overland flow (figure 1). The first two tasks of the project were to develop suitable instrumentation and to map the pipe network.

The mapping exercise proved particularly interesting, not least because the networks were much more extensive than originally supposed. Likely pipe routes were picked out mainly by collapse features, blow-holes and the sound of flowing water. These sites were marked by numbered wooden stakes , and dye tracing using different coloured proprietory dyes was carried out to prove each connection. The stakes were later surveyed at a scale of 1:1000 (figure 2). Pipe tracing was best done during the winter half-year, at least over the river terrace area, which shows characteristics typical of a landsurface unit 2(2) in the Conacher and Dalrymple (1977) classification. Here the major pipes run at depths of 40-80 cm below the surface, with their beds tending to follow the interface between the peat O horizons and the mineral AE(g) horizons, commonly of silty clay loam texture with abundant gravel, cobbles and boulders in the matrix. Horizontal saturated hydraulic conductivity generally shows a marked increase below this interface from 0.01 $mm\ s^{-1}$ in the peat to 0.61 $mm\ s^{-1}$ (means of 15 paired sample cores), although there was some evidence that the mineral horizon was less permeable than elsewhere in the immediate vicinity of the pipes.

Table 20.1 Classification of soil pipes in terms of initiating process and hydrological function

A. HYDROLOGICAL FUNCTION.

1 Regime		2 Within storm/season function		3 Connectivity with stream network	4 Internal network connectivity
Ephemeral	(E)	Primary (P)	Radical(r) Overflow(o)	Disjunct (Dj)	Discontinuous (Ds)
Seasonal	(S)	Secondary(S)	Radical(r) Overflow(o)	Connected (Cd)	Continuous (Cs)
Perennial	(P)	Tertiary (T)	Radical(r) Overflow(o)		

Explanatory notes:

Primary, secondary and tertiary describe pipe levels in order of activation during storm. For piping at a single depth, all routes would be primary.

Radical flow routes are activated to a greater or lesser degree whenever pipeflow occurs at a given level: overflow routes derive some water from radical pipes during high flows.

Hydrological connectivity may be interpreted sufficiently widely as to include pipe routes with short sections of surface, channelised flow within the 'connected' category. Such cases would be 'connected' but 'discontinuous' routes.

Suggested use of classification:

Example: PPrCdCs = a perennial, primary radical, connected to the stream network and internally continuous.

B. INITIATING PROCESS.

Initial hydrological milieu	Process leading to initial formation	Genetic classification after Parker and Jenne (1967)	Common geomorphic environment
Vadose (Cr)	flow through desiccation cracking, mass movement cracking, biotic holes	Desiccation-stress crack piping	particularly valley side slopes
Phreatic (Hv)	'boiling' AND 'confined aquifer heaving'	Variable permeability-consolidation piping AND Entrainment piping	streambanks, below bog areas, within unconsolidated sediments. Heaving in permeable lenses or layers between less permeable layers or at interfaces at base of layers of low permeability.

Complete classification: Based on both initiating process and hydrological function, for example:

EPrDjDs/Cr = an ephemeral, primary radical, not connected to the stream network and internally discontinuous, initiated by flow through cracking.

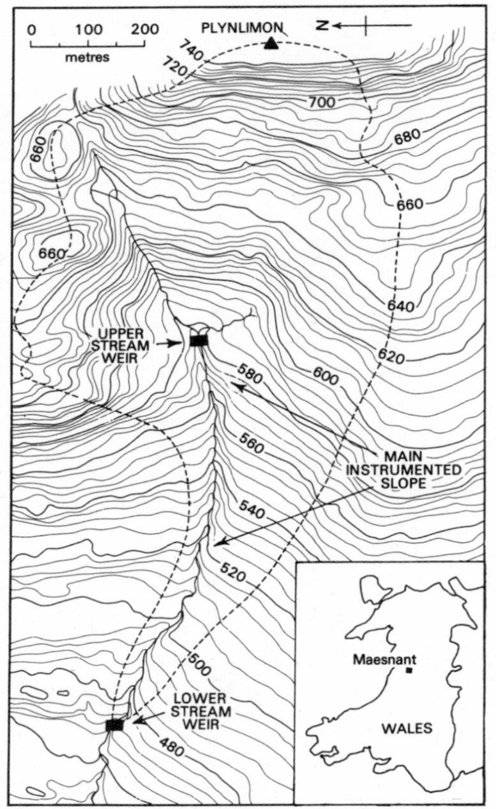

Figure 20.1 The Maesnant Experimental catchment

There was clear evidence of pipes flowing at various depths across the terrace area. At one site, minor dry pipes occurred at c. 15 cm depth in the peat, linked with desiccation cracks, whilst a perennial pipe continued to flow at a depth of 50 cm at the base of the O horizon. In other cases high-flow overflow routes were traced during storms, which could never have been located in drier weather. This was the case at the outfall of the pipe system now monitored by pipeflow stations 2 and 2B (figure 2). Dye tracing indicated that what was normally a single perennial outlet split into four outlet points during heavy storms. This meant that the monitoring site originally planned had to be shifted upslope to a point where flow was thought to be in a single channel, apart from a surface storm overflow which could easily be directed into the measuring site. In the event, a single extra highflow route remained which had to be monitored by an extra station, 2B. Monitoring has revealed that the pipe at site 2 shows flattened hydrograph peaks as the system reaches capacity and that the overflow route is activated by flows generally in excess of 1.0 l s^{-1} in the main pipe. The operational lessons for establishing meaningful pipeflow monitoring sites are obvious. In fact, it also became evident that the pipe used in the original

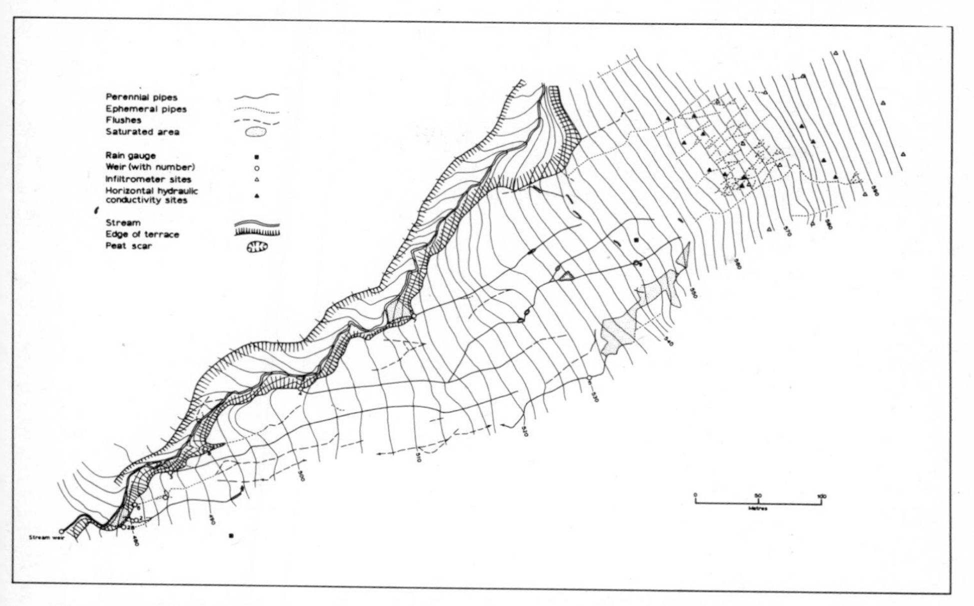

Figure 20.2 Soil pipe networks in the Maesnant basin

pilot experiment (station 1) was itself a high level limb of another, deeper system discharging at station 6. Hence, although there is an overall similarity in response between this pipe and its deeper counterpart, the total volume of stormflow was being underestimated at the original site.

A number of points were encountered where pipes discharged onto the surface, causing overland flow through saturated areas of *Sphagnum spp.* moss, which later disappeared into a continuation of the pipe system. McCaig (1979) observed the same features in Yorkshire and applied the term 'secondary source areas' to them. These were areas of perennial saturation. However, other areas with little or no *Sphagnum* growth were found to be flooded by temporary overland flow rising from overflowing pipes during certain storms and draining back into pipe blow-holes. Overland flow gauges, consisting of 20 mm O.D. perspex tubes fitted with dipsticks and filled with a small quantity of powdered charcoal, have revealed peak flows up to 6 cm in depth in these areas.

One set of these 'secondary source areas' appears to be a particularly significant point in the Maesnant networks. This is the one which occurs at the change of slope between the steeper valley side slopes covered in peaty podzols of the Hiraethog Series with slopes of c. 15° and peat covers of c. 15-20cm and the thicker peat soils developed on the soliflucted drift material of the terrace with slopes of c. 6° (figure 2). These areas appear to form a collecting ground for drainage much of which arrives, at least during storms, from extensive networks of shallow, ephemeral pipes. In contrast, the pipes draining this collecting ground are mainly perennial, perhaps deriving their extended flow from a combination of long term seepage from the upper slope and any ephemeral pipe discharge held up in the *Sphagnum* reservoir. The exact nature of flow generation here is a point of active investigation at the present time, including monitoring of flow events within the ephemeral pipe networks.

It is important to note here that the *ephemeral* pipes are very similar to those studied by Weyman and others in the Mendips and to those described and monitored in the nearby Nant Gerig basin by Gilman and Newson (1980). They are shallow, c. 15cm down, generally running in the basal section of the peaty O horizon of the podzols with diameters approaching 10cm and they can be traced with relative ease even during the summer, once the observer has developed an eye for slight linear surface depressions, 5-10cm deep, and the marker moss *Polytrichum* (cp. Newson, 1976). Proving linkages by dye tracing is, however, more problematical. In the cases of the East Twin pipes and those of the Nant Gerig the ephemeral systems remain essentially disjunct from the stream channels. On East Twins the pipes tend to feed slight topographic depressions which form linear extensions of overland flow headwards from the stream channel, whereas on Nant Gerig they tend to feed valley bottom bogs near the stream. Gilman and Newson (1980) believe that this causes a marked delay in the contribution from the ephemeral pipes to the stream.

In the case of the Maesnant pipes, the ephemeral networks on the upper slope guide water towards the heads of the perennial pipes, even if direct connexions are not always present, whilst the ephemeral pipes on the terrace slope are either high-flow routes or blow-hole inlets directing local overland flow on the main pipe system. Stagg (1978) has described yet another variation in which ephemeral pipe networks in the Black Mountains of South Wales extend through the soil from an outcrop of porous sandstone and issue into ephemeral rill systems.

The major differences in pipe network morphology between the few drainage basins so far studied suggest some major differences in the sources of pipeflow and in the role of piping in the overall drainage system between catchments affected by soil piping. The following sections briefly describe some of the preliminary results of the current Maesnant experiment and analyse their implications in terms of the significance of pipe discharge and of the sources of pipeflow.

Some preliminary results

Completion of the pipe network survey during 1980 has enabled deployment of water level recorders and a set of custom-built 2-channel data loggers across the hillslopes draining into the study reach. The instrumentation currently comprises 7 2-channel tape loggers and one 10-channel logger recording water levels at weir plates set in excavated pipe outlets, together with 2 tipping-bucket raingauges and a snowmelt gauge (Waring and Jones, 1980). This is complemented by 8 Munro chart water level recorders, 2 siphon raingauges, a class A evaporation pan, a 22-tensiometer scanivalve system (after Anderson and Burt, 1977) and a number of crest-stage gauges for overland flow and phreatic levels. The instrumentation and data processing system are described in greater detail in Jones and Crane (1981). All pipeflow is currently monitored at the final outfalls from the pipe systems which occur on the river terrace scar (figure 2), with the exception of some exploratory monitoring near the source areas. All the pipes currently under discussion are therefore perennially-flowing with the exception of two, weirs 2B and 8, which are ephemeral despite being located on the terrace scar.

The rapid response of pipeflow to rainfall indicated by a mean peak lag time of 3.9 h in the pilot experiment has been confirmed, with peak lag times ranging up to 8.7 h. In all cases the distribution of lag times is positively skewed with modes 1-2 h shorter than the mean. Comparison with mean peak lag times of 7.4 h and 10.5 h on the upper and lower stream weirs suggests that the peak pipeflow contribution to streamflow occurs on the rising limb of the stream hydrograph. This is well illustrated by the plots of pipeflow and streamflow for a series of storms in the wet, early winter period of 1979 (figure 3). These events occurred during the normal annual peak rainfall period. However, although absolute peak contribution occurs before the stream hydrograph peak, instantaneous percentage contributions appear greatest during the stream recession (figure 4). It seems likely that other sources, such as overland flow or seepage, especially seepage from a number of seepage hollows adjacent to the stream, make significant contributions to the stream peak. Whereas the pipe hydrographs tend to have a recession slope at least as gentle as the associated stream hydrograph, however, it appears that these other sources of discharge make a rather brief, flashy contribution. Attempts are now being made to define these sources more specifically. Monitoring of the flow from two seepage hollows since April 1980 has in fact suggested that their response is very similar to pipe response, but a number of small pipes have now been identified feeding these hollows. Work has now begun on instrumenting supposedly pipe-free hollows.

Estimates of the percentage contribution of flow from the 3 major pipes monitored during November 1979 to the direct runoff streamflow increment over the study reach varied from 28% to 38% in five storm periods. Very approximate allowances for flows from other pipes, of which the principal ones are now monitored, suggests that total pipeflow contributions may have amounted to over 60% of direct runoff. In fact, this level of contribution was reached in actual measured flow from 5 monitored pipes in early spring 1980, just prior to the onset of the spring drought.

Monitoring during the early part of 1980 also provided the first substantial confirmation of a three-way link between percentage pipe contribution, storm severity and antecedent rainfall (figure 5) suggested by the earlier, pilot experiment (Jones,1978). Given the reasonable assumption that the unmonitored pipes react in a similar overall fashion to the monitored pipes, the percentage contribution from pipeflow is lower in storms with high antecedent rainfall totals during the previous week and in the heavier storms. Peak percentage contribution occurs in moderate storms in moderately wet antecedent rainfall conditions. The percentage

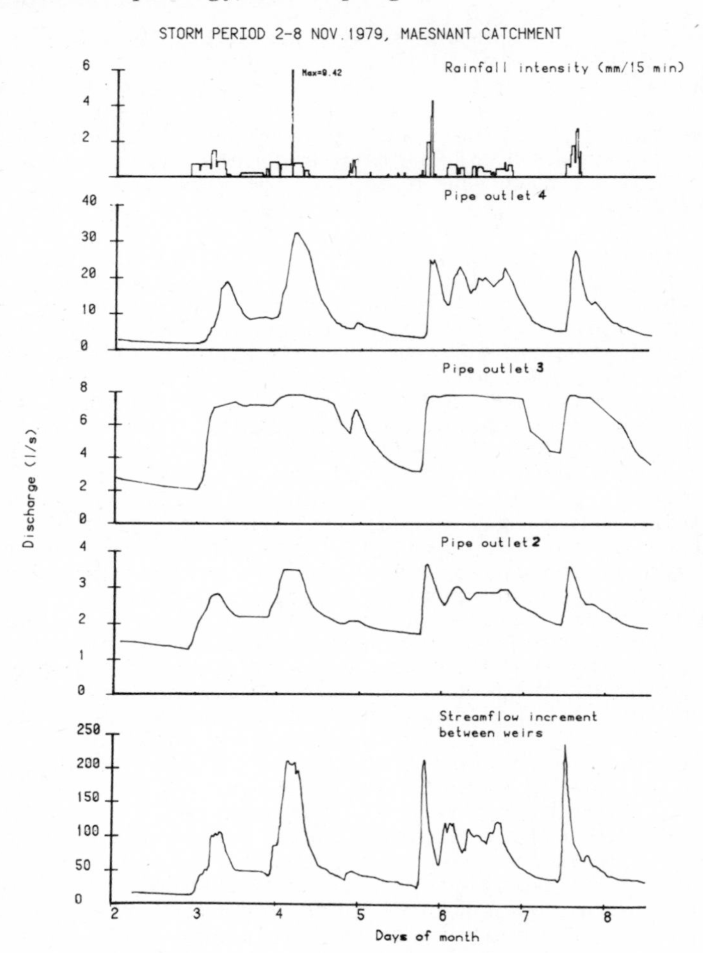

Figure 20.3 Rainfall, pipeflow and streamflow during part of November, 1979

contribution falls off again in drier antecedent conditions and in lighter storms. Quantitative definition of these relationships must clearly await the accumulation of more data. Nevertheless, it is interesting that the same general relationships have now been found on both perennial and seasonal/ephemeral pipes. They suggest that in wetter conditions more water drains to the stream by other routes. This may be for a number of reasons, for example, in wet surface conditions the infiltration capacity of the O horizon is low and not infrequently exceeded by rainfall intensity, resulting in saturation overland flow: saturated infiltration capacity measurements taken with a constant head double-ring infiltrometer indicate values of c. 0.01 mm s^{-1}. Saturation overland flow from streamside bogs is also likely to be a more important contributor in these conditions, as would subsurface seepage generally. Again, heavy rainfall is

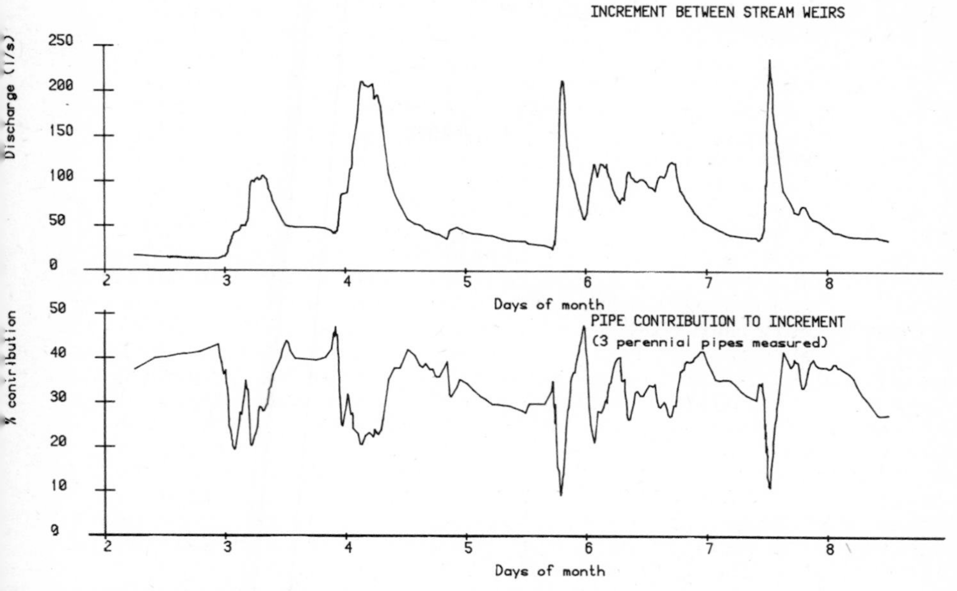

Figure 20.4 Instantaneous percentage contribution to streamflow increment between upper and lower weirs from three perennial pipe outlets.

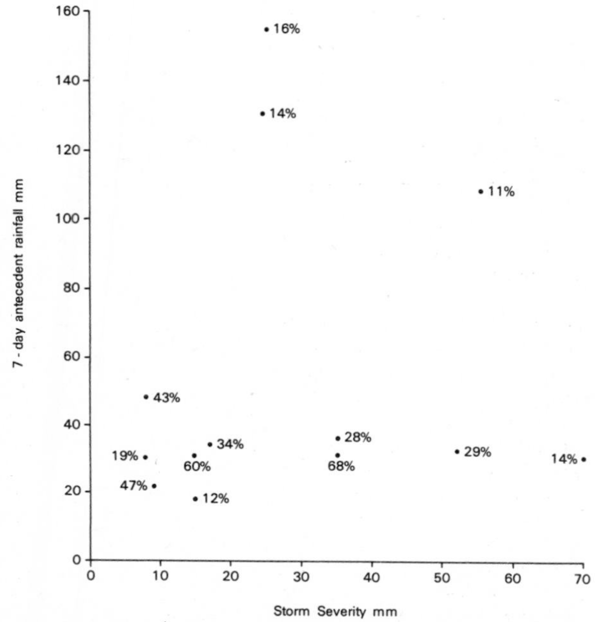

Figure 20.5 Percentage contribution to streamflow increment from three major pipe outlets during 1979-80 plotted against total 7-day antecedent rainfall and total storm rainfall (storm severity)

likely to be a good stimulant for all these alternative drainage pathways. It is notable that in light storms and dry conditions the overall percentage of rainfall contributing to the stream hydrograph is also low, due to 'losses' to soil moisture and groundwater recharge.

Another significant feature which has become evident in the current monitoring programme is the wide variation in response between pipes. Pipes 2 and 3 in figure 3 tend to reach a limiting capacity flow. In the case of 2, this then overflows into pipe 2B. In the case of pipe 3 there may be other overflows. One certainly occurs through a small conical mound built up around a small outlet (2-3 cm diameter) which is capable of producing a 6 cm high fountain in heavy storms. There may also be a linkage to outlet 8 (figure 2) which is ephemeral, despite the fact that it occupies a deep dell whereas the perennial pipe 3 has no surface expression whatsoever: perhaps an indication of former subsurface drainage capture? Once it starts flowing, however, pipe 8 can deliver 400 m^3 of stormflow. Limited pipeflow capacities may be another reason for the fall off in the proportionate contributions of pipeflow in wetter conditions. Indeed, Gilman and Newson (1980) have pointed out the relative advantage of unlimited capacity for overland flow, and also noted that whilst velocity increases linearly with discharge for overland flow the rate of transmission for pipeflow varies with the square root of the discharge. In contrast to the overflow pipes and the pipes with limited capacities, pipes 4 and 7 rarely, if ever, reach capacity and show remarkably flashy responses.

THE SOURCES OF PIPEFLOW

The sources of pipeflow are likely to be at least as varied as the different genetic types of pipe network and, perhaps, as the different flow regimes encountered. Moreover, it is probable that different sources are tapped in events of different magnitude and even at different stages within the same runoff event. In general, it appears that the sources of pipe discharge are multiple in any given case of stormflow, but we are still some way from being able to make categorical statements. In the case of the small ephemeral pipes on the upper slopes on Maesnant direct capture of overland flow through pipe inlets may be a major source in some storms, causing peak lag times of 2 h or less. However, field observations suggest that the main ephemeral pipes also collect both surface and subsurface drainage in shallow bowl-shaped collecting areas of c 5 m diameter at their heads. Again, in the perennial and seasonal pipe networks some of the rapid response may be due to direct influx through surface inlets, but the perennially saturated areas also clearly form a key source. These bogs provide areas for 'direct precipitation' and saturation overland flow as well as collecting areas for water derived from upslope, either through the ephemeral pipes, diffuse seepage or overland flow. They also maintain baseflow.

The role of the phreatic surface is interesting. Flow is not normally encountered in pipes running through unsaturated soil : this was so even in the case of small ephemeral pipes in lessivé brown earths in the English Peak District (Jones, 1978). However, it is possible to observe cases of influent seepage from soil pipes during both storm and baseflow periods. This situation was observed around small (4 cm diameter) pipe outlets in streambanks in the Peak District, and recent monitoring of phreatic levels in piezometers around 10 cm diameter perennial pipes in the peaty gleyed podzols of the Maesnant river terrace has indicated phreatic surface levels declining by 20 cm in 2 m away from the pipe which seems to suggest that the pipe is partly maintaining phreatic levels. Newson and Harrison (1978) reported significant losses of pipeflow during experiments using artificially pumped water in natural ephemeral pipes and surmised that this is a normal situation. The same opinion was expressed by Rodda *et al.* (1976). However, the peaty podzol soils appear to have been in a relatively dry state when these experiments were conducted, and although it is certainly a likely pattern of flow in some cases, dye dilution experiments on Maesnant suggest that the pipes generally derive more in the form of effluent seepage than they lose by influent seepage so that discharge continues to increase downslope. Admittedly, the evidence to date from Maesnant is based mainly on the perennial network.

The phreatic surface on the Maesnant river terrace appears to react jointly with pipeflow levels and it is suggested that rapid recharge of the phreatic surface may occur through surface cracks, areas of mass slumping and bogs, and even possibly through pipe inlets where water entering at an early stage of the storm may be largely lost by influent seepage, perhaps to return again to the pipes sometime later. A number of observed features seem to support this view. One obvious 'problem' is to explain why lag times in the pipes are neither longer nor shorter. Dye tracing experiments have indicated periods of 1-1.5 h for transmission of water from the secondary source areas to the outfall of the pipes, so that peak lag times somewhat less than the observed 4-8 h might be expected if the main source of flow were 'direct precipitation' and overland flow from small contributing areas. On the other hand, infiltration capacity and hydraulic conductivity measurements suggest that water will take of the order of 12 h to infiltrate from the surface to pipe level by diffuse seepage through the O horizon. This order of peak lag time would be expected even if piston flow were operating, in the presence of such low conductivities in the peaty horizon.

In fact, field observations show that even on the ephemeral network peak runoff can be generated 3-4 h after peak rainfall with no visible direct input from overland flow. In this case the lag is commensurate with the time needed for diffuse seepage to reach pipes 15 cm below the surface, but the surface catchment area that would drain directly to these pipes does not seem sufficient to be the sole cause of the runoff. More likely, the pipes are fed largely by phreatic surface water which has risen due to rapid recharge through fissures and more permeable areas. Multiple correlation and regression analyses based on 21 storm events for pipes 2 and 4 indicated that the only significant variable affecting the percentage of rainfall passing through these pipes was the level of low flow immediately prior to the storm hydrograph ($r = 0.72$, $\alpha = 0.001$) and that neither antecedent rainfall totals for 1, 4 and 7-days nor total storm or pre-peak rainfall were significant. This seems to suggest that the pipes are generally more efficient transmitters of rainfall within their own catchment areas when the phreatic surface is higher.

However, peak discharge levels in pipes 2 and 4 are best predicted by a combination of total rainfall prior to the hydrograph peak and the one-day antecedent rainfall total ($R = 0.88$ and 0.83 respectively), whilst 'peakedness' of the hydrograph (peak/mean discharge) on pipe 4 is best related to pre-peak rainfall ($r = 0.53$, $\alpha = 0.01$, $n = 30$). The relationships are not so clear for pipe 2 because of its tendency to reach a peak capacity and then overflow. The latter results underline the close relationship between the form of the pipe hydrographs and the pattern of storm rainfall, but give little insight into processes with the possible exception that the positive relationship between peak discharge levels and rainfall totals for the 24-hours prior to the storm may indicate that the rainwater is transmitted more readily to these pipes when the O horizon is wetter.

It seems clear, therefore, that these pipes receive drainage more quickly and in greater volumes than would be expected simply from diffuse seepage through the overburden. The statement of Whipkey and Kirkby (1978, 124) that the main difference between pipe and stream response lies in the time taken for the water to reach pipe depth clearly now requires some revision in the light of the field evidence.

THE ROLE OF SOIL PIPES IN THE DRAINAGE SYSTEM

The function and significance of piping in an individual drainage basin depend largely on three properties of the networks : the proportions of ephemeral and perennial pipes, pipe frequencies/densities, and the spatial relations of the pipe and stream networks. The results obtained to date from the current Maesnant experiment suggest that the significance of pipeflow as a drainage process in the basin is at least as great as previously supposed : a figure of 39% of storm-generated streamflow (including delayed flow) has been suggested from multiplying the discharge of the pipe monitored in the first Maesnant experiment by the ratio of the estimated total pipe-drained area derived from air photo analysis to the area of the supposed catchment of the gauged pipe (Dovey, 1976). With a far sounder basis for estimation using direct measurements on pipes with higher discharges, it now seems likely that the final estimate of mean pipe contribution will be in excess of 40%, with figures of 60% or over under certain conditions. As outlined above, the proportions of rainfall following different pathways within the drainage system vary considerably in a manner which we are only just beginning to be able to define.

Unfortunately, there is little data available for comparison from elsewhere. The Nant Gerig experiments described by Gilman and Newson (1980) were concerned only with the ephemeral network, the perennial and seasonal pipes being regarded 'as part of the main drainage net' (*ibid*, 25). Nevertheless, even the ephemeral network alone was found to be capable of transmitting large volumes of rainwater. Gilman (1971) estimated that between 4% and 72% of rain falling on a slope passed through the ephemeral pipe network during three experimental storm periods. He noted that lower percentages occurred when large amounts of overland flow were observed. Ward (1975) quotes an estimate that the Nant Gerig pipes 'carry about 20 per cent of the total rainfall measured in the contributing catchment area'. However, Ward's conclusion 'that pipeflow represents a *dominant storm runoff process*' (my italics), was perhaps a little premature since stream discharge hydrographs were not established for the Nant Gerig sites (Gilman, 1971) and the relationships between pipeflow and streamflow have not therefore been studied (cf. Gilman and Newson, 1980). To date the only other quantitative evidence of the importance of pipeflow comes from a brief, 32-day, undergraduate study by Howells (1980) conducted on farmland at an altitude of 150 m O.D. on a tributary of the River Llynfi in South Wales. This suggested daily average contributions ranging from 10% to 81% on a stream with a maximum recorded discharge of 56 l s^{-1}. Here too, percentage contributions were lower at higher stream discharges, but pipe response times appear to be rather greater than those of the stream (Jones, 1981).

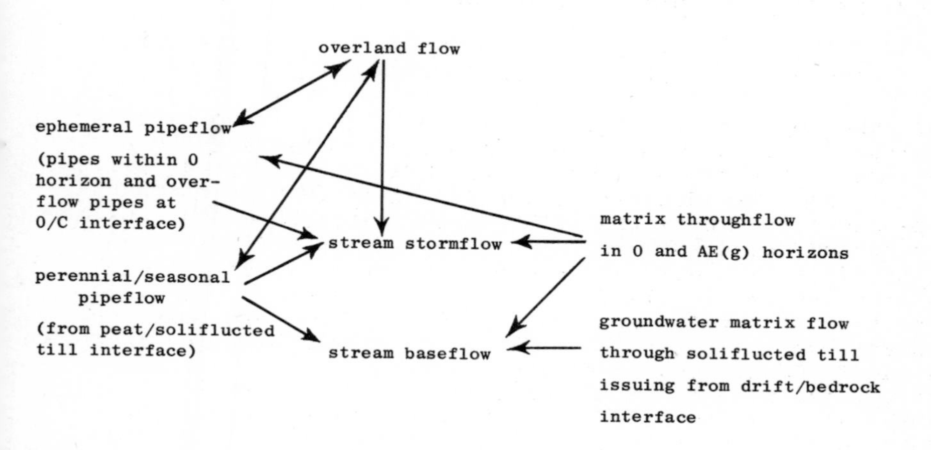

Figure 20.6 The main systematic relationships between drainage pathways in the Maesnant experimental basin

Figure 6 is an attempt to outline the functional relationships between drainage pathways in the Maesnant system. It is hoped that the next 18 months of research will enable a more detailed verification and quantification of these processes. In addition to this hydrological work, monitoring of current pipe erosion began in 1979 using sediment traps set in the pipe beds behind the weir plates. Results to date suggest that previous estimates of pipe erosion rates in Britain may be rather low (Jones, 1981). Weekly average loads have been measured in the range 0.21 - 0.38 kg, mainly of gravel and coarse sand derived from the pipe bed. However, pipe 4, which is capable of peak discharges in excess of 50 $l\ s^{-1}$, has delivered an average load of 3.8 kg per week during 14 months of monitoring, with a peak rate of 6.2 kg per week in the winter high-flow period between November 1979 and March 1980.

We are just beginning to understand the function of the pipe networks in this one basin and we are certainly a long way from being able to extrapolate to other piped basins. It is inevitable that the variety of network configurations encountered will lead to different response patterns and, indeed, different conclusions from workers in different areas, perhaps even more so than for stream networks because of the great variety of channel type. But it is important that like is compared to like, that differences in flow regime especially between ephemeral and perennial are recognised and networks sampled accordingly, and that complete systems are studied. It is understandable that Cryer (1980) working on the Maesnant should argue against a rigorous distinction between ephemeral and perennial pipes in view of the configuration of the Maesnant networks. On the other hand Wilson (1977) working in the Brecon Beacons, and Gilman and Newson (1980) on Nant Gerig, both encountered such marked topographic separation between ephemeral and perennial pipes that they did not regard the latter as part of the pipe system to be monitored. Wilson (1977) in fact concluded that the maximum discharge from ephemeral pipes as estimated from artificial pumping experiments was several orders of magnitude too low to be of any significance to streamflow. Gilman and Newson's experiments seem to suggest otherwise for Nant Gerig and there is no question of the significance of pipeflow on Maesnant.

Acknowledgements

Many thanks are due to the Natural Environment Research Council for funding, to my research assistant, F.G. Crane, to the Department of Geography, Aberystwyth, for technical support, to R.L. Collin and N.W.T. Chisholm for surveying and photogrammetry and to the students who have helped in pipe-tracing exercises.

References

Abramowitz, M. and Stegun, I.A., 1964. *Handbook of mathematical functions,* (United States National Bureau of Standards, Applied Mathematics Series, 55).

Alexander, D.E., 1977. *Transient channel forms in the Basento catchment, southern Italy.* Unpublished Ph.D. thesis, (University of London, London).

Alexander, D.E., 1980. *i calanchi* - accelerated erosion in Italy, *Geography,* 65(2), 95-100.

Alexandre, J. and Alexandre, S., 1964. Action linéaire ou en surface du ruissellement dans une région de savane (Katanga méridional), *Publications de l'Université Officielle du Congo à Elisabethville,* 7, 105-114.

Almagià, R., 1907, 1910. Studi geografici sopra le frane in Italia, *Memorie della Società Geografica Italiana,* 13-14.

Almagro Basch, M. and Arribas, A., 1963. El poblado y la necropólis megalíticas de Los Millares (Santa Fe de Mondújar, Almería), *Bibliotheca Praehistorica Hispania,* vol. 3, Madrid.

Almagro Gorbea, M.J., 1973. El poblado y la necropolis de El Barranquete (Almeria), *Acta Arqueológico Hispanica,* 6.

Almagro Gorbea, M.J., 1976. Memoria de las excavaciones efectuadas en el yacimiento de Tarajal (Almeria), *Noticiario Arqueológico Hispanico, Praehistoria,* 5, 195-198, Madrid.

Almagro Corbea, M.J., 1979. *C-14 y Prehistoria de la Peninsula Iberica,* Madrid: Fundación Juan March.

Amerman, C.R. and McGuiness, J.L., 1967. Plot and small watershed runoff and its relation to larger areas, *Transactions of the American Society of Agricultural Engineers,* 464-66.

Anderson, M.G. and Burt, T.P., 1977. Automatic monitoring of soil moisture conditions in a hillslope spur and hollow, *Journal of Hydrology,* 33, 383-390.

Anderson, M.G. and Burt, T.P., 1978a. Experimental investigations concerning the topographical control of soil water movement on hillslopes, *Zeitschrift für Geomorphologie,* 29, 52-63.

Anderson, M.G. and Burt, T.P., 1978b. Time synchronised stage recorders for the monitoring of incremental discharge inputs in small streams, *Journal of Hydrology,* 34, 101-109.

Anderson, M.G. and Burt, T.P., 1978c. The role of topography in controlling throughflow generation, *Earth Surface Processes,* 3, 331-344.

Anderson, M.G. and Burt, T.P., 1980. The role of topography in controlling throughflow generation: a reply, *Earth Surface Processes*, 5, 193-195.

Anderson, M.G. and Kneale, P.E., 1980a. Topography and hillslope soil water relationships in a catchment of low relief, *Journal of Hydrology*, 47, 115-128.

Anderson, M.G. and Kneale, P.E., 1980b. Pore water pressure and stability conditions on a motorway embankment, *Earth Surface Processes*, 5, 37-46.

Andolfato, U. and Zucchi, F., 1971. The physical setting, in: *Arts of Cappadocia*, ed Giovannini, L., (Nagel Publishers, Geneva).

Andrews, J.T., Carrara, P.E., King, F.B. and Stuckenrath, R., 1975. Holocene environmental changes in the alpine zone, northern San Juan Mountains, Colorado: Evidence from bog stratigraphy and palynology, *Quaternary Research*, 5, 173-197.

Anon, 1962. Flooding and salt problems in the wheatbelt, *Western Australia Department of Agriculture Bulletin*, 3051.

Arribas, A., 1968. Las bases economicas del Neolítico al Bronce, in: *Estudios de Economía Antigua de la Península Ibérica*, ed Taradell, M., (Vicnes-Vives, Barcelona), 33-60.

Arribas, A. and Molina, F., 1979. *El poblado de 'Los Castillejos' en Las Peñas de Los Gitanos (Montefrio, Granada)*, (University of Granada, Granada).

Arribas, A., Molina Gonzalez, F., Arteaga Matute, O., and Molina Fajardo, F., 1974. Excavaciones en el poblado de la Edad del Bronce 'Cerro de la Encina', Monachil (Granada) (El corte estatigrafico no. 1), *Excavaciones Arqueologicas en España*, 81.

Arulanandan, K., Loganathan, P., and Krone, R.B., 1975. Pore and eroding fluid influences on surface erosion of soil. *American Society of Civil Engineers, Journal of Geotechnical Engineering Division*, 101 (GT1), 51-66.

Arulanandan, K. and Heinzen, R.T., 1977. Factors influencing erosion in dispersive clays and methods of identification, *International Association of Scientific Hydrology, Publication*, 122, 75-81.

Ashida, K. and Michuie, M., 1971. An investigation of bed degradation downstream of a dam, *International Association Hydraulic Research, 14th Congress Paris*, Paper C-30, 1-9.

Atkinson, T.C., 1978. Techniques for measuring subsurface flow on hillslopes, in: *Hillslope hydrology*, ed Kirkby, M.J., (Wiley, London), 73-120.

Atmospheric Environment Service, 1978. Rainfall intensity, duration and frequency values for Brooks horticultural station, Alberta: Hydrometeorology Division, Canadian Climate Centre.

Atmospheric Environment Service 1969-1976. *Monthly record: Meteorologic observations in Canada*, Downsview, Ontario.

Australian Bureau of Statistics Western Australian Office, 1977. *Western Australian Year Book*, (Government Printer, Western Australia).

Barendregt, R.W. and Ongley, E.O., 1977. Piping in the Milk River Canyon, southeastern Alberta - a contemporary dryland geomorphic process, in: *Erosion and solid material transport in inland water, International Association of Scientific Hydrology, Publication*, 122, 233-243.

Barendregt, R.W. and Ongley, E.D., 1979. Slope recession in the One-Four Badlands, Alberta, Canada: an initial appraisal of contrasted moisture regimes, *Canadian Journal of Earth Sciences*, 16, 224-229.

Batini, F.E. and Selkirk, A.B., 1978. Salinity sampling in the Helena catchment, Western Australia, *Forests Department of Western Australia Research Paper*, 45.

Batini, F.E., Hatch, A.B. and Selkirk, A.B., 1977. Variations in level and salinity of perched and semi-confined groundwater tables, Hutt and Wellbucket experimental catchments, *Forests Department of Western Australia Research Paper*, 33.

Batini, F.E., Selkirk, A.B. and Hatch, A.B., 1976. Salt content of soil profiles in the Helena catchment, Western Australia, *Forests Department of Western Australia Research Paper*, 23.

Baur, A.J., 1952. Soil and water conservation glossary, *Journal of Soil and Water Conservation*, 7, 41-52, 93-104, 144-156.

Baver, L.D., Gardner, W.H. and Gardner, W.B., 1972. *Soil Physics*. (Wiley, New York).

Begin, Z.B., 1978. *Aspects of degradation of alluvial streams in response to base-level lowering*. Unpublished Ph.D. thesis, (Colorado State University, Fort Collins).

Begin, Z.B., Meyer, D.F., Schumm, S.A., 1980. Knickpoint migration due to base-level lowering, *American Society Civil Engineers, Journal of Waterway, Port, Coastal and Ocean Division*, v. 106, pp 369-388.

Begin, Z.B., Meyer, D.F., Schumm, S.A., 1981a. Development of longitudinal profiles of alluvial channels in response to base-level lowering, *Earth Surface Processes and Landforms*, v. 6, 49-68.

Begin, Z.B., Meyer, D.F., Schumm, S.A., 1981b. Sediment production of alluvial channels in response to base-level lowering, *Transactions American Society Agricultural Engineers*, (in press).

Benjamin, B.T., 1957. Wave formation in laminar flow down an inclined plane, *Journal of Fluid Mechanics*, 2, 554-574.

Bennett, H.H., 1960. Soil erosion in Spain, *Geographical Review*, 50, 59-72.

Bentor, Y.K., 1966. The clays of Israel. *Guide book to the excursions. The International clay conference, Jerusalem, Israel*, 123.

Betancourt, J.L. and Van Devender, T.R., 1980. *Holocene environments in Chaco Canyon, New Mexico: the pack-rat midden record*. (Report submitted to the Division of Cultural Research, Southwest Cultural Resources Center, National Park Service).

Betson, R.P., 1964. What is watershed runoff? *Journal of Geophysical Research*, 69, 1541-52.

Betson, R.P. and Marius, J.B., 1969. Source areas of storm runoff. *Water Resources Research*, 5(3), 574-582.

Bettenay, E., Blackmore, A.V. and Hingston, F.J., 1964. Aspects of the hydrologic cycle and related salinity in the Belka valley, Western Australia, *Australian Journal of Soil Research*, 2, 187-210.

Beven, K., 1976. Hillslope hydrolographs by the finite element method, *University of Leeds, Department of Geography Working Paper*, 143.

Beven, K., 1979. *Soil water flow in a combined micropore/ macropore system - ideas for a preliminary model*, unpublished manuscript.

Birot, P., 1964. *La Méditerranée et le Moyen-Orient, Vol.1. Généralités: Péninsule Iberique, Italie*, (Presses Universitaires de France, Paris), 550 pp.

Birot, P. and Solé Sabaris, 1959. La morphologie due sud-est de l'Espagne, *Revue de Géographique des Pyrénnées et du Sud-Quest*, XX, 119-184.

Blance, B., 1971. Die Anfänge der Metallurgie auf der Iberischen Halbinsel, *Studien zu den Anfängen der Metallurgie*, 4.

Blench, T., 1969. *Mobile bed fluviology*, (University of Alberta Press, Edmonton).

Bloom, A.L., 1978. *Geomorphology: A systematic analysis of late Cenozoic landforms*. (Prentice-Hall, Inc., New Jersey).

Boenzi, F., Radina, B., Ricchetti, G. and Valduga, A., 1971. *Note illustrative per Foglio 201 (Matera)*, (Servizio Geologico d'Italia: Carta Geologica d'Italia).

Bombicci, C., 1881, Geologia dell'Appennino bolognese, in: *Appennino Bolognese*, (C.A.I., Bologna).

Bork, H.R. and Rohdenburg, H., 1979. The behaviour of overland flow and infiltration under simulated rainfall, *Symposium on agricultural soil erosion in temperate climates, Strasbourg 20-23 September, 1978*. 225-237.

Bower, C.A. and Hatcher, J.T., 1962. Characterisation of salt-affected soils with respect to sodium, *Soil Science*, 93, 275-280.

Bradley, W.H., 1940. Pediments and pedestals in miniature, *Journal of Geomorphology*, 3, 244-255.

Branson, F.A. and Owen, J.B., 1970. Plant cover, runoff and sediment yield relationships on Mancos Shale in Western Colorado. *Water Resources Research*, 6(3), 783-790.

Brice, J.C., 1966. Erosion and deposition in the loess mantled Great Plains, Medicine Creek basin, Nebraska, *United States Geological Survey Professional Paper*, 352 H.

Bridges, E.M. and Harding, D.M., 1971. Micro-erosion processes and factors affecting slope development in the Lower Swansea Valley, *Institute of British Geographers Special Publication*, 3, 65-79.

Brock, R.B., 1969. Development of roll wave trains in open channels, *Journal of the Hydraulics Division, American Society of Civil Engineers*, 95, 1401-1427.

Brown, G.W., 1962. Piping erosion in Colorado, *Journal of Soil and Water Conservation*, 17, 220-222.

Brunsden, D., 1974. The degradation of a coastal slope, Dorset, *Institute of British Geographers Special Publication* 7, 79-98.

Bryan, R.B., 1973. Surface crusts formed under simulated rainfall on Canadian soils, *Consiglio Nazionale Delle Richerche, Laboratorio per la Chemica del Terreno, Piza,* Conference 2.

Bryan, R.B. and Campbell, I.A., 1980. Sediment entrainment and transport during local rainstorms in the Steveville Badlands Alberta, *Catena*, 7, 51-65.

Bryan, R.B., Yair, A. and Hodges, W.K., 1978. Factors controlling the initiation of runoff and piping in Dinosaur Provincial Park badlands, Alberta, Canada, *Zeitschrift für Geomorphologie, Supplement Band*, 29, 151-168.

Bucciante, M., 1922. Sulla distribuzione geografica dei calanchi in Italia, *L'Universo*, 3(9), 585-605.

Buchanan, R.G., 1977. *Landforms and observed hazard mapping, South Thompson Valley, British Columbia*. Geotechnical and Materials Branch Report, B.C. Ministry of Highways and Public Works.

Buchanan, R.G. and Evans, S.G., 1977. Piping and related processes. Appendix 3 in: *Geological hazards and urban development of silt deposits in the Penticton area* (eds D. Nyland and G.E. Miller).

Buckman, A.F. and Cockfield W.E., 1950. Gullies formed by sinking of the ground. *American Journal of Science*, v. 248, 137-141.

Bull, W.B., 1964. Geomorphology of segmented alluvial fans in western Fresno County, California, *U.S. Geological Survey Professional Paper*, 352-E.

Bull, W.B., 1975. Allometric change of landforms, *Geological Society of American Bulletin*, 86, 1489-1498.

Burt, T.P., 1978. An automatic fluid scanning switch tensiometer system, *British Geomorphological Research Group, Technical Bulletin*, 21.

Burt, T.P., 1979. Diurnal variations in stream discharge and throughflow during a period of low flow, *Journal of Hydrology*, 41, 291-301.

Burt, T.P., 1980. Rainfall in the southern Pennines, *Huddersfield Polytechnic Department of Geography Occasional Paper*, 8.

Burt, T.P. and Anderson, M.G., 1980. Soil moisture conditions on an instrumented slope March-October 1976, in: *Atlas of Drought in Great Britain*, ed Doornkamp, J.C., and Gregory, K.J. (Institute of British Geographers, London), 44.

Burvill, G.H., 1947. Soil salinity in the agricultural area of Western Australia, *Journal of Agriculture Western Australia*, 13, 9-19.

Burvill, G.H., 1956. Salt land survey 1955, *Journal of the Australian Institute of Agricultural Science*, 5, 113-119.

Butzer, K.W., 1961. Remarks on soil erosion in Spain (Abstract), *Annals, Association of American Geographers*, 52, 405.

Butzer, K.W., 1980. Holocene alluvial sequences: Problems of dating and correlation, in: *Timescales in Geomorphology*, ed Cullingford, R.A., Davidson, D.A. and Lewin, J. (Wiley, Chichester), 130-142.

Byers, P.N., 1969. Mineralogy and origin of the Upper Eastend and Whitemud formations of South-central and Southeastern Saskatchewan and Southeastern Alberta. *Canadian Journal of Earth Sciences*, 6(2), 317-334.

Calzecchi-Onesti, A., 1957. *Sistemazione in collina*, (Ramo Editoriale degli Agricoltori, Rome).

Campbell, I.A., 1970a. Micro-relief measurements on unvegetated shale slopes, *Professional Geographer*, 22, 215-220.

Campbell, I.A., 1970b. Erosion rates in the Steveville badlands, Alberta, *The Canadian Geographer*, 14, 202-216.

Campbell, I.A., 1974. Measurements of erosion on badlands surfaces, *Zeitschrift für Geomorphologie, Supplement Band*, 21, 122-137.

Campbell, I.A., 1977a. Sediment origin and sediment load in a semi-arid drainage basin. pp 165-186 in: *Geomorphology in Arid Regions*, ed D.O. Doehring. Publications in Geomorphology, Fort Collins.

Campbell, I.A., 1977b. Stream discharge, suspended sediment and erosion rates in the Red Deer River basin, Alberta, Canada. *International Association of Hydrological Sciences*, 122, 244-259.

Carman, M.A., 1958. Formation of badland topography, *Bulletin of the Geological Society of America*, 69, 789-790.

Carrara, P.E. and Andrews, J.T., 1973. Holocene deposits in the alpine of the San Juan Mountains, Southwest Colorado, *Geological Society of America Abstracts with Programs*, 5, 469-470.

Carson, M.A. and Kirkby, M.J., 1972. *Hillslope form and process* (Cambridge University Press).

Carter, C.S., and Chorley, R.J., 1961. Early slope development in expanding stream system, *Geological Magazine*, v. 98, pp 117-130.

Castelvecchi, A. and Vittorini, S., 1967. Osservazioni preliminari per uno studio sull'erosione in Val d'Orcia, *Atti del XX Congresso Geografico Italiano*, III, 151-168.

Chapman, R.W., 1978. The evidence of prehistoric water control in southeast Spain, *Journal of Arid Environments*, 1, 261-277.

Chisci, G.C., 1978. The management of clay soils on hill land: a model for a Mediterranean environment, *Outlook on Agriculture*, 13 (I.C.I. Plant Protection Division), 5-12.

Chorley, R.J., 1978. The hillslope hydrological cycle, in: *Hillslope hydrology*, ed Kirkby, M.J., (Wiley, London).

Chow, Ven Te, 1959. *Open channel hydraulics*, (McGraw-Hill, New York).

Cockfield, W.E. and Buckham, A.F., 1946. Sink-hole erosion in the white silts at Kamloops. *Transactions of the Royal Society of Canada*, Series 3, Volume 40, section 4, pp 1-13.

Coles, J.M. and Harding, A.F., 1979. *The Bronze Age in Europe*, (Methuen, London).

Conacher, A.J., 1970. Processes of salt-scald formation in the Western Australian wheatbelt, in: *Arid Zone Newsletter 1970*, (CSIRO, Canberra), 50-52.

Conacher, A.J., 1974. Rehabilitation of salt scalds in the Western Australian wheatbelt by the interception and diversion of overland flow and throughflow on valley-side slopes, *Science and Australian Technology*, 11, 14-16.

Conacher, A.J., 1975. Throughflow as a mechanism responsible for excessive soil salinisation in non-irrigated, previously arable lands in the Western Australian wheatbelt: a field study, *Catena*, 2, 31-36.

Conacher, A.J., 1979a. Salinity problems in southwestern Australia, *Search*, 10, 162-164.

Conacher, A.J., 1979b. Water quality and forests in south-western Australia: review and evaluation, *Australian Geographer*, 14, 150-159.

Conacher, A.J. and Dalrymple, J.B., 1977. The nine unit land surface model: an approach to pedogeomorphic research, *Geoderma*, 18(1/2), 1-154.

Conacher, A.J. and Dalrymple, J.B., 1978. Identification, measurement and interpretation of some pedogeomorphic processes, *Zeitschrift für Geomorphologie Supplement Band*, 29, 1-9.

Conacher, A.J. and Murray, I.D., 1973. Implications and causes of salinity problems in the Western Australian wheatbelt: the York-Mawson area, *Australian Geographical Studies*, 11, 40-61.

Conacher, A.J., Brine, A.E. and Glassford, D.K., 1972. *Causes of salt scalding in the Dalwallinu area: a summary report prepared for the Dalwallinu Shire Council*, Department of Geography, University of Western Australia.

Cooke, R.U. and Reeves, R.W., 1977. *Arroyos and environmental change in the American South-west*, (Clarendon Press, Oxford).

Cori, B. and Vittorini, S., 1974. Ricerche sui fenomeni di erosione accelerata in Val d'Era (Toscana), *L'erosione del suolo in Italia e i suoi fattori*, I, (Consiglio Nazionale delle Ricerche, Istituto di Geografia del-l'Universita di Pisà).

Corriere della Sera, 1979. Sono costati cinquantamila miliardi i danni delle frane in trenta anni, (Notizie dall'Interno, Lunedì 26 novembre, 1979).

Crouch, R.J., 1976. Field tunnel erosion - a review, *Journal Soil Conservation Service NSW*, 32, 98-111.

Crouch, R.J., 1978. Variation in the structural stability of soil in a tunnel-eroding area, in: *Modification of Soil Structure*, ed Emerson, W.W., Bond, R.D. and Dexter, A.R., (Wiley, Chichester), 267-274.

Cryer, R., 1980. The chemical quality of some pipeflow waters in upland mid-Wales and its implications, *Cambria*, 6(2), 1-19.

Culling, W.E.H., 1960. Analytical theory of erosion, *Journal of Geology*, 68, 336-344.

Curtis, W.R., and Cole, W.D., 1972. Micro-topographic profile gage, *Agriculture Engineering*, 53, 17.

Dalrymple, J.B., Blaug, R.J. and Conacher, A.J., 1968. An hypothetical nine-unit landsurface model, *Zeitschrift für Geomorphologie*, 12, 60-76.

Daly, R.A., 1915. A geological reconnaissance between Golden and Kamloops, British Columbia, along the Canadian Pacific Railway. *Canadian Geological Survey Memoir*, 68.

Daniels, R.B., 1960. Entrenchment of the Willow Drainage Ditch, Harrison Country, Iowa, *American Journal of Science*, 258, 161-176.

Davidson, D.A., 1980. Erosion in Greece during the first and second millenia B.C., in: *Timescales in Geomorphology*, ed Cullingford, R.A., Davidson, D.A. and Lewin, J., (Wiley, Chichester), 143-152.

Davies, T.R.H. and Samad, M.F.A., 1978. Fluid dynamic lift on a bed particle, *Proceedings of the American Society of Civil Engineers, Hydraulics Division*, 104, HY 8, 1171-1182.

Davis, W.M., 1909. *Geographical Essays*, Boston, Ginn and Company, (reprinted 1954 by Dover Publications, New York) 777 pp.

Dawson, G.M., 1978. On the surficial geology of British Columbia. *Quarterly Journal of the Geological Society of London*, 34, 89-123.

Dawson, G.M., 1879. Preliminary report on the physical and geological features of the southern portion of the interior of British Columbia, *Canadian Geological Survey Report of Progress*, 1877-1878.

Dawson, G.M., 1891. On the later physiographical geology of the Rocky Mountain region of Canada, *Proceedings and Transactions of the Royal Society of Canada*, v. 8, section 4, pp 3-74.

Dawson, G.M., 1895. Report on the area of the Kamloops map-sheet, British Columbia, *Canadian Geological Survey Annual Report*, 1894 (1896), v. 7, part B.

Del Prete, M. and Valentini, G., 1971. Le caratteristiche geotecniche delle argille azzure dell'Italia sud-orientale in relazione alle differenti situazioni stratigrafiche e tettoniche, *Geologia Applicata e Idrogeologia* (Bari), 6, 197-215.

De Mas, P. and Jungerius, P.D., 1980. (in press) Marginale landbouw in semi-aride Marokko - een geintegreerd fysisch en sociaal geografisch ondersoek op perceelsniveau naar de 'decision environment van de Riffijnse fellah', *Geofrafisch Tijdschrift*.

Denny, C.S., 1965. Alluvial fans in the Death Valley region, California and Nevada, *U.S. Geological Survey Professional Paper*, 466.

Denny, C.S., 1967. Fans and pediments, *American Journal of Science*, 265, 81-105.

De Ploey, J., 1964. Cartographie géomorphologique et morphongénèse aux environs du Stanley-Pool (Congo), *Acta geographica Lovaniensia*, 3, 431-441.

De Ploey, J., 1971. Liquefaction and rainwash, *Zeitschrift für Geomorphologie*, 15, 491-496.

De Ploey, J., 1980. (in press) Crusting and time-dependent rainwash mechanisms on loamy soils, *Proceedings of 'Conservation 80', at the National College of Agricultural Engineering Silsoe* (England).

De Ploey, J., Savat, J. and Moeyersons, J., 1976. The differential impact of some soil loss factors on flow, runoff creep and rainwash, *Earth Surface Processes*, 1, 151-161.

Dewever, J., 1980. *De consistentie-index en erosieverschijnselen op zandige tot lemige akkers*, Thesis, Laboratory for Experimental Geomorphology, (University of Leuven).

Dickinson, W.T., Holland, M.E. and Smith, G.L., 1967. *An experimental rainfall-runoff facility*, Fort Collins, Colorado State University Hydrology Paper 25, 81 p.

Dimmock, G.M., Bettenay, E. and Mulcahy, M.J., 1974. Salt content of lateritic profiles in the Darling Range, Western Australia, *Australian Journal of Soil Research*, 12, 63-69.

Dortignac, E.J., 1951. Design and operation of the Rocky Mountain infiltrometer. *U.S.D.A. Forest Service. Rocky Mountain Forest and Range Experiment Station, Station Paper*, 5, 68 p.

Driesch, A. von den, 1973. Fauna, Klima und Landschaft in Süden der Iberischen Halbinsel während der Metallzeit, in: *Domestikations-forschung und Geschichte der Haustiere*, ed Matólsci, J. (Akademiai Kiado, Budapest), 245-254.

Dudley, J.H., 1970. Review of collapsing soils, *Journal of the Soil Mechanics and Foundations Division, Proceedings of the American Society of Civil Engineers*, 925-947.

Duley, F.L., 1939. Surface factors affecting the rate of intake of water by soils. *Soil Science Society of America, Proceedings*, 4, 60-64.

Dunin, F.X., 1976. Infiltration: its simulation for field conditions, in: *Facets of Hydrology*, ed Rodda, J.C. (J. Wiley, Chichester), 199-227.

Dunne, T., 1978. Field studies of hillslope flow processes, in: *Hillslope Hydrology*, ed Kirkby, M.J. (Wiley, London), 227-293.

Dunne, T., 1980. Formation and controls of channel networks, *Progress in Physical Geography*, 4(2), 211-239.

Dunne, T. and Dietrich, W.E., 1980. Experimental investigation of Horton overland flow on tropical hillslopes. 2. Hydraulic characteristics and hillslope hydrographs, *Zeitschrift für Geomorphologie, Supplementband*, 35, 60-80.

Dunne, T., Moore, T.R. and Taylor, C.H., 1975. Recognition and prediction of runoff producing zones in humid regions, *Hydrological Sciences Bulletin*, 20, 305-327.

Einstein, H.A., 1950. The bed-load function for sediment transportation in open channel flows, *U.S. Department of Agriculture, Soil Conservation Service, Technical Bulletin*, 1026.

Elias, F., 1963. Precipitaciones maximas en España: Regimen de intensidades y frecuencias, *Servicio de Conservacion de Suelos*, (Madrid).

Elias, F. and Giminez Oritz, R., 1965. Evapotranspiraciones potenciales y balances de Agua en España, *Spain, Direccion General de Agricultura* (Madrid).

Ellison, W.D., 1947. Soil erosion studies, *Journal of Agricultural Engineering*, 28, 145-146, 197-201, 245-248, 297-300, 349-351.

Emerson, W.W., 1967. A classification of soil aggregates based on their coherence in water, *Australian Journal of Soil Research*, 5, 47-57.

Emmett, W.W., 1970. The hydraulics of overland flow on hillslopes, *United States Geological Survey Professional Paper*, 662-A.

Emmett, W.W., 1974. Channel aggradation in western US as indicated by observations at Vigil Network sites, *Zeitschrift für Geomorphologie, Supplementband*, 21, 52-62.

Emmett, W.W., 1978. Overland flow, in: *Hillslope hydrology*, ed Kirkby, M.j. (Wiley, Chichester), 145-176.

Engelen, G.B., 1973. Runoff processes and slope development in Badlands National Monument, South Dakota, *Journal of Hydrology*, 18, 55-79.

Environment Canada, 1972. *Temperature and precipitation: 1941-1970*. (Atmospheric Environment Service, Downsview, Ontario).

Eyles, R.J., 1968. Stream net ratio in West Malaysia, *Bulletin of the Geological Society of America*, 79, 701-712.

Farres, P., 1978. The role of time and aggregate size in the crusting process. *Earth Surface Processes*, 3, 243-254.

Faulkner, P.H., 1970. *Aspects of channel and basin morphology in the Steveville badlands, Alberta*, unpublished M.Sc. thesis, (Department of Geography, University of Alberta.

Fenneman, N.M., 1931. *Physiography of the Western United States*. McGraw-Hill Co. N.Y.

Fisher, D.J., Erdmann, C.E. and Reeside, J.B. Jr., 1961. Cretaceous and Tertiary formations of the Book Cliffs. Carbon, Emery and Grand Counties, Utah and Garfield and Mesa Counties, Colorado. *U.S. Geological Survey Professional Paper*, 332.

Flint, R.F., 1935. White silt deposits in the Okanagan Valley, British Columbia. *Transactions of the Royal Society of Canada*, series 3, v. 29, 107-114.

FAO, 1967. *Guidelines for soil description; first edition*, (FAO, Rome).

FAO-UNESCO, 1974. *Soil map of the world, 1:5 000 000,* (UNESCO, Vol. 1, Legend, Paris).

Foster, G.R. and Meyer, L.D., 1972. Transport of soil particles by shallow flow, *Transactions of the American Society of Agricultural Engineers,* 15, 99-102.

Fournier, F., 1960. *Climat et érosion: la relation entre l'érosion du sol par l'eau et les précipitations atmosphériques,* (Paris).

Freeze, R.A., 1972. Role of subsurface flow in generating surface runoff 1. Base flow contributions to channel flow, *Water Resources Research,* 8, 1272-1283.

Freeze, R.A., 1978. Mathematical models of hillslope hydrology, in: *Hillslope hydrology,* ed Kirkby, M.J. (Wiley, London), 177-225.

Freitag, H., 1971. Die natürliche Vegetation des Südostspanischen Trockengebietes, *Botanisches Jahrbücher,* 91, 147-308.

Fulton, R.J., 1965. Silt deposition in late-glacial lakes of British Columbia, *American Journal of Science,* 263, 553-570.

Fulton, R.J., 1967. Deglaciation in the Kamloops region, British Columbia, *Geological Survey of Canada Bulletin,* 154.

Fulton, R.J., 1975. Quaternary geology and geomorphology, Nicola-Vernon area, British Columbia, *Geological Survey of Canada Memoir,* 380.

Gabriels, D., Pauwels, J.M. and De Boodt, M., 1977. A quantitative rill erosion study on a loamy sand in the hilly region of Flanders, *Earth Surface Processes,* 2-3, 257-260.

García Sanches, M., 1963. El poblado argárico del Cerro del Culantrillo, en Gorafe (Granada), *Archivo de Prehistoria Levantina,* 10, 97-164.

García Sanchez, M. and Spahni, J.C., 1959. Sepulcros megalíticos de la región de Gorafe (Granada), *Archivo de Prehistoria Levantina,* 8, 43-114.

Garfunkel, Z., 1978. The Negev, Regional synthesis of sedimentary Basins, in: *Excursion guidebook Part I. Tenth International Congress on sedimentology, Jerusalem, Israel,* 34-110.

Geiger, F., 1970. Die ariditat in Sudostspanien, *Stuttgarter Geographische Studien, Band,* 77.

Geiger, F., 1971. El Sureste español y los problemas de la aridez, *Estudio Geografico.*

Gerlach, T., 1967. Gerlach troughs - overland flow traps. *Revue de Géomorphologie Dynamique,* 17, 170.

Gerson, R., 1972. *Geomorphic processes of Mt. Sdom,* Ph.D. thesis, (Hebrew University, Jerusalem).

Gerson, R., 1977. Sediment transport for desert watersheds in erodible materials, *Earth Surface Processes,* 2, 343-361.

Gessler, J., 1971. Aggradation and degradation. in: *River Mechanics*, ed Shen, H.W., (Fort Collins, Colorado), vol. 1, pp. 8-1 - 8-24.

Gilbert, G.K., 1877. Report on the Geology of the Henry Mountains. *U.S. Geography and Geology Survey of the Rocky Mountain Region*, 160p.

Gilman, A., 1976. Bronze Age dynamics in southeast Spain, *Dialectical Anthropology*, 1, 307-319.

Gilman, K., 1971. A semi-quantitative study of the flow of natural pipes in the Nant Gerig sub-catchment, *Natural Environment Research Council, Institute of Hydrology, Subsurface Section, Wallingford, Internal Report*, 36.

Gilman, K. and Newson, M.D., 1980. *Soil pipes and pipeflow - a hydrological study in upland Wales*, (British Geomorphological Research Group Research Monograph No. 1, Geo Books, Norwich).

Glass, L.J. and Smerdon, E.T., 1967. The effect of rainfall on the velocity distribution in shallow channel flow, *Transactions of American Society of Agricultural Engineers*, 10, 330-332, 336.

Glock, W.S., 1931. The development of drainage systems: A synoptic view, *Geographical Review*, v. 21, pp 475-482.

Goldberg, P., 1976. Upper Pleistocene Geology of the Avdat/Aqev area, in: *Prehistory and Paleoenvironments in the Central Negev, Israel*, v.1, ed Marks, A.E. (SMU. Press, Dallas), 25-53.

Goldberg, P., (in press). Late Quaternary stratigraphy of Israel: an eclectic view, *C.N.R.S. Colloquium Prehistoire du Levant: Chronologie et organisation de l'espace depuis les origines Jusqu'au VI Millenaire, Lyon*, 1980.

Graf, W.H., 1971. *Hydraulics of sediment transport*, (McGraw-Hill, New York).

Graf, E.D. and Arora, H.S., 1972. Comments in *Proceedings of the Speciality Conference on the Performance of Earth and Earth-Supported Structures, A.S.C.E.*, Vol. 3, 105-106.

Green, W.H. and Ampt, G.A., 1911. Studies on soil physics. 1. The flow of air and water through soils, *Journal of Agricultural Science*, 4(1), 1-24.

Gregory, K.J. and Walling, D.E., 1974. The geomorphologist's approach to instrumented watersheds in the British Isles, in: *Fluvial Processes in Instrumented Watersheds*, ed Gregory K.J. and Walling, D.E. (Institute of British Geographers Special Publication No. 6, London), 1-6.

Grissinger, E.H., 1966. Resistance of selected clay systems to erosion by water, *Water Resources Research*, 2(1), 131-138.

Grissinger, E.H., 1972. Laboratory studies of the erodibility of cohesive materials, *Mississippi Water Resources Conference, Proceedings*, 19-36.

Grissinger, E.H. and Asmussen, L.E., 1963. Channel stability in undisturbed cohesive soils: a discussion, *American Society of Civil Engineers, Journal Hydraulics Division*, 3708 (HY6), 259-261.

Gusi Jener, F., 1975. La aldea neolítica de Terrera Ventura (Tabernas, Almeria), *XIII Congreso Nacional de Arquelogía, Huelva, 1973*, 331-314.

Gutierrez, A.A., 1980. *Channel and hillslope geomorphology of badlands in the San Juan Basin, New Mexico*, unpublished M.S. thesis, (University of New Mexico, Albuquerque).

Hack, J.T., 1960. Interpretation of erosional topography in humid temperate regions, *American Journal of Science*, 258a, 80-97.

Hack, J.T., 1965. Post glacial drainage evolution and geometry in the Ontonagon Area, Michigan, *U.S. Geological Survey Professional Paper*, 504B, 40p.

Hadley, R.F. and Schumm, S.A., 1961. Sediment sources and drainage basin characteristics in upper Cheyenne River Basin, *U.S. Geological Survey Water Supply Paper*, 1531, 137-198.

Hadley, R.F. and Lusby, G.C., 1967. Runoff and hillslope erosion resulting from a high intensity thunderstorm near Mack, Western Colorado. *Water Resources Research*, 3(1), 134-143.

Haigh, M.J., 1977. The use of erosion pins in the study of slope evolution, in: *British Geomorphological Research Group Technical Bulletin No. 18, Shorter Technical Methods*, ed Finlayson, B.L.

Haigh, M.J., 1978. Micro-rills and desiccation cracks: some observations, *Zeitschrift für Geomorphologie*, 22(4), 457-461.

Hall, S.A., 1977. Late quaternary sedimentation and paleoecologic history of Chaco Canyon, New Mexico, *Geological Society of American Bulletin*, 88, 1593-1618.

Hall, S.A., 1980. *Geology of archeologic sites and associated sand dunes, San Juan County, New Mexico*, unpublished report to Navajo Cultural Resource Management Program.

Hallam, S.J., 1975. *Fire and hearth: a study of Aboriginal usage and European usurpation in south-western Australia*, (Australian Institute of Aboriginal Studies, Canberra).

Harr, R.D., 1977. Water flux in soil and subsoil on a steep forested slope, *Journal of Hydrology*, 33, 37-58.

Harrison, R.J., 1974. A reconsideration of the Iberian background to Beaker metallurgy, *Palaeohistoria*, 16, 63-105.

Harvey, A.M., 1977. Event frequency in sediment production and channel change, in: *River Channel Changes*, ed Gregory, K.J., (John Wiley and Sons), 301-315.

Harvey, A.M., 1978. Dissected alluvial fans in southeast Spain, *Catena*, 5, 177-211.

Heede, B.H., 1971. Characteristics and processes of soil piping in gullies, *United States Department of Agriculture Forest Service Research Paper* RM-68, 1-13.

Heede, P.H., 1974. Stages in development of gullies in western United States of America, *Zeitschrift für Geomorphologie*, 18, 260-271.

Heede, B.H., 1976. Gully development and control, the status of our knowledge, *United States Department of Agriculture Forest Service Research Paper*, RM 196.

Henderson, F.M. and Wooding, R.A., 1964. Overland flow and groundwater flow from a steady rainfall of finite duration, *Journal of Geophysical Research*, 69, 1531-1540.

Henderson, F.M., 1966. *Open channel flow*. (MacMillan, New York).

Henschke, C.J., 1980. The extent of the dry land salinity problem, *Land and Stream Salinity Seminar, Western Australia*, Government Printer, Perth, 19.1-19.2.

Herbert, E.J., Shea, S.R. and Hatch, A.B., 1978. Salt content of lateritic profiles in the Yarragil catchment, Western Australia, *Forests Department of Western Australia Research Paper*, 32.

Hernández Hernández, F. and Dug Godoy, I., 1977. Escavaciones en el poblado de 'El Pichacho', *Excavaciones Arqueologicas en España*, 95.

Hewlett, J.D. and Hibbert, A.R., 1967. Factors affecting the response of small watersheds to precipitation in humid areas, *Proceedings of International Symposium of Forest Hydrology*, ed Sopper, W.E. and Lull, H.W., 275-290.

Hewlett, J.H. and Nutter, W.L., 1970. The varying source area of streamflow from upland basins, in: *Proceedings of the Symposium on Interdisciplinary Aspects of Watershed Management*, (Montana State University, Boseman, American Society of Civil Engineers), 65-83.

Heywood, B.H.J., 1961. Studies in frost-heave cycles at Schefferville, *McGill Sub-Arctic Research Report*, 11, 6-10.

Heywood, R.T., 1977. Probability of rainfall levels in Alberta, *Manuscript, Irrigation Division, Alberta Agriculture, Lethbridge*.

Higgins, C.G., 1953. Miniature pediments near Calistoga, California, *Journal of Geology*, 61, 461-465.

Hillel, D., 1960. Crust formation in loessial soils, *Proceedings of the 7th International Congress of Soil Science*, 330-339.

Hillel, D. and Gardner, W.R., 1970. Transient infiltration into crust-topped profiles, *Soil Science*, 109, 69-76.

Hills, R.C., 1970. The determination of the infiltration capacity of field soils using the cylinder infiltrometer, *British Geomorphological Research Group, Technical Bulletin*, 3.

Hingston, F.J. and Galaitis, V., 1976. The geographical variation of salt precipitated over Western Australia, *Australian Journal of Soil Research*, 14, 319-335.

Hogg, S.E., 1978. *The near surface hydrology of the Steveville badlands, Alberta:* Unpublished M.Sc. thesis, University of Alberta.

Holland, M.E., 1969. *Colorado State University experimental rainfall-runoff facility - Design and testing of rainfall system.* Fort Collins, Colorado State University. Engineering Research Center Publication, CER 69-70-MEH21.

Holmes, J.W., 1971. Salinity and the hydrologic cycle, in Salinity and Water Use: a National Symposium on Hydrology, ed Talsma, T. and Philip, J.R., (Macmillan, Melbourne), 25-40.

Holmes, J.W., 1979. *The Whittington interceptor drain trial: report to the Public Work Department, Western Australia,* (School of Earth Sciences, The Flinders University of South Australia).

Hooke, R. le B., 1968. Steady-state relationships on arid-region alluvial fans in closed basins, *American Journal of Science*, 266, 609-629.

Horton, R.E., 1938. Rain wave-trains, *Transactions American Geophysical Union*, 19, 368-374.

Horton, R.E., 1945. Erosional development of streams and their drainage basins: hydrophysical approach to quantitative morphology, *Bulletin Geological Society of America*, 56, 275-370.

Horton, R.E., Leach, H.R. and Van Vliet, R., 1934. Laminar sheet flow, *American Geophysical Union Transactions*, 15(2), 393-404.

Houldsworth, E., 1941. The Big Muddy Valley of Southern Saskatchewan, *Canadian Geographical Journal*, 23, 116-131.

Howard, A.D., 1971. Simulation of stream networks by headward growth and branching, *Geographical Analysis*, 3, 29-50.

Howells, K.A., 1980. *Pipeflow and its contribution to streamflow in a small upland catchment*, Unpublished B.Sc. dissertation, (University College, Swansea).

Hunt, C.B., Averitt, P., and Miller, R.L., 1953. Geology and geography of the Henry Mountains, *U.S. Geological Survey Professional Paper 228*.

Hunt, C.B., 1974. *Natural regions of the United States and Canada,* (Freeman, San Francisco).

Imeson, A.C. and Kwaad, F.J.P.M., 1980a. Gully types and gully prediction, *Geografisch Tijdschrift,* 14, 430-441.

Imeson, A.C. and Kwaad, F.J.P.M., 1980b. (in press) Field measurements of infiltration in a mountainous area, *Studia geomorphologica Carpatho-Balcanica.*

Imeson, A.C. and Verstraten, J.M., 1981. Suspended solids concentrations and river water chemistry, *Earth Surface Processes and Landforms,* 6, 251-263.

Inglis, C.C., 1968. Discussion of 'Systematic evaluation of river regime', *Proceedings of the American Society of Civil Engineers, Waterways and Harbors Division,* 94, WW1, 109-114.

ICONA, 1969. *Institute National Para la Conservacion de la Naturaleza,* Projecto de restauracion hydrological - forestal del embalse de Beninar, Granada, Spain.

ICONA, 1979. *Precipitaciones maximas en España,* Instituto Nacional para la Conservacion de la Naturealeza, Ministerie de Agricultura, Monografia 21.

Instituto Geologico y Minero de España, 1971. *Mapa geologica de España,* Escala 1:200 000 sintesis de la carografia existente, No. 78 (Baza), No. 84/85 (Almeria-Garrucha), No. 83, (Granada-Malaga), No. 77 (Jaen), Instituto Geologico y Minero de España, Madrid).

Iorns, W.V., Hembree, C.H. and Oakland, G.L., 1965. Water resources of the Upper Colorado River Basin--Technical Report. *U.S. Geology Survey Professional Paper 441.*

Ireland, H.A., Sharpe, C.F.S. and Eargle, D.H., 1939. Principles of gully erosion in the Piedmont of South Carolina. *United States Department of Agriculture Technical Bulletin,* 63.

Ishihara, T., Iwagaki, Y. and Ishihara, Y., 1953. On the rollwave-trains appearing in the water flow on a steep slope surface, *Memoirs of the Faculty of Engineering, Kyoto University,* Japan, 14-2, 83-91.

Izzard, C.F., 1944. The surface profile of overland flow, *Transactions, American Geophysical Union,* 25, 959-968.

James, D.M.D., 1971. The Nant-y-Moch formation, Plynlimon inlier, west central Wales. *Quarterly Journal of Geological Society,* 127, 177-181.

Jansen, J.M.L. and Painter, R.B., 1974. Predicting sediment yield from climate and topography, *Journal of Hydrology,* 21, 371-380.

Janssens, V., 1979. *Experimenten betreffende korrelselectiviteit in runoff,* Thesis (Laboratory for Experimental Geomorphology, University of Leuven).

Jeffreys, H., 1925. Flow of water in inclined channels of rectangular section, *Philosophical Magazine*, 49, 793-808.

Johnson, D., 1932. Miniature rock fans and pediments, *Science*, 76, 546.

Jones, J.A.A., 1971. Soil piping and stream channel initiation, *Water Resources Research*, 7(3), 602-610.

Jones, J.A.A., 1976. *Soil piping and the subsurface initiation of stream channel networks*, Unpublished Ph.D. thesis, (University of Cambridge, Cambridge).

Jones, J.A.A., 1978. Soil pipe networks: distribution and discharge, *Cambria*, 5(1), 1-21.

Jones, J.A.A., 1979. Extending the Hewlett model of stream runoff generation. *Area*, 11(2), 110-114.

Jones, J.A.A., 1981. *The nature of soil piping - a review of research*, (British Geomorphological Research Group Research Monograph No. 3, Geo Books, Norwich), 301pp.

Jones, J.A.A. and Crane, F.G., 1980. New evidence of rapid interflow contributions to the streamflood hydrograph, *Beitrage für Hydrologie, Sonderheft* 3.

Jopling, A.V. and Forbes, D.L., 1979. Flume study of silt transportation and deposition, *Geografiska Annaler*, 61A, 67-85.

Jungerius, P.D. and Wusten, H.H. van der, 1980. A rapid method for the determination of soil dispersion and its application to soil erosion problems in the Rif Mountains Morocco, *Studia geomorphologica Carpatho-Balcanica*, (in press).

Kälin, M., 1977. Hydraulic piping: theoretical and experimental findings, *Canadian Geotechnical Journal*, 14, 107-124.

Kalinske, A.A., 1947. Movement of sediment as bed-load in rivers, *Transactions of the American Geophysical Union*, 28-4.

Kamphorst, A. and Bolt, G.H., 1976. Saline and sodic soils. in: *Soil chemistry A. Basic elements*, ed Bolt, G.H. and Bruggenwert, M.G.M., (Elsevier, Amsterdam), 171-191.

Karki, K.S., Chander, S. and Malhotra, R.C., 1972. Supercritical flow over sills at incipient jump conditions, *Proceedings of the American Society of Civil Engineers, Hydraulics Division*, 98, HY 10, 1753-1764.

Karcz, I. and Kersey, D., 1980. Experimental study of free-surface flow instability and bedforms in shallow flows, *Sedimentary Geology*, 27, 263-300.

Kayser, B., 1964. *Studi sui terreni e sull'erosione del suolo in Lucania*, (Edizione Montemurro, Matera).

Kelley, W.P., 1964. Review of investigations on cation exchange and semi-arid soils, *Soil Science*, 97, 80-88.

Kendrick, G.W., 1977. Middle Holocene marine molluscs from near Guildford, Western Australia, and evidence for climatic change, *Journal of the Royal Society of Western Australia,* 59, 97-104.

Kharaka, Y.K. and Barnes, I., 1973. SOLMNEQ: Solution-Mineral equilibrium computations, *United States Geological Survey, National Technical Information Service,* PB 215899, (U.S. Department of Commerce, Springfield).

Kilinc, M., and Richardson, E.V., 1973. Mechanics of soil erosion from overland flow generated by simulated rainfall, *Colorado State University Hydrology Paper,* 63.

Kimber, P.C., 1974. The root system of jarrah (Eucalyptus marginata), *Forests Department of Western Australia Research Paper,* 10.

King, C.A.M., 1966. *Techniques in geomorphology,* (Edward Arnold Ltd., London), 342 pp.

Kirkby, M.J., (ed), 1978. *Hillslope hydrology:* (Wiley Interscience, Chichester).

Kirkby, M.J., 1978. Implications for sediment transport, in: *Hillslope hydrology,* ed Kirkby, M.J. (Wiley, Interscience, Chichester), 325-363.

Kirkby, M.J., 1980. The stream head as a significant geomorphological threshold, in: *Thresholds in Geomorphology,* eds Coates, D.R. and Vitek, J.D., (George Allen and Unwin).

Kirkby, M.J. and Chorley, R.J., 1967. Throughflow, overland flow and erosion, *International Association of Scientific Hydrology Bulletin,* 12, 5-21.

Kirkby, M.J., Callan, J., Weyman, D.R. and Wood, J., 1976. Measurement and modelling of dynamic contributing areas in very small catchments, *University of Leeds, Department of Geography Working Paper,* 167.

Knapp, B.J., 1973. A system for the field measurement of soil water movement, *British Geomorphological Research Group Technical Bulletin,* 9.

Knobel, E.V., Dansdill, R.K. and Richardson, M.L., 1955. Soil survey of the Grand Junction area, Colorado. *U.S.D.A. Soil Survey Series 1940,* 19, 118p.

Koons, D., 1955. Cliff retreat in the southwestern United States, *American Journal of Science,* v. 253, pp 44-52.

Kostiakov, A.M., 1932. Reported in R.M. Dixon, 1977. Air-earth interface concept for wide-range control of infiltration, *American Society of Agricultural Engineers Paper,* 77-2062.

Kutiel, H., 1978. *The distribution of rain intensities in Israel,* M.Sc. thesis, (The Hebrew University, Jerusalem).

Lam, K.C., 1977. Patterns and rates of slopewash on the badlands of Hong Kong, *Earth Surface Processes*, 2, 319-332.

Lane, E.W., 1955. Design of stable channels, *Transactions of the American Society of Civil Engineers*, 120, 1234-1279.

Langbein, W.B. and Schumm, S.A., 1956. Yield of sediment in relation to mean annual precipitation, *Transaction of the American Geophysical Union*, 1076-1084.

Laronne, J.B., 1977. *Dissolution potential of surficial Mancos Shale and alluvium*, unpublished Ph.D. thesis, (Department of Earth Resources, Colorado State University).

Laronne and Shen, 1982. *Journal of Hydrology*, (in Press).

Laws, J.O., and Parsons, D.A., 1943. Relation of raindrop size to intensity, *Transactions of the American Geophysical Union*, 24, 452-460.

Leopold, L.B., 1978. El asunto del arroyo, in: *Geomorphology: present problems and future prospects*, ed Embleton, C., Brunsden, D. and Jones D.K.C., (Oxford University Press, Oxford).

Leopold, L.B., Emmett, W.W. and Myrick, R.M., 1966. Channel and hillslope processes in a semi-arid area, *United States Geological Survey, Professional Paper*, 252.

Leopold, L.B., and Langbein, W.B., 1962. The concept of entropy in landscape evolution, *U.S. Geological Survey Professional Paper* 500A, 20p.

Livesly, B., 1976. The sedimentary influence of a tributary stream growth of the Niabrara Delta, *Proceedings of the 3rd Federal Inter Agency Symposium on Sedimentation. Sedimentation Committee, Water Resources Council*, 4-127-137.

Longley, R.W., 1968. *Climatic maps for Alberta:* Department of Geography, University of Alberta, 8pp.

Longley, R.W., 1972. The climate of the Prairie Provinces, *Environment Canada Climatological Studies*, 13.

Lopez Bermudez, F., 1973. *La Vega alta del Segura* (Deparamento de Geographia, Universidad de Murcia).

López García, P., 1978. Resultados polínicos del Holoceno en la Peninsula Ibérica, *Trabajos de Prehistoria*, 33, 9-44.

Love, D.W., 1980. *Quaternary geology of Chaco Canyon, northwestern New Mexico*, unpublished Ph.D. dissertation, (University of New Mexico, Albuquerque).

Loveday, J. and Pyle, J., 1973. The Emerson dispersion test and its relationship to hydraulic conductivity, *Division of Soils Technical Paper*, 15 (CSIRO, Australia), 1-7.

Low, A.J., 1954. The study of soil structure in the field and in the laboratory, *Journal of Soil Science*, 5, 57-74.

Luckman, B., 1978. Measurements of debris movement on alpine talus slopes, *Zeitschrift für Geomorphologie, Supplement Band*, 29, 117-129.

Lulli, L., 1974. Una ipotesi sulla formazione dei calanchi della Valle dell'Era (Toscana), *Bollettino della Società Italiana della Scienza del Suolo*, 8, 3-6.

Lulli, L. and Ronchetti, G., 1973. Prime osservazioni sulla crepacciatura dei suoli nelle argille plioceniche marine della Valle dell'Era (Toscana), *Annali dell' Istituto Sperimentale per lo Studio e la Difesa del Suolo* (Florence), IV, 143-9.

Lusby, G.C., 1979. Effects of grazing on runoff and sediment yield from desert rangeland at Badger Wash in Western Colorado, 1953-1973. *U.S. Geological Survey Water Supply Paper 1532-I.*

Lusby, G.C., Reid, V.H. and Knipe, O.D., 1971. Effects of grazing on the hydrology and geology of the Badger Wash Basin in Western Colorado, 1953-1966. *U.S. Geological Survey Water Supply Paper 1532-D.*

Lvovitch, M.I., 1980. Soil trend in hydrology. *Hydrological Sciences Bulletin*, 25(1), 33-46.

Mabbutt, J.A., 1977. *Desert landforms*, (The MIT Press, Cambridge, Massachusetts).

Macaulay, H.A., Hobson, G.B. and Fulton, R.J., 1972. Bedrock topography of the north Okanagan Valley and stratigraphy of the unconsolidated valley fill, *Geological Survey of Canada Paper*, 72-8.

Machemehl, J.L., 1968. Sediment transport in shallow supercritical flow disturbed by simulated rainfall, *Water Resources Institute, Texas A & M University, Technical Report*, 14.

Marks, A.E., 1977. Introduction: A preliminary overview of central Negev Prehistory, 3-31, *Prehistory and Palaeoenvironments in the Central Negev, Israel*. Vol. II, ed Marks, E.A., (SMU Press, Dallas).

Marks, A., Phillips, H., Crea, H. and Ferring, R., 1971. Prehistoric sites near Ein Avdat in the Negev, *Israel Exploration Journal*, 21, 13-124.

Martin, L., 1979. Accelerated soil erosion from tractor wheelings: a case study in mid-Bedfordshire, England, in: *Symposium on Agricultural Soil Erosion in Temperate Climates*, Strasbourg 20-23 September 1978, 157-162.

Masannat, Y.M., 1980. Development of piping erosion conditions in the Benson area, Arizona, U.S.A., *Quarterly Journal of Engineering Geology*, 13, 53-61.

Massy, M., 1980. *Bijdrage tot het onderzoek naar de oorzaken en de mechanismen van de differentiëele verslemping en de geulerosie op akkers*, unpublished thesis (Laboratory for Experimental Geomorphology, University of Leuven).

Mayer, P.G., 1959. Roll waves and slug flows in inclined open channels, *Journal of the Hydraulics Division, American Society of Civil Engineers*, 85, 99-140.

McArthur, W.M. and Bettenay, E., 1979. The land, in: *Environment and science*, ed O'Brien, B.J., (University of Western Australia Press for the Education Committee of the 150th Anniversary Celebrations Perth), 22-52.

McArthur, W.M. and Clifton, A.L., 1975. Forestry and agriculture in relation to soils in the Pemberton area of Western Australia, *CSIRO Soils and Land Use Series*, 54.

McCaig, M., 1979. The pipeflow streamhead - a type description. *University of Leeds, School of Geography Working Paper*, 242.

McIntyre, D.S., 1958a. Permeability measurements of soil crusts formed by raindrop impact. *Soil Science*, 85, 185-189.

McIntyre, D.S., 1958b. Soil splash and the formation of surface crusts by raindrop impact. *Soil Science*, 85, 261-266.

McNeal, B.L., 1970. Prediction of interlayer swelling of clays in mixed-salt solutions, *Soil Science Society of America Proceedings*, 34, 201-206.

McNeal, B.L., Norvell, W.A. and Coleman, N.T., 1966. Effect of solute composition on the swelling of extracted soil clays, *Soil Science Society of America Proceedings*, 30, 313-317.

Mears, B., 1963. Karst-like features in badlands of the Arizona Painted Desert, *Wyoming University Contributions in Geology*, 2(1), 7-11.

Mears, B., 1968. Piping, in: *Encyclopedia of Geomorphology*, ed Fairbridge, R.W., (Reinhold Book Company, New York), 849-850.

Mein, R.G. and Larson, C.L., 1973. Modelling infiltration during a steady rain, *Water Resources Research*, 9(2), 384-394.

Melton, M.A., 1960. Intravalley variation in slope angles related to microclimate and erosional environment, *Geological Society of America Bulletin*, 71, 133-144.

Mendoza, A., Molino, F., Aguayo, P., Carrasco, J. and Nájera, T., 1975. El poblado del 'Cerro de los Castellones' (Laborcillas, Granada), *XIII Congreso Nacional de Arqueología, Huelva*, 315-322. Zaragosa.

Meyer, C. and Yenne, K., 1940. Notes on the mineral assemblage of the white silt terraces in the Okanagan Valley, British Columbia, *Journal of Sedimentary Petrology*, 10, 8-11.

Meyer-Peter, E. and Muller, R., 1948. Formulae for bed-load transport, *International Association for Hydraulic Research,* 2nd Meeting Stockholm 1948, 39-64.

Miller, J.P. and Leopold, L.B., 1963. Simple measurements of morphological changes in river channels and hill slopes, *Changes of Climate, UNESCO Arid Zone Research Series XX,* 421-427.

Mitchell, J.K., 1976. *Fundamentals of soil behavior,* (Wiley, New York).

Mock, S.J., 1971. A classification of channel links in stream networks, *Water Resources Research,* 7, 1558-1566.

Molina González, F. and Pareja López, E., 1975. Excavaciones en la Cuesta del Negro (Purullena, Grandada), Excavaciones Arqueólogicas en España, 86.

Mollard, J.D., 1974. *Landforms and surface materials of Canada,* (J.D. Mollard Association Limited, Regina).

Monnon, J.L., 1978. Observations préliminaires sur l'érosion des sols en Outre-Forêt (Alsace), *Recherches Geographiques a Strasbourg,* 9, 47-52.

Morgan, R.P.C., 1972. Observations on factors affecting the behaviour of a first-order stream, *Transactions of the Institute of British Geographers,* 56, 171-185.

Morgan, R.P.C., 1977. Soil erosion in the United Kingdom: field studies in the Silsoe area 1973-1975, *National College of Agricultural Engineering Silsoe, Occasional Paper* 5.

Morgan, R.P.C., 1979. *Topics in applied geography: soil erosion,* (Longman, London).

Mori, A., 1968. Considerazioni sull'erosione accelerata del suolo in Abruzzo, *Bollettino della Società Geografica Italiana,* (Rome), Serie IX, Vol. IX, 67-78.

Morin, J. and Benyamini, Y., 1977. Rainfall infiltration into bare soils, *Water Resources Research,* 13(5), 813-817.

Moss. A.J., Walker, P.H. and Hutka, J., 1979. Raindrop-simulated transportation in shallow water flows: an experimental study, *Sedimentary Geology,* 22, 165-184.

Moss, A.J., Walker, P.H. and Hutka, J., 1980. Movement of loose, sandy detritus by shallow water flows: an experimental study, *Sedimentary Geology,* 25, 43-66.

Mücher, H.J. and De Ploey, J., 1977. Experimental and micromorphological investigations of erosion and redeposition of loess by water, *Earth Surface Processes,* 2, 117-124.

Mulcahy, M.J., 1971. Landscapes, laterites and soils in south Western Australia, in: *Landform Studies from Australia and New Guinea,* ed Jennings, J.N. and Mabbutt, J.A., (A.N.U. Press, Canberra), 211-230.

Mulcahy, M.J., 1978. Salinisation in the southwest of Western Australia, *Search*, 9, 269-272.

Mundorff, J.D., 1972. Reconnaissance of chemical quality of surface water and fluvial sediment in the Price River Basin, Utah. *State of Utah Department of National Research Technical Publication 39*, 55p.

Nasmith, H., 1962. Late glacial history and surficial deposits of the Okanagan Valley, British Columbia, B.C., *Department of Mines and Petroleum Resources Bulletin*, 46.

Newson, M.D., 1976. Soil piping in upland Wales: a call for more information, *Cambria*, 3(1), 33-39.

Newson, M.D. and Harrison, J.G., 1978. Channel studies in the Plynlimon experimental catchments, *Natural Environment Research Council Institute of Hydrology, Wallingford, Berkshire, Report No. 47*, 61pp.

Nordin, C.F., 1964. Study of channel erosion and sediment transport, *American Society of Civil Engineers, Journal of Hydraulics Division*, 90, 173-192.

Nortcliff, S., Thornes, J.B. and Waylen, M.J., 1979. Tropical forest systems: a hydrological approach, *Amazonia*, 6, 557-568.

Northcote, K.H., 1971. *A factual key for the recognition of Australian soils*, 3rd ed., (Rellim Technical Publications, Glenside, South Australia).

Nulsen, R.A., 1978. Water movement through soil, *Journal of Agriculture Western Australia*, 19, 106-107.

Nyland, D. and Miller, G.E., (ed) 1977. *Geological hazards and urban development of silt deposits in the Penticton area*, (Geotechnical and Materials Branch Report, B.C. Ministry of Highways and Public Works).

Oster, J.D. and Sposito, G., 1980. The Gapon coefficient and the exchangeable sodium percentage - sodium adsorption ratio relation, *Soil Science Society of American Proceedings*, 44, 258-260.

Otvos, E.G., Jr., 1965. Types of rhomboid beach surface patterns, *American Journal of Science*, 263, 271-276.

Panicucci, M., 1972. Ricerche orientative sui fenomeni erosivi nei terreni argillosi, *Atti delle Giornate di Studio della Prima Sezione C.I.G.R., Firenze, 12-16 Settembre, 1972*, 260-280.

Parizek, R.R., 1964. *Geology of the Willow Bunch Lake Area, Saskatchewan*, (Saskatchewan Research Council, Geology Division, Report 4).

Parker, G.G., 1965. Piping: a geomorphic agent in landform development of the drylands, *International Association of Scientific Hydrology, Special Publication*, 65, 103-113.

Parker, G., 1978. Self-formed straight rivers with equilibrium banks and mobile bed. Part 2. The gravel river, *Journal of Fluid Mechanics*, 89-1, 127-146.

Parker, G., 1979. Hydraulic geometry of active gravel rivers, *Proceedings of the American Society of Civil Engineers, Journal of the Hydraulics Division*, 105, HY 9, 1185-1201.

Parker, R.S., 1977. *Experimental study of drainage basin evolution and its hydrologic implications*, unpublished Ph.D. thesis, (Fort Collins, Colorado State University).

Parker, R.S., 1977. Experimental study of drainage basin evolution and its hydrologic implications. Colorado State University, *Hydrology Paper*, 90.

Parsons, D.A., 1949. Depths of overland flow, *Soil Conservation Service Technical Paper*, 82.

Passerini, G., 1957. La degradazione idrometeorica dei terreni argillosi italiani, *Atti del Primo Simposio Internazionale de Agrochimica, Pisa, 7-8 maggio, 1957*, 85-107.

Peck, A.J., 1976. Interactions between vegetation and stream water quality in Australia, in: *Watershed management on range and forest lands, Proceedings of the 5th workshop of the U.S./Australia Rangelands Panel, Boise, Idaho, June 15-22, 1975*, ed Heady, H.F., Falkenborg, D.H. and Riley, J.P., (Utah Water Research Laboratory, College of Engineering, Utah State University, Logan), 149-155.

Peck, A.J., 1977. Development and reclamation of secondary salinity, in: *Soil factors in crop production in a semi-arid environment*, ed Russell, J.S. and Greacen, E.L., (University of Queensland Press, Brisbane), 301-319.

Peck, A.J., 1978a. Salinisation of non-irrigated soils and associated streams: a review, *Australian Journal of Soil Research*, 16, 157-168.

Peck, A.J., 1978b. Note on the role of a shallow dryland aquifer in dryland salinity, *Australian Journal of Soil Research*, 16, 237-240.

Peck, A.J. and Hurle, D.H., 1973. Chloride balance of some farmed and forested catchments in southwestern Australia, *Water Resources Research*, 9, 648-757.

Peck, A.J. and Watson, J.D., 1979. Hydraulic conductivity and flow in non-uniform soil, in: *Proceedings of Workshop on Soil Physics and Field Heterogeneity, CSIRO Division of Environmental Mechanics, Canberra, Australia*, February 1979.

Peck, A.J., Height, M.I., Hurle, D.H. and Hendle, P.A., 1979. Changes in groundwater systems after clearing for agriculture, *Proceedings of agriculture and the environment in Western Australia, Perth, 10th October 1979*, (Western Australian Institute of Technology, Bentley), 95-97.

Phelps, H.O., 1975. Shallow laminar flows over rough granular surfaces, *Proceedings of the American Society of Civil Engineers, Journal of the Hydraulics Division*, 101, HY 3, 367-384.

Philip, J.R., 1957. The theory of infiltration, 4. Sorptivity and algebraic infiltration equations, *Soil Science*, 84(3), 257-264.

Pickup, G., 1975. Downstream variations in morphology, flow conditions and sediment transport in an eroding channel, *Zeitschrift für Geomorphologie*, 19, 443-459.

Pike, R.J. and Wilson, S.E., 1971. Elevation-relief ratio, hypsometric integral, and geomorphic area-altitude analysis, *Geological Society of America Bulletin*, 82, 1079-1084.

Pilgrim, A.T., 1979. Landforms, in: *Western landscapes*, ed Gentilli, J., (University of Western Australia, Press for the Education Committee of the 150th Anniversary Celebrations, Perth), 49-87.

Pinczès, Z., 1972. Die Formen der Bodenerosion und der Kampf gegen Sie im Weingebiet des Tokajer Berges, *Abhandlungen aus dem geographischen Institut der Kossuth Universität in Debrecen (Hungary)*, 10, 63-70.

Polemio, M. and Rhoades, J.D., 1977. Determining cation exchange capacity: A new procedure for calcareous and gypsiferous soils, *Soil Science Society of America Proceedings*, 41, 524-527.

Ponce, S.L., 1975. *Examination of a non-point source loading function for the Mancos Shale wildlands of the Price River Basin, Utah*, unpublished Ph.D. thesis, Department of Civil and Environmental Engineering, Utah State University.

Ponce, S.L. and Hawkins, R.H., 1978. Salt pickup by overland flow in the Price River Basin, Utah. *Water Resources Bulletin*, 14(5), 1187-1200.

Pond, S.F., 1971. Qualitative investigation into the nature and distribution of flow processes in Nant Gerig, *Institute of Hydrology, Wallingford, Report*, 28.

Poulovassilis, A. and Tzimas, E., 1974. The hysteresis in the relationship between hydraulic conductivity and suction, *Soil Science*, 117, 250-256.

Public Works Department, 1979. *Investigations of the Whittington interceptor system of salinity control: Progress Report May 1979*, (Water Resources Information Note, Water Resources Section, Planning Design and Investigation Branch, Public Works Department, Perth, Western Australia).

Puvaneswaran, P., 1981. *Soil-slope relationships in the Wugong Brook catchment, Western Australia*, unpublished M.A. thesis, Department of Geography, University of Western Australia.

Quigley, R.M., 1977. *Mineralogy, chemistry and structure. Penticton and South Thompson silt deposits*, Research Report to B.C. Department of Highways.

Quinlan, J.R., 1974. *Karst, pseudokarst and dolines: a classification and review*, unpublished Ph.D. thesis, (University of Texas, Austin, Texas).

Quirk, J.P., 1978. Some physico-chemical aspects of soil structural stability - a review, in: *Modification of Soil Structure*, ed Emerson, W.W., Bond, R.D. and Dexter, A.R., (Wiley, Chichester), 3-16.

Quirk, J.P. and Schofield, R.K., 1955. The effect of electrolyte concentration on soil permeability, *Journal of Soil Science*, 6, 163-178.

Rapetti, F. and Vittorini, S., 1972. I venti piovosi a Legoli (Toscana) in relazione ai processi di erosione del suolo, *Atti della Società Toscana di Scienze Naturali*, Serie A, Vol. LXXIX, 150-175 (Pisa).

Rapetti, F. and Vittorini, S., 1975. La temperatura del suolo in due versanti contrapposti del preappennino argilloso toscano, *Bollettino della Società Italiana di Scienze del Suolo*, 9, 9-25.

Rapp, A., 1961. Recent development of mountain slopes in Kärkevagge and surroundings, Northern Scandinavia, *Geografiska Annaler*, 42, 72-200.

Rapp, A., Murray-Rust, D.H., Christiansson, C. and Berry, L., 1972. Soil erosion and sedimentation in four catchments near Dodoma, Tanzania, *Geografiska Annaler*, 54-A, 3-4, 255-318.

Rendell, H.M., 1975. *Clay hillslope erosion in a semi-arid area: the Basento Valley, southern Italy*, unpublished Ph.D. thesis, (University of London, London).

Rhoades, J.D. and Merrill, S.D., 1976. Assessing the suitability of water for irrigation: theoretical and empirical approaches, in: *Prognosis of salinity and alkalinity, FAO Soils Bulletin*, 31, (FAO, Rome), 69-109.

Richards, L.A. (ed) 1954. Diagnosis and improvement of saline and alkali soils, *USDA, Washington, Handbook*, 60.

Richter, G., 1965. Bodenerosion, Schäden und gefährdete Gebiete in der Bundesrepublik Deutschland, *Forschungen zur Deutschen Landeskunde*, 152, 592 p.

Roberts, F.J. and Carbon, B.A., 1971. Water repellence in sandy soils of south-western Australia. I. Some studies related to field occurrence, *Field Station Records Division of Plant Industries CSIRO*, 10, 13-20.

Roberts, F.J. and Carbon, B.A., 1972. Water repellence in sandy soils of south-western Australia. II. Some chemical characteristics of the hydrophobic skins, *Australian Journal of Soil Research*, 10, 35-42.

Rodda, J.C., Downing, R.A. and Law, F.M., 1976. *Systematic hydrology*, (Butterworth, London).

Rolfe, B.N., Miller, R.F. and McQueen, I.S., 1960. Dispersion characteristics of montmorillonite, kaolinite and illite clays in waters of varying quality and their control with phosphate dispersants, *United States Geological Survey Professional Paper*, 334-G, 229-271.

Roose, E.J. and Lelong, F., 1976. Les facteurs de l'érosion hydrique en Afrique tropicale: études sur petites parcelles expérimentales de sol, *Revue de Géographie Physique et de Géologie Dynamique*, 18, 365-374.

Rottner, J., 1959. A formula for bed-load transportation, *La Houille Blanche*, 4-3, 301-307.

Rowell, D.L., 1963. Effect of electrolyte concentration on the swelling of orientated aggregates of montmorillonite. *Soil Science*, 96, 368-374.

Rowell, D.L., 1965. Influence of positive charge on the inter- and intra- crystalline swelling of orientated aggregates of Na-montmorillonite in NaCl Solutions, *Soil Science*, 100, 340-347.

Rowell, D.L., Payne, D. and Ahmad, N., 1969. The effect of the concentration and movement of solutions on the swelling, dispersion and movement of clay in saline and alkali soils, *Journal of Soil Science*, 20, 176-188.

Rubey, W.W., 1928. Gullies in the great plains formed by sinking of the ground, *American Journal of Science*, v. 15, pp 417-422.

Rubin, J. and Steinhardt, R., 1964. Soil water relations during rain infiltration, 3. Water uptake at incipient ponding. *Proceedings of the Soil Science Society of America*, 28(5), 614-619.

Rudeforth, C.C., 1970. *Soils of North Cardiganshire*. (Memoirs of the Soil Survey of Great Britain).

Ruhe, R.V., 1950. Graphic analysis of drift topographies, *American Journal of Science*, 248, pp 435-443.

Ruh-Ming Li, Simons, D.B. and Stevens, M.A., 1976. Morphology of cobble streams in small watersheds, *Proceedings of the American Society of Civil Engineers, Journal of the Hydraulics Division*, 100, HY 8, 1101-1117.

Russell, E.W., 1973. *Soil conditions and plant growth. 10th ed.* (Longman, London).

Saba El-Ghossain, T., 1978. Erosion et caractéristiques analytiques des formations superficielles par un exemple du vignoble alsacien, *Recherches Géographiques à Strasbourg*, 9, 79-94.

Savat, J., 1975. Some morphological and hydraulic characteristics of river patterns in the Zaire Basin, *Catena*, 2, 161-180.

Savat, J., 1976. Discharge velocities and total erosion of a calcareous loess: a comparison between pluvial and terminal runoff, *Revue de Géomorphologie Dynamique*, XXIV, 113-122.

Savat, J., 1977. The hydraulics of sheetflow on a smooth surface and the effect of simulated rainfall, *Earth Surface Processes*, 2, 125-140.

Savat, J., 1979. Laboratory experiments on erosion and deposition of loess by laminar sheetflow and turbulent rillflow, *Proceedings on the Seminar Agricultural Soil Erosion in Temperate Non-Mediterranean Climate: Strasbourg-Colmar*, 20-23, September 1979, 139-143.

Savat, J., 1980. Resistance to flow in rough supercritical sheet flow, *Earth Surface Processes*, 5, 103-122.

Saxton, K.E., Spomer, R.G. and Kramer, L.A., 1971. Hydrology and erosion of loessial watersheds, *Proceedings of the American Society of Civil Engineers, Journal of the Hydraulics Division*, 97, HY 11, 1835-1851.

Scategni, P., 1971. *Esperienze di correzione dei torrenti dissestati di tipo alpino* (Ministero dell'Agricoltura e delle Foreste, Direzione Generale per l'Economia Montana e per le Foreste, Rome).

Schiedegger, A.E., Schumm, S.A. and Fairbridge, R.W., 1968. Badlands, in: *The Encyclopedia of Geomorphology*, ed R.W. Fairbridge, (Reinhold Book Corporation, New York), 43-48.

Schlichting, H., 1979. Boundary layer theory, (McGraw Hill, New York).

Schmidt, R.G., 1979. Probleme der Erfassung und Quatifizierung von Ausmass und Prozessen der aktuellen Bodenerosion (Abspülung) auf Ackerflächen, *Physiogeographica, Basel*, 1.

Schubart, H., 1975. Cronología relativa de la cerámica sepulcral de El Argar, *Trabajos de Prehistoria*, 32, 79-92.

Schuldenrein, J. and Goldberg, P., (in press). Late Quaternary Paleoenvironments and Prehistoric site distributions in the Lower Jordan Valley: A preliminary report, in: *Paleorient*.

Schüle, W., 1967. Feldbewässerung in Alt-Europa, *Madrider Mitteilungen*, 8, 79-99.

Schüle, W. and Pellicer, M., 1966. El Cerro de la Virgen (Orce, Granada) I. *Excavaciones Arqueologicos en España*, 46.

Schultz, J.D., 1980. *Geomorphology, sedimentology, and Quaternary history of the eolian deposits, West-central San Juan Basin, northwest New Mexico*, unpublished M.S. thesis, (University of New Mexico, Albuquerque).

Schultz, J.D. and Wells, S.G., 1981. Geomorphology of the Chaco dune field, northwestern New Mexico, *Geological Society of America Abstracts with Programs*, Cordilleran Section, 13, 105.

Schumm, S.A., 1956a. Evolution of drainage systems and slopes in badlands at Perth Amboy, New Jersey, *Bulletin of the Geological Society of America*, 67, 597-646.

Schumm, S.A., 1956b. The role of creep and rainwash on the retreat of badlands slopes, *American Journal of Science*, 254, 693-706.

Schumm, S.A., 1962. Erosion on miniature pediments in Badlands National Monument, South Dakota, *Geological Society of American Bulletin*, 73, 719-724.

Schumm, S.A., 1964. Seasonal variations of erosion rates and processes on hillslopes in western Colorado. *Zeitschrift für Geomorphologie*, 5, 215-238.

Schumm, S.A., 1967. Rates of surficial rock creep on hillslopes in Western Colorado, *Science*, 155, 560-561.

Schumm, S.A., 1973. Geomorphic thresholds and complex response of drainage systems, in: *Fluvial geomorphology*, ed Morisawa, M., (Publications in Geomorphology, State University of New York, Binghampton), 299-310.

Schumm, S.A., 1975. Episodic erosion: A modification of the geomorphic cycles, in: *Theories of landform development*, ed Melhorn, W.N. and Flenal, R.C., (Publication in Geomorphology, State University of New York, Binghampton, 69-85).

Schumm, S.A., 1977. *The fluvial system*, Wiley-Interscience Publication, New York.

Schumm, S.A. and Hadley, R.F., 1957. Arroyos and the semiarid cycle of erosion, *American Journal of Science*, 255, 161-174.

Schumm, S.A. and Hadley, R.F., 1961. Progress in the application of landform analysis in studies of semiariel erosion, *U.S. Geological Survey Circular*, 473, 13.

Schumm, S.A. and Lichty, R.W., 1965. Time, space and causality in geomorphology, *American Journal of Science*, 263, 110-119.

Schumm, S.A. and Lusby, G.C., 1963. Seasonal variations in infiltration capacity and runoff on hillslopes of Western Colorado, *Journal of Geophysical Research*, 68, 3655-3666.

Scoging, H.M., 1976. A stochastic model of daily rainfall simulation in a semi-arid environment, *LSE Discussion Paper*, 59, 1-22.

Scoging, H.M. and Thornes, J.B. 1980. Infiltration characteristics in a semi-arid environment, *Proceedings of Symposium on the hydrology of areas of low precipitation*, Canberra, December 1979, *International Association of Scientific Hydrology Publication*, 128, 159-168.

Scott, G.R., O'Sullivan, R.B. and Heller, J.S., 1979. Preliminary geologic map of the Burnham Trading Post Quadrangle, San Juan County, New Mexico, *U.S. Geological Survey Miscellaneous Field Studies Map*, MF-1076.

Servizio Idrografico Italiano, 1953. *Le sorgenti italiane*, (Pubblicazione No. 14, Ministero dei Lavori Pubblici, Rome).

Seymour, J., 1980. Soil erosion: can we dam the flood? *Ecosystem*, 25, 3-11.

Sfalanga, M. and Rizzo, V., 1974. Caratteristiche tecniche delle argille plioceniche e pleistoceniche in relazione al loro assetto morfologico, *Annali dell' Istituto Sperimentale per lo Studio e la Difesa del Suolo* (Florence), 5, 255-306.

Sharma, M.L., Gander, G.A. and Hunt, C.G., 1980. Spatial variability of infiltration in a watershed, *Journal of Hydrology*, 45, 101-122.

Shaw, J., 1975. Sedimentary successions in Pleistocene ice-marginal lakes, in: *Glaciofluvial and glaciolacustrine sediments*, ed Jopling, A.V. and MacDonald, B.C., (Society of Economic Paleontologists and Mineralogists, Special Publication 23).

Shea, S.R. and Hatch, A.B., 1976. Stream and groundwater salinity levels in the South Dandalup catchment of Western Australia, *Forests Department of Western Australia Research Paper*, 22.

Shen, H.W. and Ruh-Ming Li, 1973. Rainfall effect on sheet flow over smooth surfaces, *Proceedings of the American Society of Civil Engineers, Journal of the Hydraulics Division*, 97, HY 5, 771-792.

Shen, H.W., Enck, E., Sunday, G. and Laronee, J.B., 1979. Salt loading from hillslopes. *International Association of Hydraulic Research Processes*, 5, 99-105.

Sherard, J.L., Ryker, N.L. and Decker, R.S., 1972. Piping in earth dams of dispersive clay, in: *Proceedings of the Special Conference on the Performance of Earth and Earth-Supported Structures, A.S.C.E.*, 1, 150-161.

Sherwood, M.J., 1969. *An examination of the source of a stream in a semi-arid environment*, unpublished M.A. Dissertation (University of Western Australia).

Simons, D.B. and Albertson, M.L., 1961. Flume studies using medium sand (0.45 mm), *U.S. Geological Survey, Water Supply Paper*, 1498-A.

Singh, V.P., 1976. A distributed converging overland flow model: III Application to natural watersheds, *Water Resources Research*, 12, 902-908.

Siret, H. and Siret, L., 1887. *Les Premiers Age du Métal dans le Sud-Est de l'Espagne* (Antwerp).

Smart, I.S. and Moruzzi, V.L., 1971. Random-walk model of stream network development, *Journal of Research and Development*, IBM, v. 15, No. 3, pp 197-203.

Smart, I.S. and Moruzzi, V.L., 1971. Computer simulation of Clinch Mountain drainage Networks, *Journal of Geology*, 79, 572-584.

Smerdon, E.T., 1964. Effect of rainfall on critical tractive forces in channels with shallow flow, *Transactions of American Agricultural Engineers*, 7, 287-290.

Smith, D.D. and Wischmeier, W.H., 1957. Factors affecting sheet and rill erosion, *Transactions of the American Geophysical Union*, 38, 889-896.

Smith, D.I. and Stopp, P., 1979. *The river basin*, (Cambridge University Press).

Smith, K.G., 1958. Erosional processes and landforms in Badlands National Monument, South Dakota, *Geological Society of America Bulletin*, 69, 975-1008.

Smith, R.E., 1972. The infiltration envelope: results from a theoretical infiltrometer. *Journal of Hydrology*, 17, 1-21.

Smith, R.E. and Parlange, J.Y., 1978. A parameter-efficient hydrologic infiltration model, *Water Resources Research*, 14, 533-538.

Smith, S.T., 1962. *Some aspects of soil salinity in Western Australia*, unpublished M.Sc thesis, (University of Western Australia).

Smith, S.T., 1966. The relationship of flooding and saline water tables, *Journal of Agriculture Western Australia*, 7, 334-340.

Southard, J.B., 1970. Impossibility of Bernouilli pressure forces on particles suspended in boundary layers: discussion of a paper by R.V. Fisher and J.M. Mattinson, *Journal of Sedimentary Petrology*, 40-1, 518-519.

Southern, R.L., 1979. The atmosphere, in: *Environment and science*, ed O'Brien, B.J., (University of Western Australia Press for the Education Committee of the 150th Anniversary Celebrations, Perth), 183-226.

Sposito, G., 1977. The Gapon and the Vanselow selectivity coefficients, *Soil Science Society of America Proceedings*, 41, 1205-1206.

Sposito, G. and Mattigod, S.V., 1977. On the chemical foundation of the sodium adsorption ratio, *Soil Science Sociology of America Proceedings*, 41, 323-329.

Springer, M.E., 1958. Desert pavement and vesicular layer of some soils of the desert of the Lahontan Basin, Nevada, *Soil Science Society of America, Proceedings*, 22, 63-66.

Stagg, M.J., 1978. Rill patterns derived from air photographs of the Grwyne Fechan catchment, Black Mountains, *Cambria*, 5(1), 22-36.

Stauffer, M.R., Hajnal, Z. and Gendzwill, D.J., 1976. Rhomboidal lattice structure: a common feature on sandy beaches, *Canadian Journal of Earth Sciences*, 13, 1667-1677.

Steering Committee, 1978a. *Report by the Steering Committee on research into the effects of bauxite mining on the water resources of the Darling Range*, Western Australia Department of Industrial Development Perth).

Steering Committee, 1978b. Report by the Steering Committee on research into the effects of the woodchip industry on water resources in south western Australia, *Western Australia Department of Conservation and Environment Bulletin*, 31.

Stefanini, G., 1914. Sulle 'biancane' del Volterrano e del Senese, *Rivista Geografica Italiana*, XXI, 657-667.

Stocking, M.A., 1977. *The erosion of soils on Karroo sands in central Rhodesia with particular reference to gully form and process*, unpublished Ph.D. thesis, (University of London, London).

Strahler, A.N., 1950. Equilibrium theory of erosional slopes approached by frequency distribution analysis, *American Journal of Science*, 248, 673-696, 800-814.

Strahler, A.N., 1952. Hypsometric (area-altitude) analysis of erosional topography, *Geological Society of America Bulletin*, 63, 1117-1142.

Strahler, A.N., 1957. Quantitative analysis of watershed geomorphology, *Transactions of the American Geophysical Union*, 38, 913-920.

Strahler, A.N., 1958. The nature of induced erosion and aggradation, in: *Man's role in changing the face of the earth*, ed Thomas, W.L., (University of Chicago Press, Chicago), 621-638.

Swenson, J., Erickson, D.T., Donaldson, K.M. and Shiozaki, J.J., 1970. Soil survey of Carbon-Energy area, Utah. *U.S.D.A. Soil Conservation Services*, Washington, D.C., 78p.

Tackett, J.L. and Pearson, R.W., 1965. Some characteristics of soil crusts formed by simulated rainfall. *Soil Science*, 99, 407-413.

Talsma, T., 1969. *In situ* measurements of sorptivity, *Australian Journal of Soil Research*, 7, 277-284.

Teakle, L.J.H., 1938. Soil salinity in Western Australia, *Journal of Agriculture Western Australia*, 15, 434-452.

Teakle, L.J.H. and Burvill, G.H., 1938. The movement of soluble salts in soils under light rainfall conditions, *Journal of Agriculture Western Australia*, 15, 218-245.

Terzaghi, K. and Peck, R.B., 1966. *Soil Mechanics in Engineering Practice*, (Wiley, New York).

Thrones, J.B., 1976. Semi-arid erosional systems, London School of Economics, *Geographical Paper*, 7.

Thornethwaite, C.W., 1948. An approach toward a rational classification of climate, *Geographical Review*, 38, 55-94.

Torre Pena, F. and Aguayo de Hoyos, P., 1976. Materiales argáricos procedentes del 'Cerro del Gallo' de Fonelas (Granada), *Cuadernos de Prehistoria Granadina*, 1, 159-197.

Trotman, C.H. (ed), 1974. The influence of land use on stream salinity in the Manjimup area, Western Australia, *Western Australia Department of Agriculture Technical Bulletin*, 27.

University of London Computer Centre, 1975. SYMAP: Geographic mapping on lineprinter, in: *Users' guide*, ed Gilbert, J.C.

Valentine, P.S., 1976. A preliminary investigation into the effects of clear cutting and burning on selected soil properties in the Pemberton area of Western Australia, *Geowest*, 8 (Department of Geography, University of Western Australia).

Vanmaercke-Gottigny, M.C., 1967. De geomorfologische kaart van het Zwalmbekken, *Verhandelingen van de Koninklijke Vlaamse Akademie, Klasse der Wetenschappen*, 99, 93p.

Vanmaercke-Gottigny, M.C., 1977. Hellingsmorfometrie en morfodynamische toestand der loesshellingen in Zuid-Oost Vlaanderen, *Verslag van de Contactgroepen Nationaal Fonds voor Wetenschappelijk Onderzoek NFWO*, 261-273.

Vanoni, V.A. (ed), 1975. Sedimentation Engineering, *American Society of Civil Engineers*, New York, pp 405-406.

Vaux, A. and Morony, D., 1977. *General review of salt problems and applications of interceptor banks in salt affected areas of Western Australia*, unpublished 3rd year Honours Pedogeomorphology Project Report, (Department of Geography, University of Western Australia).

Vera, J.A., 1970. Estudio estratigrafico de la Depresion de Guadix-Baza, *Bol. Geol. Min.*, LXXXI-V, 429-462.

Verstraten, J.M., 1980. *A new procedure for the determination of exchangeable bases in calcareous, gypsiferous and slightly saline soils*, (Internal Report Laboratory for Physical Geography and Soil Science, University of Amsterdam).

Viessman, W. Jr., Knapp, J.W., Lewis, G.L. and Harbaugh, T.E., 1977. *Introduction to hydrology*, second edition, (Intext Educational Publishers- Dun-Donnelley, New York).

Vita-Finzi, C., 1969. *The Mediterranean valleys*. (Cambridge University Press).

Vita-Finzi, C., 1976. Diachronism of Old World alluvial sequences, *Nature*, 263(5574), 218-219.

Vittorini, S., 1974. La valutazione quantitativa dell'erosione nei suoli argillosi pliocenici della Val d'Era, *L'erosione del suolo in Italia e i suoi fattori*, I, (Consiglio Nazionale delle Ricerche, Istituto Geografico dell'Universita di Pisa), 23-40.

Vittorini, S., 1977. Osservazioni sull-origine e sul ruole di due forme di erosione nelle argille: calanchi e biancane, *Bollettino della Società Geografica Italiana*, Serie X, Vol. VI, 25-54.

Vittorini, S., 1979. Ruscellimento, deflusso ipodermico ed erosione nelle argille plastiche, *Rivista Geografica Italiana*, 86(3), 338-346.

Vries, M. de., 1975. A morphological time-scale for rivers, *International Association Hydraulic Research 16th Congress*, B-3, 17-23.

Ward, R.C., 1975. *Principles of hydrology*, second edition (McGraw Hill, London).

Waring, E.A. and Jones, J.A.A. 1980. A snowmelt and water equivalent gauge for British conditions, *Hydrological Sciences Bulletin*, 25(2), 129-134.

Watson, E., 1970. The Cardigan Bay area, in: *The glaciations of Wales and adjoining regions*, ed Lewis, C.A., 125-145.

Wein, R.W. and West, N.E., 1973. Soil changes caused by erosion control treatments on a salt desert area. *Soil Science Society of American Proceedings*, 37, 98-103.

Wells, S.G., 1977. Geomorphic controls of alluvial fan deposition in the Sonoran Desert, southwestern Arizona, *Geomorphology in Arid Regions*, ed Doehring, D.O., 27-50.

Wells, S.G., 1978. Geomorphic framework of an open drainage basin in the Basin and Range Province of southwestern Arizona, *Geological Society of America Abstracts with Program, Cordilleran Section*, 10, 153.

Wells, S.G. and Rose, D.E., 1981. Applications of geomorphology to surface coal-mining reclamation, northwestern New Mexico, *Environmental Geology and Hydrology in New Mexico*, eds S.G. Wells and L. Lambert, New Mexico Geological Society Special Publication 10, 69-83.

Wells, S.G., Smith, L.N., Bullard, T.F. and Schultz, J.D., 1981. Quaternary landscape evolution in the southeastern Colorado Plateau, *Geological Society of America Abstracts with Programs*, Cordilleran Section, 13, 113.

Weyman, D.R., 1970. Throughflow on hillslopes and its relation to the stream hydrograph, *Bulletin International Association Scientific Hydrology*, 15, 25-33.

Weyman, D.R., 1973. Measurements of the downslope flow of water in the soil. *Journal of Hydrology*, 20, 267-288.

Weyman, D.R., 1975. *Runoff processes and streamflow modelling*, (Oxford University Press).

Weyman, D.R., 1978. The process which lead to streamflow, *Teaching Geography*, 4(2), 71-76.

Weyman, D.R. and Weyman, V., 1977. *Landscape processes: an introduction to geomorphology*, (George Allen and Unwin, London).

Whipkey, R.Z., 1965. Subsurface stormflow from forested slopes, *Bulletin International Association Scientific Hydrology*, 10, 74-85.

Whipkey, R.Z. and Kirkby, M.J., 1978. Flow within the soil, in: *Hillslope Hydrology*, ed Kirkby M.J., (Wiley, London).

White, W.D. and Wells, S.G., 1979. Forest-fire devegetation and drainage basin adjustments in mountainous terrain, *Adjustments of the Fluvial System*, ed D.D. Rhodes and G.P. Williams, (Kendall/Hunt, Dubuque, Iowa), 199-224.

Whittington, H.S., 1975. *A battle for survival against salt encroachment at 'Springhill', Brookton, Western Australia*, (P.O. Box 9, Brookton, W.A. 6306).

Williams, T.H.L., 1978. An automatic scanning and recording tensiometer system, *Journal of Hydrology*, 39, 175-183.

Williamson, D.R. and Bettenay, E., 1979. Agricultural land use and its effect on catchment output of salt and water - evidence from southern Australia, *Progress in Water Technology*, 11, 463-480.

Williamson, D.R. and Johnston, C.D., 1979. Effects of land use changes on salt and water transport, *Proceedings of Agriculture and the Environment in Western Australia, 10 October 1979*, (Western Australian Institute of Technology, Bentley), 87-93.

Wilson, A.G. and Kirkby, M.J., 1975. *Mathematics for geographers and planners*. (Clarendon Press, Oxford).

Wilson, C.M., 1977. *The generation of storm runoff in an upland catchment*, unpublished Ph.D. thesis, (University of Bristol).

Wilson, T.V. and Ligon, J.T., 1973. The interflow process on sloping watershed areas, *Water Resources Research Institute, Clemson University, South Carolina, Report*, 38.

Wischmeier, W.H. and Smith, D.D., 1978. Predicting rainfall erosion losses - guide to conservation planning. *United States Department of Agriculture, Agricultural Handbook*, 537.

Wischmeier, W.H., Johnson, C.B. and Cross, B.V., 1971. A soil erodibility nomograph for farmland and construction sites. *Journal of Soil and Water Conservation*, 26, 189-193.

Wolman, M.G. and Miller, J.P., 1960. Frequency and magnitude of forces in geomorphic processes, *Journal of Geology*, 68, 54-74.

Wood, W.E., 1924. Increase of salt in soil and streams following the destruction of the native vegetation, *Journal of the Royal Society of Western Australia*, 10, 35-47.

Wright, A.C.S. and Kelley, C.C., 1959. *Soil erosion in the Penticton Series, Wester Bench Irrigation District, B.C.*, (Soil Survey Branch, Department of Agriculture, Kelowna, B.C.).

Yaalon, D.H. and Kalmar, D., 1978. Dynamics of cracking and swelling clay soils: displacement of skeletal grains, optimum depth of slickensides, and rate of intrapedonic turbation. *Earth Surface Processes*, 3, 31-42.

Yair, A., 1972. Observations sur les effets d'un ruisellement dirigé selon la pente des interfluves dans une région semi-aride d' Israel, *Revue de Géographie Physique et de Géologie Dynamique*, 14, 537-548.

Yair, A., 1974. Sources of runoff and sediment supplied by the slopes of a first order drainage basin in an arid environment, Northern Negev, Israel, in: *Report of the International Geographical Union Commission on Present-day Geomorphological Processes, No. 24*, 403-417.

Yair, A. and Klein, M., 1973. The influence of surface properties on flow and erosion processes on debris covered slopes in an arid area, *Catena*, 1, 1-18.

Yair, A. and Lavee, H., 1974. Areal contribution to runoff on scree slopes in an extreme arid environment, *Zeitschrift für Geomorphologie, Supplement Band*, 21, 106-121.

Yair, A., Sharon, D. and Lavee, H., 1978. An instrumented watershed for the study of partial area contribution of runoff in the arid zone, *Zeitschrift für Geomorphologie, Supplement Band*, 29, 71-82.

Yair, A., Sharon, D. and Lavee, H., 1980b. Trends in runoff and erosion processes over an arid limestone hillside, northern Negev, Israel. *Hydrological Science Bulletin*, 25(3), 243-255.

Yair, A., Bryan, R.B., Lavee, H. and Adar, E., 1980a. Runoff and erosion processes and rates in the Zin Valley badlands, Northern Negev, Israel, *Earth Surface Processes*, 5, 205-225.

Yair, A., Goldberg, P., Bryan, R.B. and Lavee, H., 1980b. Present and past geomorphic evidences in the development of a badlands landscape: Zin Valley northern Negev, Israel, in: *Palaeoecology of Africa and the surrounding islands*, ed M. Sarnthein, Seibold, E. and Rognon, P., (Balkema Press, Vol. 12), 125-135.

Yalin, M.S., 1977. *Mechanics of sediment transport*, 2nd edition, (Pergamon Press, New York).

Yang, C.T. and Stall, J.B., 1971. Note on the map scale effect in the study of stream morphology, *Water Resources Research*, 7, 311-322.

Yong Nam Yoon and Wenzel, H.G., 1971. Mechanics of sheet flow under simulated rainfall, *Proceedings of the American Society of Civil Engineers, Hydraulics Division*, 95, HY 9, 1367-1386.

Yoder, R.E., 1936. A direct method of aggregate analysis of soils and a study of the physical nature of erosion losses. *Journal of the American Society of Agronomy*, 28, 337-351.